Quantum Entanglement

by
Trevor G. Underwood

2nd Edition (August 6, 2025)

By the same author:

Quantum Electrodynamics – annotated sources. Volumes I and II. (April 2023);

Special Relativity. (June 2023);

General Relativity. (November 2023);

Gravity. (March 2024);

Electricity & Magnetism. (May 2024);

Quantum Entanglement. (June 2024);

The Standard Model. (September 2024);

New Physics. (October 2024);

The Cosmological Redshift of Light. (November 2024);

Cosmic Microwave Background Radiation. (January 2025);

Fundamental Physics. (May 2025).

all distributed by Lulu.com.

Dedicated to my father, Alfred Ernest Underwood.

Published by Trevor G. Underwood
18 SE 10th Ave.
Fort Lauderdale, FL 33301

ISBN: 979-8-218-45794-5 (hardcover)
Library of Congress Control Number: 2024912713

Printed and distributed by Lulu Press, Inc.

700 Park Offices Dr
Ste 250
Durham, NC, 27709
http://www.lulu.com/shop

CONTENTS.

Page no.

Pauli, decided to pursue instead the alternative *non-relativistic* theory to the problem of *completion of electron groups in an atom*, in order to draw conclusions only about the *number of possible stationary states* of an *atom* when several equivalent *electrons* are present. But this did not address the position and relative order of the term values. On the basis of these results, Pauli obtained a general classification of every *electron* in the *atom* by the principal quantum number n and two auxiliary quantum numbers k_1 and k_2 to which he added a further quantum number m_1 in the presence of an external field, in agreement with experiments. In particular, his rule explained Stoner's result in a natural way and with it the period lengths 2, 8 18, 32.

76 **George Eugene Uhlenbeck (December 6, 1900–October 31, 1988).**

78 **Samuel Abraham Goudsmit (July 11, 1902–December 4, 1978).**

80 **Uhlenbeck, G. E. & Goudsmit, S. (November, 1925). Ersetzung der Hypothese vom unmechanischen Zwang durch eine Forderung bezuglich des inneren Verhaltens jedes einzelnen Elektrons. (Replacement of the hypothesis of unmechanical coercion by a requirement regarding the internal behavior of each individual electron.)** *Naturw.*, 13, 47, 953-4 (in German); https://doi.org/10.1007/ BF01558878; translation by T. G. Underwood; also in Underwood, T. G. (2023). *Quantum Electrodynamics - annotated sources*, Volume I, pp. 282-6; without being aware of Compton's suggestion Uhlenbeck and Goudsmit noted doublets in the alkali spectra that did not conform to current models of the atom. They proposed applying the model of the *spinning electron* to interpret a number of features of the quantum theory of the *anomalous Zeeman effect*, and applied the classical formula for spherical rotating electron with finite radius and surface charge.

84 **Uhlenbeck, G. E. & Goudsmit, S. (February 20, 1926). Spinning Electrons and the Structure of Spectra.** *Nature*, 117, 264-5; https://doi.org/10.1038/117264a0.

85 **Paul Adrien Maurice Dirac (August 8, 1902–October 20, 1984).**

93 **Dirac, P. A. M. (February, 1928). The Quantum Theory of the Electron.** *Roy. Soc. Proc., A*, 117, 778, 610–24; https://doi.org/10.1098/rspa.1928.0023; Dirac noted that the new quantum mechanics applied to the problem of the structure of the atom with *point-charge electrons* did not give results in agreement with experiment. The discrepancies consisted of "duplexity" phenomena; the observed number of stationary states for an electron in an atom being twice the number given by the theory. Goudsmit and Uhlenbeck introduced the idea of an electron with a *spin*. Previous *relativity* treatments by Gordon and Klein obtained the operator of

the wave equation by the same procedure as in the *non-relativity* theory; they substituted classical *quantum differential operators* for the *momentum vector* in the amended *relativistic Hamiltonian equation* and applied the resulting differential operator to the *wave function* to obtain the *Klein-Gordon equation.* Dirac noted that Gordon and Klein's treatments gave rise to two difficulties. The *first difficulty* was in the physical interpretation of solutions of ψ as the *charge* and the *current.* This was satisfactory for emission and absorption of radiation, but only provided the probability of any dynamical variable at any specific time having a value between specified limits if they referred to the *position* of the electron, but, unlike the *non-relativity* theory, *not if they refer to its momentum or any other dynamical variable.* The *second difficulty* was that the conjugate imaginary of the wave equation was the same as that for an electron with charge – e and negative energy. *This paper only addressed the removal of the first of difficulties.* The resulting theory was only an approximation but appeared sufficient to address the duplexity problems without further assumptions. Dirac applied the method of *q-numbers* and using non-commutative algebra exhibited the properties of a free electron and of an electron in a central field of electric force. He showed that simplest Hamiltonian for a *point charge electron satisfying requirements of both relativity and the general transformation theory* of quantum mechanics led to an explanation of all duplexity phenomena of number of stationary states being twice the observed value *without further assumption about spin.* In contrast to the Schrödinger equation which described wave functions of only one complex value, Dirac introduced *vectors of four complex numbers* (known as bispinors). This resulted in a *relativistic equation of motion* for the *wave function of the electron* referred to as the *Dirac equation*, $\{p_0 + \rho_1 (\boldsymbol{\sigma}, \mathbf{p}) + \rho_3 mc\} \psi = 0$, where **p** is the *momentum* vector, and $\boldsymbol{\sigma}$ denotes the vector $(\sigma_1, \sigma_2, \sigma_3)$. This included a term equal to the spin correction given by Darwin and Pauli. It described all spin-½ particles with mass, but did not address the second class of solutions of the wave equation in which *charge of the electron is positive* and *energy of a free electron is negative.*

107 **Dirac, P. A. M. (March, 1928). The quantum theory of the Electron. Part II.** *Roy. Soc. Proc., A*, 118, 779, 351-61; https://doi.org/10.1098/rspa.1928.0056; application of the *Dirac equation* to the conservation theorem, the selection principle, the relative intensities of the lines of a multiplet, and the Zeeman effect.

119 **Pauli, W. (1940). The Connection Between Spin and Statistics.** *Phys. Rev.*, 58, 8, 716–22; doi:10.1103/PhysRev.58.716; Pauli derived that for a *relativistically invariant wave equation* for free particles: from postulate (I), *according to which the energy must be positive, the necessity of Fermi-Dirac statistics for particles*

with arbitrary half-integral spin; from postulate (II), *according to which observables on different space-time points with a space-like distance are commutable, the necessity of Einstein-Bose statistics for particles with arbitrary integral spin.* Postulate I was introduced because "in case of *half-integral spin, …, a positive definite energy density, as well as a positive definite total energy, is impossible*". Similarly, postulate II was introduced so that "all physical quantities at finite distances exterior to the light cone … are commutable. … The justification for our postulate lies in the fact that measurements at two space points with a space-like distance can never disturb each other, *since no signals can be transmitted with velocities greater than that of light*".

130 PART II Exchange Interaction.

139 **Dirac, P. A. M. (October, 1926). On the Theory of Quantum Mechanics.** *Roy. Soc. Proc., A*, 112, 762, 661-77; https://doi.org/10.1098/rspa.1926.0133.JSTOR 94692; Dirac developed a *relativistic* treatment of Schrodinger's wave theory from a more general point of view in which the time t and its conjugate momentum – W were treated from the beginning on the same footing as the other variables. He applied his *relativistic formulation* to a system containing an atom with two electrons and found that *if the positions of the two electrons were interchanged the new state of the atom was physically indistinguishable from the original one*. In order that that the theory only enabled calculation of *observable quantities* it was necessary to treat (*mn*) and (*nm*) as only one *state*. *Unsymmetrical* functions of the co-ordinates (and momenta) of the two electrons could not be represented by matrices. *Symmetrical* functions such as the total *polarizations* of the atom could be considered to be represented by matrices without inconsistency. These matrices were by themselves sufficient to determine all the physical properties of the system. The *theory of uniformizing variables* introduced by the author *could no longer apply*. The new theory allowed two solutions satisfying the necessary conditions; one led to Pauli's principle that not more than one electron can be in any given orbit, and the other, when applied to the analogous problem of the ideal gas, led to the Einstein-Bose statistical mechanics. *With neglect of relativity mechanics* this accounted for the absorption and stimulated emission of radiation and showed that the elements of the matrices representing the total polarization determined the transition probabilities. *This could not be applied to spontaneous emission.*

156 **Walter Heinrich Heitler (January 2, 1904 – November 15, 1981).**

159 **Fritz Wolfgang London (March 7, 1900–March 30, 1954).**

161 **Heitler, W. & London, F. (June, 1927). Wechselwirkung neutraler Atome und homöopolare Bindung nach der Quantenmechanik. (Interaction of neutral atoms and homeopolar bonding according to quantum mechanics.)** *Zeit. Phys.*, 44, 455–72. https://doi.org/10.1007/BF01397394; also at http://quantum-chemistry-history.com/Heitler_London_Dat/WechselWirk1927/WechselWirk1927.htm (in German); translation by T. G. Underwood; in this paper, Heitler and London examined the interaction between *neutral atoms* which resulted in non-polar valance bonds. They applied quantum mechanics to the calculation of the *interaction energy* of the atoms when they move closer together. Due to *quantum entanglement*, it was found that two neutral atoms could interact with each other in two ways. *The problem was twofold degenerate, corresponding to the two ways of assigning the electrons to the neutral atoms*. Examination of the different cases of two H atoms and two He atoms showed that by applying the *Pauli principle*, the selected eigenfunctions of the system changed or maintained their sign, respectively, when two electrons were swapped if the two electrons compared had the same, or different, *spin*. It was found that in the case of He there was only one solution, which yields about the right size of the He gas kinetic-radius, *due to the fact that 2 He atoms (and the same applied to all noble gases) could not differ in their spin* – in contrast to hydrogen (and all atoms with unfinished shells) – so that 2 He atoms had only one possible mode of behaving.

176 **Frenking, G.* (2021). The Chemical Bond – an Entrance Door of Chemistry to the Neighboring Sciences and to Philosophy.** *Israel Journal of Chemistry*; doi.org/10.1002/ijch.202100070.

183 **Dirac, P. A. M. (April, 1929). Quantum Mechanics of Many-Electron Systems.** *Roy. Soc. Proc., A*, 123, 792, 714-33; https://doi.org/10.1098/rspa.1929.0094; also in Underwood, T. G. (2023). *Quantum Electrodynamics - annotated sources*, Volume I, pp. 565-79; in this paper Dirac introduced the term *exchange interaction*. He noted that the general theory of quantum mechanics was now almost complete, and that the imperfections that still remained were in connection with the exact fitting in of the theory with *relativity ideas*, which only gave rise to difficulties when high-speed particles were involved and were therefore *of no importance in the consideration of atomic and molecular structure and ordinary chemical reactions*. The difficulty was only that the exact application of these laws led to equations much too complicated to be soluble. He noted that it was desirable that approximate practical methods of applying quantum mechanics should be developed which could lead to an explanation of the main features of complex atomic systems without too much computation. Current *non-relativistic* quantum theory could not give an explanation of *multiplet structure* without an extraneous

assumption of *large forces coupling the spin vectors of the electrons in an atom*. The explanation was provided by ***quantum entanglement*** through *exchange interaction* arising from electrons being indistinguishable one from another resulted in large *exchange energies* between electrons in different atoms. This accounted for homopolar valency bonds, in which, for each *stationary state* of the atom there was one magnitude of the *total spin vector*. He also noted that developments of the *theory of exchange* made by Heitler & London and Heisenberg made extensive use of *group theory*, which was a theory of certain quantities that did not satisfy the commutative law of multiplication and should thus form a part of quantum mechanics, and then translated the methods and results of *group theory* into the language of *quantum mechanics*. He demonstrated that *exchange interaction* equal to a constant *perturbation energy*, together with *coupling energy* between spin vectors, determined *energy* levels; and showed that in the first approximation the *exchange interaction* between the electrons could be replaced *by a coupling between their spins*, the energy of this coupling for each pair of electrons being equal to the scalar product of their *spin vectors* multiplied by a numerical coefficient given by the *exchange energy*.

199 PART III Susceptibility of materials to a magnetic field.

203 **Feynman, R. P. (1963). The Magnetism of Matter.** *The Feynman Lectures on Physics*, Volume II, Chapter 34. Basic Books, New York, 34-7 to 34-8. https://www.feynmanlectures.caltech.edu/II_34.html.

210 **Werner Karl Heisenberg (December 5, 1901–February 1, 1976).**

217 **Heisenberg, W. (September, 1928). Zur Theory of Ferromagnetismus. (On the theory of ferromagnetism.)** *Zeit. Phys.*, 49, 619–36 (in German); https://doi.org/10.1007/BF01328601; translation by D. H. Delphenich; https://neo-classical-physics.info/uploads/3/0/6/5/3065888/Heisenberg_-_on_the_theory_of_ferromagnetism.pdf; in another brilliant paper, Heisenberg noted that empirical results exhibited *ferromagnetism* as an entirely similar state of affairs to what was previously observed in the spectrum of the helium atom. It seemed to follow from the levels in the helium atoms that *a powerful interaction prevailed between the spin directions of two electrons* that led to the splitting of the level structure into systems of singlets and triplets. He also noted that this was closely related to explaining ferromagnetic phenomena as being implied by the *exchange phenomenon* (resulting from *quantum entanglement*). He proceeded to show that the *Coulomb interaction*, together with the *Pauli principle*, succeeded in evoking the same effects as the molecular field that Weiss had postulated, noting that it was

only in recent times that mathematical methods were developed for the treatment of such a complicated problem in the important investigations of Wigner, Hund, and Heitler & London. Heisenberg assumed as a first approximation that the lattice separations were very large, and that every electron belonged to its own atom, and applied Heitler-London's calculations to the case of 2n electrons in a state of interaction, finding 2n electrons in 2n positionally different quantum cells. Due to their smallness, he was able to leave the magnetic interactions outside of consideration, and showed that *the spin moments of all electrons become partly parallel and partly anti-parallel as a result of the exchange processes*. By adding the fundamental Pauli principle to this, viz., that the *eigenfunctions* of the total system should be *anti-symmetric* in all electrons, he showed that an entirely well-defined *total magnetic moment* belonged to each level value of the perturbed system, and there were (2n)! levels in the unperturbed system (ignoring the Pauli principle and spin). He then showed that a *statistical treatment of ferromagnetism was possible when all energy values had been calculated.* Heisenberg concluded that *an atom in a lattice could only be exchanged with its "neighbors"*; exchanges with atoms that lie further away that the "neighboring atoms" could be neglected. The *number of "neighbors" of an atom* was, e.g., 1 in a molecular lattice of diatomic molecules, 2 in a linear chain, 4 in a quadratic surface lattice, 6 in a simple cubic lattice, 8 in a cubic, space-centered lattice, and 12 in a cubic, face-centered lattice. By assuming a distribution of energy values about the mean had the approximate form of a Gaussian error curve, Heisenberg showed that small or negative values of the constant β [= zJ_0/kT)] result in *paramagnetism*; and that *ferromagnetism is only possible for lattice types for which an atom had at least eight neighbors*, which was the case for Fe, Co, Ni, whose lattices are all cubic, some of which were space-centered ($z = 8$) and some of which were face-centered ($z = 12$). He concluded that two conditions were necessary for the appearance of *ferromagnetism*: the crystal lattice must be a type such that *any atom had at least 8 neighbors*; and the *principal quantum number* of the electrons that were responsible for magnetism must be $n \geq 3$.

228 **John Hasbrouck Van Vleck (March 13, 1899–October 27, 1980).**

230 **Van Vleck, J. H. (1932). The Theory of Electric and Magnetic Susceptibilities.** Clarendon Press, Oxford; https://dn790000.ca.archive.org/0/items/theoryofelectric031070mbp/theoryofelectric031070mbp.pdf; extracts are provided from John Van Vleck's 400-page book, published in 1932, which provides an extremely comprehensive explanation of the *non-relativistic* quantum theory of the *electric and magnetic susceptibities of materials*, including the quantum theory of *diamagnetism* and *paramagnetism*, and detailed explanations of the *exchange*

effect and Heisenberg's *theory of ferromagnetism.* As Van Vleck noted in the Preface, "the analysis of experimental *magnetic susceptibilities* cannot be attempted until the quantum chapters, since *the numerical values of magnetic susceptibilities are inextricably connected with the quantization of angular momentum*". Van Vleck also noted that *the theory of the electron spin may be presented in two ways,* viz. by means of what he calls a *semi-mechanical model* (the Uhlenbeck-Goudsmit model) or by means of *Dirac's 'quantum theory of the electron'.* In the *semi-mechanical model*, matrix expressions for the *spin angular momentum* are written down by analogy with the *orbital angular momentum* matrices, with certain postulates regarding the occurrence of a half-quantum of *spin* per electron and *the ratio of spin magnetic moment to spin angular momentum,* in order to bring the theory into line with experiments. He noted that the interaction of the *spin* with *external magnetic fields* was handled perfectly well by the *semi-mechanical model*, but Einstein's *special theory of relativity* is inconsistent with this. He noted, on the other hand, that *the extension of Dirac's relativistic theory to many electron systems was at that time in a rather unsettled state* resulting in a nonvanishing probability of the mass of the electron changing sign, *an obvious absurdity.* Consequently, Van Vleck decided to present the quantitative aspects of the spin entirely with the aid of the older *semi-mechanical model.* In considering *diamagnetism,* [the property of materials that are repelled by a magnetic field; which creates an induced magnetic field in them in the opposite direction] Van Vleck supposed that the atoms were in singlet S states [in which all electrons in the S orbital are paired], as otherwise there was an overwhelming *paramagnetism* [a form of magnetism whereby some materials are weakly attracted by an externally applied magnetic field, and form internal induced magnetic fields in the direction of the applied magnetic field]. In such states there cannot be even an instantaneous *magnetic moment* in the absence of external fields. In other cases, Van Vleck distinguished different cases, including where the *inter-atomic forces* were so small that the magnetism could be calculated by treating the atoms of the solid to be free (exemplified by *rare earth salts*); cases where *the orbital and spin magnetic effects were both largely destroyed*, resulting in feeble *paramagnetism* (which most elements exhibit in the solid state), or even *diamagnetism*; and solids in which *the Heisenberg exchange forces tended to align the spins parallel and so create ferromagnetism* (as for iron, nickel, and cobalt). Van Vleck noted that the *exchange forces had the effect of introducing a very strong coupling between the spins of paramagnetic atoms or ions. Diamagnetic* atoms or ions have no resultant spin and so do not give rise to any *exchange forces* tending to orient the *spins* of other atoms. Moreover, these *exchange forces* became of subordinate importance in media in which the density of *paramagnetic* atoms or ions was low because the great majority of the atoms are *diamagnetic*, and in most salts involving the iron group which are consequently only *paramagnetic*. Van Vleck then provided a detailed account of how entanglement between electrons with the same *quantum spin states* creates an *exchange effect* which results in a strong coupling between their *spins*. He concluded by explaining how Heisenberg applied this in his theory of *ferromagnetism*.

260 **The Nobel Prize in Physics 1977, Press release, October 11, 1977.** The Nobel Prize in Physics 1977 was awarded to Van Vleck, jointly with Philip Warren Anderson and Sir Nevill Francis Mott, "for their fundamental theoretical investigations of the electronic structure of magnetic and disordered systems". The electrical and magnetic properties of different materials are determined by how the electrons move about in relation to the atomic nucleus. When an atom from a foreign substance is inserted into a crystalline structure, the crystal's properties can be altered. During the 1930s John Van Vleck developed theories about how electrical fields in a crystal affect a foreign atom and how such an atom can be bound to nearby atoms through its electrons. He also showed how the interaction between the electron's movements can create local magnetic moments in crystals.

262 **John H. Van Vleck – 1977 Nobel Lecture, December 8, 1977.** *Quantum Mechanics the key to understanding Magnetism*. [https://www.nobelprize.org/uploads/2018/06/vleck-lecture.pdf]; in his Nobel Prize lecture Van Vleck provided a brief history of the evolution of the quantum theory of *magnetic materials*.

268 **Jaščur, M. (October, 2013). Quantum Theory of Magnetism.** http://www.upjs.sk/pracoviska/univerzitna-kniznica/e-publikacia/#pf (Slovak); English translation; https://www.upjs.sk/public/media/5596/Qauntum-Theory-of-Magnetism.pdf.

283 PART IV Quantum Entanglement.

292 **Einstein, A., Podolsky, B. & Rosen, N. (May, 1935). Can Quantum-Mechanical Description of Physical Reality Be Considered Complete?** *Phys. Rev.*, 47, 10, 777-80; https://doi.org/10.1103/PhysRev.47.777; also, at https://journals.aps.org/pr/pdf/10.1103/PhysRev.47.777; "in a complete theory there is an element corresponding to each element of reality. *A sufficient condition for the reality of a physical quantity is the possibility of predicting it with certainty, without disturbing the system*. In quantum mechanics in the case of two physical quantities described by non-commuting operators, *the knowledge of one precludes the knowledge of the other*. Then either (1) the description of reality given by the *wave function* in

quantum mechanics is not complete or (2) these two quantities cannot have simultaneous reality. Consideration of the problem of making predictions concerning a system on the basis of measurements made on another system that had previously interacted with it leads to the result that if (1) is false then (2) is also false. One is thus led to conclude that the description of reality as given by a wave function is not complete." [Einstein was wrong, again.]

302 **Erwin Rudolf Josef Alexander Schrödinger (August 12, 1887–January 4, 1961).**

309 **Schrödinger, E. (October, 1935). Discussion of probability relations between separated systems.** *Mathematical Proceedings of the Cambridge Philosophical Society*, 31, 4, 555–63; https://doi.org/10.1017/S0305004100013554; also at https://sci-hub.se/10.1017/S0305004100013554; "when two systems, of which we know the states by their respective representatives, enter into temporary physical interaction due to known forces between them, and when after a time of mutual influence, the systems separate again, then they can no longer be described in the same way as before, viz. by endowing each of them with a representative of its own. *I would not call that one but rather the characteristic trait of quantum mechanics*, the one that enforces its entire departure from classical lines of thought. *By the interaction the two representatives (or ψ-functions) have become entangled.*" [A brilliant rebuttal to Einstein, Podolsky, & Rosen. (May, 1935).]

318 **Bell, J. S. (1964). On the Einstein-Podolsky-Rosen Paradox.** *Physics*, 1, 195-200; https://journals.aps.org/ppf/pdf/10.1103/PhysicsPhysiqueFizika.1.195; in this paper, John Stewart Bell, a physicist from Northern Ireland, building upon the Einstein–Podolsky–Rosen paradox, determined that *quantum mechanics was incompatible with local hidden-variable theories,* given some basic assumptions about the nature of measurement, subsequently known as *Bell's theorem*. "Local" here referred to the principle of locality, the idea that a particle can only be influenced by its immediate surroundings, and that interactions mediated by physical fields could not propagate faster than the speed of light. "Hidden variables" were putative properties of quantum particles that were not included in quantum theory but nevertheless affected the outcome of experiments. Bell deduced that if measurements are performed independently on the two separated particles of an entangled pair, then the assumption that the outcomes depended upon hidden variables within each half implied a mathematical constraint on how the outcomes on the two measurements were correlated. This constraint was subsequently called a *Bell inequality*. Bell then showed that quantum physics predicted correlations that violated this inequality. The first rudimentary experiment designed to test *Bell's theorem* was performed in 1972 by John Clauser and Stuart Freedman.

323 **John Francis Clauser (born December 1, 1942).**

324 **First Experimental Proof That Quantum Entanglement Is Real (1972).**

325 **Freedman, S. J. & Clauser, J. F. (April, 1972). Experimental Test of Local Hidden-Variable Theories.** *Physical Review Letters*, 28, 14, 938–41; https://journals.aps.org/prl/pdf/10.1103/ PhysRevLett.28.938; "in the present work we measured the correlation in linear polarization of two *photons* emitted in an atomic cascade. The decaying atoms were viewed by two symmetrically placed optical systems, each consisting of two lenses, a wavelength filter, a rotatable and removable polarizer, and a single-photon detector. We made the following assumptions for any *local hidden-variable* theory: (1) The two *photons* propagate as separated localized particles. (2) A binary selection process occurs for each *photon* at each polarizer (transmission or no-transmission). This selection does not depend upon the orientation of the distant polarizer. In addition, we made the following assumption to allow a comparison of the *generalization* of *Bell's inequality* without experiment: (3) All *photons* incident on a detector have a probability of detection that is independent of whether or not the *photon* has passed through a polarizer. It has been shown by this generalization of *Bell's inequality* that the existence of local hidden variables imposes restrictions on this correlation in conflict with the predictions of quantum mechanics. Our data, in agreement with quantum mechanics, violate these restrictions to high statistical accuracy, *thus providing strong evidence against local hidden-variable theories*."

328 **The Nobel Prize in Physics 2022, Press release, October 4, 2022.** The Nobel Prize in Physics 2022 was awarded to Clauser, jointly with Anton Zeilinger, University of Vienna, Austria, and Alain Aspect, Institut d'Optique Graduate School – Université Paris-Saclay and École Polytechnique, Palaiseau, France "for experiments with *entangled photons*, establishing the violation of Bell inequalities and pioneering quantum information science". One of the most remarkable traits of quantum mechanics is that it allows two or more particles to exist in what is called an entangled state. What happens to one of the particles in an entangled pair determines what happens to the other particle, even if they are far apart. In 1972, John Clauser conducted groundbreaking experiments using entangled light particles, photons. This and other experiments confirm that quantum mechanics is correct and pave the way for quantum computers, quantum networks and quantum encrypted communication.

331 **Experiments testing macroscopic quantum superpositions must be slow. (September, 2015). Mari, A., De Palma, D. & Giovannetti, V.** (Addition: June 23, 2025). *Nature*, Scientific Reports, 6, 22777 (2016); arXiv:1509.02408v1 [quant-ph]; we consider a thought experiment where the preparation of a macroscopically massive or charged particle in a quantum superposition and the associated dynamics of a distant test particle apparently allow for superluminal communication. We give a solution to the paradox which is based on the following fundamental principle: any local experiment, discriminating a coherent superposition from an incoherent statistical mixture, necessarily requires a minimum time proportional to the mass (or charge) of the system. For a charged particle, we consider two examples of such experiments, and show that they are both consistent with the previous limitation. In the first, the measurement requires to accelerate the charge, that can entangle with the emitted photons. In the second, the limitation can be ascribed to the quantum vacuum fluctuations of the electromagnetic field. On the other hand, when applied to massive particles our result provides indirect evidence for the existence of gravitational vacuum fluctuations and for the possibility of entangling a particle with quantum gravitational radiation.

PREFACE

In my last book, "Electricity & Magnetism", two conclusions stood out; the *rotary* nature of magnetism compared with the *linear* translation of electricity; and the dependence of *ferromagnetism* on *entanglement* of *electrons* with the same *quantum spin states*. This book takes a closer look at *quantum entanglement*. *Entanglement* occurs between *valence electrons* with the same *quantum spin states* creating *non-polar (valence) bonds*.

> [*Valence electrons* are electrons in the outermost shell of an atom, and that can participate in the formation of a chemical bond if the outermost shell is not closed.]

Entanglement also occurs between electrons with the *same quantum spin states* in *materials* in a *magnetic field*, creating various types of *magnetism*, depending on structure of the atomic lattice of the material. It has also been demonstrated that *quantum* entanglement occurs *at a distance* between electrons, photons, top quarks, and molecules.

Part I describes the development of the current theory of the *spin of the electron*. **Part II** describes the development of the theory of how *entanglement* between electrons with the same *quantum spin states* result in *exchange interaction*. **Part III** addresses the quantum theory of the susceptibility of *materials* to a *magnetic field* and the resulting types of magnetism. **Part IV** describes *quantum entanglement at a distance* between electrons and other particles with the same *quantum spin states*.

Part I Electron Spin starts with Arthur Compton's paper, "(1921). The Magnetic Electron", in which the idea of a *quantized spinning of the electron* was put forward for the first time. Compton hypothesized that the electron *magnetic moment* was intrinsically connected to the electron's *spin*.

This is followed by Wolfgang Pauli,'s paper "(1925). Über den Zusammenhang des Abschlusses der Elektronengruppen im Atom mit der Komplexstruktur der Spektren. (On the connection between the closure of the electron groups in the atom and the complex structure of the spectra.)" in which Pauli first reviewed the established theories for the energy differences *of the triplet levels of the alkaline earths*, based respectively, on *the anomaly of the relativity correction* of the *optically active electron*, and *the dependence of the interaction between the electron and the atom core on the relative orientation of these two systems*. He noted a serious difficulty with the former is the connection of these ideas with the *correspondence principle*, which was well known to be a necessary means to explain the selection rules for the *quantum numbers* k_1, j, and m and the polarization of the Zeeman components, in particular that *it was necessary that the totality of the stationary states of an atom corresponded to a collection (class) of orbits with a definite type of periodicity properties*. The dynamic explanation of this kind of motion of the *optically active electron*, which was based upon the assumption of deviations of the forces between the *atom* core and the *electron* from central symmetry, *seemed to be incompatible with the possibility to represent the alkali doublet (and thus also the magnitude of the corresponding precession frequency) by relativistic formulae*. Consequently, Pauli, decided to apply instead the alternative *non-relativistic* theory to the problem of *completion of electron groups in an atom*, in order to draw conclusions about the *number of possible stationary states* of an *atom* when several equivalent *electrons* are present. But this did not address the position and relative order of the term values. On the basis of these results, Pauli obtained a general classification of every *electron* in the *atom* by the principal quantum number n and two auxiliary quantum numbers k_1 and k_2 to which he added a further quantum number m_1 in the presence of an external field, in agreement with experiments. In particular, his rule explained Stoner's result in a natural way and with it the period lengths 2, 8 18, 32.

Then, without being aware of Compton's suggestion, Uhlenbeck and Goudsmit noted in their paper, "(1925). Ersetzung der Hypothese vom unmechanischen Zwang durch eine Forderung bezuglich des inneren Verhaltens jedes einzelnen Elektrons", that doublets in the alkali spectra did not conform to current models of the atom. They applied the model of the *spinning electron* to interpret a number of features of the quantum theory of the *anomalous Zeeman effect*, using the classical formula for spherical rotating electron with finite radius and surface charge.

This is followed by Paul Dirac's paper, "(February, 1928). The Quantum Theory of the Electron", in which Dirac noted that the new quantum mechanics applied to the problem of the structure of the atom with *point-charge electrons* did not give results in agreement with experiment. The discrepancies consisted of "duplexity" phenomena; the observed number of stationary states for an electron in an atom being twice the number given by the theory. Goudsmit and Uhlenbeck had introduced the idea of an electron with a *spin.* Previous *relativity* treatments by Gordon and Klein obtained the operator of the wave equation by the same procedure as in the *non-relativity* theory; they substituted classical *quantum differential operators* for the *momentum vector* in the amended *relativistic Hamiltonian equation* and applied the resulting differential operator to the *wave function* to obtain the *Klein-Gordon equation.* Dirac noted that Gordon and Klein's treatments gave rise to two difficulties. The *first difficulty* was in the physical interpretation of wave-mechanical expressions for the *charge* and the *current.* This was satisfactory for emission and absorption of radiation, but only provided the probability of any dynamical variable at any specific time having a value between specified limits if they referred to the *position* of the electron, but, unlike the *non-relativity* theory, *not if they refer to its momentum or any other dynamical variable*. The *second difficulty* was that the conjugate imaginary of the *wave equation* was the same as that for an electron with charge – e and negative energy.

This paper only addressed the removal of the first of difficulties. The resulting theory was only an approximation but appeared sufficient to address the duplexity problems without further assumptions. Dirac applied the method of *q-numbers* and using non-commutative algebra exhibited the properties of a free electron and of an electron in a central field of electric force. He showed that simplest Hamiltonian for a *point charge electron satisfying requirements of both relativity and the general transformation theory* of quantum mechanics led to an explanation of all duplexity phenomena of number of stationary states being twice the observed value *without further assumption about spin*. In contrast to the Schrödinger equation which described wave functions of only one complex value, Dirac introduced *vectors of four complex numbers* (known as bispinors). This resulted in a *relativistic equation of motion* for the *wave function of the electron* referred to as the *Dirac equation*, $\{p_0 + \rho_1 (\boldsymbol{\sigma}, \mathbf{p}) + \rho_3 mc\} \psi = 0$, where $\mathbf{p}$ is the *momentum* vector, and $\boldsymbol{\sigma}$ denotes the vector $(\sigma_1, \sigma_2, \sigma_3)$. This included a term equal to the spin correction given by Darwin and Pauli. It described all spin-½ particles with mass, but did not address the second class of solutions of the wave equation in which *charge of the electron is positive* and *energy of a free electron is negative*.

In the second part of this paper, "(March 1928) The Quantum Theory of the Electron. Part II", Dirac applied the *Dirac equation* to the conservation theorem, the selection principle, the relative intensities of the lines of a multiplet, and the Zeeman effect.

The last paper in this section, "Pauli, W. (1940). The Connection Between Spin and Statistics", appeared to be a rather peculiar effort to reinstate Einstein's *theory of special relativity*, possibly due to the fact that it was published at the outbreak of World War II, just prior to Pauli and Einstein both moving to Princeton. Pauli demonstrated that for a *relativistically invariant wave equation* for free particles: from postulate (I), *according to which the energy must be positive, the necessity of Fermi-Dirac statistics for particles with arbitrary half-integral spin*; from postulate (II), *according to which observables on different space-time points with a space-like distance are commutable, the necessity of Einstein-Bose statistics for particles with arbitrary integral spin*. Postulate I was introduced because "in case of *half-integral spin, …, a positive definite energy density, as well as a positive definite total energy, is impossible*". Similarly, postulate II was introduced so that "all physical quantities at finite distances exterior to the light cone … are commutable. … The justification for our postulate lies in the fact that measurements at two space points with a space-like distance can never disturb each other, *since no signals can be transmitted with velocities greater than that of light*".

In the first paper in **Part II Exchange Interaction**, ("Dirac, P. A. M. (October 1926). On the Theory of Quantum Mechanics"), Dirac developed a *relativistic* treatment of *Schrodinger's wave theory* from a more general point of view in which the time t and its conjugate momentum – W were treated from the beginning on the same footing as the other variables. He applied his *relativistic formulation* to a system containing an atom with two electrons and found that *if the positions of the two electrons were interchanged the new state of the atom was physically indistinguishable from the original one.* In order that that the theory only enabled calculation of *observable quantities* it was necessary to treat (*mn*) and (*nm*) as only one *state. Unsymmetrical* functions of the co-ordinates (and momenta) of the two electrons could not be represented by matrices. *Symmetrical* functions such as the total *polarizations* of the atom could be considered to be represented by matrices without inconsistency. These matrices were by themselves sufficient to determine all the physical properties of the system. The *theory of uniformizing variables* introduced by the author *could no longer apply.* The new theory allowed two solutions satisfying the necessary conditions; one led to Pauli's principle that not more than one electron can be in any given orbit, and the other, when applied to the analogous problem of the ideal gas, led to the Einstein-Bose statistical mechanics. *With neglect of relativity mechanics this accounted for the absorption and stimulated emission of radiation* and showed that the elements of the matrices representing the total polarization determined the transition probabilities. *This could not be applied to spontaneous emission.*

This is followed by the path-breaking paper ("Heitler & London (1927). Wechselwirkung neutraler Atome und homöopolare Bindung nach der Quantenmechanik. (Interaction of neutral atoms and homeopolar bonding according to quantum mechanics)"), in which Walter Heitler and Fritz London examined the interaction between *neutral atoms* known as *non-polar valance bonds.* They applied quantum mechanics to the calculation of the *interaction energy* of the atoms when they move closer together. Due to *quantum entanglement*, it was found that two neutral atoms could interact with each other in two ways. *The problem was twofold degenerate, corresponding to the two ways of assigning the electrons to the neutral atoms.* Examination of the different cases of two H atoms and two He atoms showed that by applying the *Pauli principle*, the selected eigenfunctions of the system changed or maintained their sign, respectively, when two electrons were swapped if the two electrons compared had the same, or different, *spin.* It was found that in the case of He there was only one solution, which yields about the right size of the He gas kinetic-radius, *due to the fact that 2 He atoms (and the same applied to all noble gases) could not differ in their spin* – in contrast to hydrogen (and all atoms with unfinished shells) – so that 2 He atoms had only one possible mode of behaving.

Then, in another paper, "Dirac, P. A. M. (1929). Quantum Mechanics of Many-Electron Systems", in which Dirac introduced the term *exchange interaction*, he noted that the general theory of quantum mechanics was now almost complete, and that the imperfections that still remained were in connection with the exact fitting in of the theory with *relativity ideas*. They only gave rise to difficulties when high-speed particles were involved and were therefore *of no importance in the consideration of atomic and molecular structure and ordinary chemical reactions*. The difficulty was only that the exact application of these laws led to equations much too complicated to be soluble. Dirac noted that it was desirable that approximate practical methods of applying quantum mechanics should be developed which could lead to an explanation of the main features of complex atomic systems without too much computation. Current *non-relativistic* quantum theory could not give an explanation of *multiplet structure* without an extraneous assumption of *large forces coupling the spin vectors of the electrons in an atom*. The explanation was provided by *exchange interaction* arising from electrons being indistinguishable one from another resulted in large *exchange energies* between electrons in different atoms. This had accounted for *homopolar valency bonds*, in which, for each *stationary state* of the atom there was one magnitude of the *total spin vector*. Dirac also noted that developments of the *theory of exchange* made by Heitler & London and Heisenberg made extensive use of *group theory*, which was a theory of certain quantities that did not satisfy the commutative law of multiplication and should thus form a part of quantum mechanics, and then translated the methods and results of *group theory* into the language of *quantum mechanics*. He demonstrated that *exchange interaction* equal to a constant *perturbation energy*, together with *coupling energy* between spin vectors, determined *energy* levels; and showed that in the first approximation the *exchange interaction* between the electrons could be replaced *by a coupling between their spins*, the energy of this coupling for each pair of electrons being equal to the scalar product of their *spin vectors* multiplied by a numerical coefficient given by the *exchange energy*.

PART III Susceptibility of materials to a magnetic field begins with another brilliant paper by Heisenberg, "(1928). Zur Theory of Ferromagnetismus (On the theory of ferromagnetism)", in which Heisenberg noted that empirical results exhibited *ferromagnetism* as an entirely similar state of affairs to what was previously observed in the spectrum of the helium atom. It seemed to follow from the levels in the helium atoms that *a powerful interaction prevailed between the spin directions of two electrons* that led to the splitting of the level structure into systems of singlets and triplets. He also noted that this was closely related to explaining ferromagnetic phenomena as being implied by the *exchange phenomenon* (resulting from *quantum entanglement*). Heisenberg proceeded to show that the *Coulomb interaction*, together with the *Pauli principle*, succeeded in evoking the same effects as the molecular field that Weiss had postulated, noting that it was only in recent times that mathematical methods were developed for the treatment of such a complicated problem in the important investigations of Wigner, Hund, and Heitler & London. Heisenberg assumed as a first approximation that the lattice separations were very large, and that every electron belonged to its own atom, and applied Heitler-London's calculations to the case of 2n electrons in a state of interaction, finding 2n electrons in 2n positionally different quantum cells. Due to their smallness, he was able to leave the magnetic interactions outside of consideration, and showed that *the spin moments of all electrons become partly parallel and partly anti-parallel as a result of the exchange processes*. By adding the fundamental Pauli principle to this, viz., that the eigenfunctions of the total system should be *anti-symmetric* in all electrons, he showed that an entirely well-defined *total magnetic moment* belonged to each level value of the perturbed system, and there were (2n)! levels in the unperturbed system.

> [An *eigenfunction* of a *linear operator* defined on some *function space* is any non-zero function in that space that, when acted upon by the linear operator, is only multiplied by some scaling factor called an *eigenvalue*. In quantum mechanics, the Schrödinger equation with the Hamiltonian operator can be solved by separation of variables if the Hamiltonian does not depend explicitly on time. In that case, the *wave function* leads to two differential equations, which are *eigenvalue* equations. The *eigenfunctions* of the Hamiltonian operator are *stationary states* of the quantum mechanical system, each with a corresponding *energy*. They represent allowable *energy states* of the system and may be constrained by boundary conditions.]

He then showed that a *statistical treatment of ferromagnetism was possible when all energy values had been calculated.* Heisenberg concluded that *an atom in a lattice could only be exchanged with its "neighbors"*; exchanges with atoms that lie further away that the "neighboring atoms" could be neglected. The *number of "neighbors" of an atom* was, e.g., 1 in a molecular lattice of diatomic molecules, 2 in a linear chain, 4 in a quadratic surface lattice, 6 in a simple cubic lattice, 8 in a cubic, space-centered lattice, and 12 in a cubic, face-centered lattice. By assuming a distribution of energy values about the mean had the approximate form of a Gaussian error curve, Heisenberg showed that small or negative values of the constant β [= zJ_0/kT)] result in *paramagnetism*; and that *ferromagnetism is only possible for lattice types for which an atom had at least eight neighbors*, which was the case for Fe, Co, Ni, whose lattices are all cubic, some of which were space-centered ($z = 8$) and some of which were face-centered ($z = 12$). He concluded that two conditions were necessary for the appearance of *ferromagnetism*: the crystal lattice must be a type such that *any atom had at least 8 neighbors*; and the *principal quantum number* of the electrons that were responsible for magnetism must be $n \geq 3$.

Extracts are then provided from John Van Vleck's 400-page book, published in 1932, which provides an extremely comprehensive account of the *non-relativistic* quantum theory of the *electric and magnetic susceptibities of materials*, including the quantum theory of *diamagnetism* and *paramagnetism* and detailed explanations of the *exchange effect* and Heisenberg's *theory of ferromagnetism*. Van Vleck noted in the Preface, "the analysis of experimental *magnetic susceptibilities* cannot be attempted until the quantum chapters, since *the numerical values of magnetic susceptibilities are inextricably connected with the quantization of angular momentum*". He also noted that *the theory of the electron spin may be presented in two ways,* viz. by means of what he calls a *semi-mechanical* (*Uhlenbeck-Goudsmit) model*) or by means of *Dirac's 'quantum theory of the electron'.* In the *semi-mechanical model*, matrix expressions for the *spin angular momentum* are written down by analogy with the *orbital angular momentum* matrices, with certain postulates regarding the occurrence of a half-quantum of *spin* per electron and *the ratio of spin magnetic moment to spin angular momentum* in order to bring the theory into line with experiments. He observed that the interaction of the *spin* with *external magnetic fields* was handled perfectly well by the *semi-mechanical model*, but Einstein's *special theory of relativity* was inconsistent with it. However, *the extension of Dirac's relativistic theory to many electron systems was at that time in a rather unsettled state* resulting in a nonvanishing probability of the mass of the electron changing sign, *an obvious absurdity*. So Van Vleck decided to present the quantitative aspects of the spin entirely with the aid of the older *non-relativistic semi-mechanical model*. In considering *diamagnetism*, [the property of materials that are repelled by a magnetic field; which creates an induced magnetic field in them in the opposite direction] he supposed that the atoms were in *singlet S states* [in which all electrons in the S orbital are paired], as otherwise there was an overwhelming *paramagnetism* [the form of magnetism whereby some materials are weakly attracted by an externally applied magnetic field, and form internal induced magnetic fields in the direction of the applied magnetic field]. In such *states* there cannot be even an instantaneous *magnetic moment* in the absence of external fields. Van Vleck distinguished other cases, including where the *inter-atomic forces* were so small that the magnetism could be calculated by treating the atoms of the solid to be free (exemplified by *rare earth salts*); where *the orbital and spin magnetic effects were both largely destroyed*, resulting in feeble *paramagnetism* (which most elements exhibit in the solid state); and solids in which *the Heisenberg exchange forces tended to align the spins parallel and so create ferromagnetism* (as for iron, nickel, and cobalt). Van Vleck noted that the *exchange forces had the effect of introducing a very strong coupling between the spins of paramagnetic atoms or ions*. *Diamagnetic* atoms or ions have no resultant spin and so do not give rise to any *exchange forces* tending to orient the *spins* of other atoms. Moreover, these *exchange forces* became of subordinate importance where the density of *paramagnetic* atoms or ions was low, because the great majority of the atoms are *diamagnetic*, and in most salts involving the iron group, which are consequently only *paramagnetic*. He then described how *entanglement* between electrons with the same *quantum spin states* creates an *exchange effect* which results in a strong coupling between their *spins*, and concluded by explaining how Heisenberg applied this in his theory of *ferromagnetism*.

Part IV. Quantum Entanglement addresses *quantum entanglement* at a distance, which Einstein referred to as "spooky action at a distance." The first paper I this section is what is known as the EPR paper, "Einstein, A., Podolsky, B. & Rosen, N. (May, 1935). Can Quantum-Mechanical Description of Physical Reality Be Considered Complete?", in which the authors argued that "in a complete theory there is an element corresponding to each element of reality. *A sufficient condition for the reality of a physical quantity is the possibility of predicting it with certainty, without disturbing the system.* In quantum mechanics in the case of two physical quantities described by non-commuting operators, *the knowledge of one precludes the knowledge of the other.* Then either (1) the description of reality given by the *wave function* in quantum mechanics is not complete or (2) these two quantities cannot have simultaneous reality. Consideration of the problem of making predictions concerning a system on the basis of measurements made on another system that had previously interacted with it leads to the result that if (1) is false then (2) is also false. One is thus led to conclude that the description of reality as given by a *wave function* is not complete." Einstein was wrong, again.

This is followed by a brilliant rebuttal by Erwin Schrodinger, "(October, 1935). Discussion of probability relations between separated systems", in which he introduced the word "*entanglement*". Schrodinger argued that when two systems, of which we know the states by their respective representatives, enter into temporary physical interaction due to known forces between them, and when the systems separate again, then they can no longer be described in the same way as before, viz. by endowing each of them with a representative of its own. "*I would not call that one but rather the characteristic trait of quantum mechanics*, the one that enforces its entire departure from classical lines of thought. *By the interaction the two representatives (or ψ-functions) have become entangled* ".

30 years later, in 1964, in "On the Einstein-Podolsky-Rosen Paradox', John Stewart Bell, a physicist from Northern Ireland, determined that *quantum mechanics was incompatible with local hidden-variable theories* given some basic assumptions about the nature of measurement, subsequently known as *Bell's theorem*. "Local" here referred to the principle of *locality*, the idea that a particle could only be influenced by its immediate surroundings, and that interactions mediated by physical fields could not propagate faster than the speed of light. "Hidden variables" were properties of quantum particles that were not included in quantum theory but nevertheless affect the outcome of experiments. Bell deduced that if measurements were performed independently on the two separated particles of an entangled pair, then the assumption that the outcomes depended upon hidden variables within each half implied a mathematical constraint on how the outcomes on the two measurements were correlated. This constraint was subsequently called a *Bell inequality*. Bell then showed that quantum physics predicts correlations that violate this inequality.

The first rudimentary experiment designed to test *Bell's theorem* was performed in 1972 by John Clauser and Stuart Freedman, and reported in "Freedman, S. J. & Clauser, J. F. (April, 1972). Experimental Test of Local Hidden-Variable Theories". "In the present work we measured the correlation in linear polarization of two *photons* emitted in an atomic cascade. The decaying atoms were viewed by two symmetrically placed optical systems, each consisting of two lenses, a wavelength filter, a rotatable and removable polarizer, and a single-photon detector. We made the following assumptions for any *local hidden-variable* theory: (1) The two *photons* propagate as separated localized particles. (2) A binary selection process occurs for each *photon* at each polarizer (transmission or no-transmission). This selection does not depend upon the orientation of the distant polarizer. In addition, we made the following assumption to allow a comparison of the *generalization* of *Bell's inequality* without experiment: (3) All *photons* incident on a detector have a probability of detection that is independent of whether or not the *photon* has passed through a polarizer. It has been shown by this generalization of *Bell's inequality* that the existence of local hidden variables imposes restrictions on this correlation in conflict with the predictions of quantum mechanics. Our data, in agreement with quantum mechanics, violated these restrictions to high statistical accuracy, *thus providing strong evidence against local hidden-variable theories*."

For this work in 1972, 50 years later, Clauser, jointly with Anton Zeilinger and Alain Aspect, was awarded the 2022 Nobel Prize in Physics, "for experiments with *entangled photons*, establishing the violation of Bell inequalities and pioneering quantum information science"… "One of the most remarkable traits of quantum mechanics is that it allows two or more particles to exist in what is called an *entangled state*. What happens to one of the particles in an *entangled* pair determines what happens to the other particle, even if they are far apart. In 1972, John Clauser conducted groundbreaking experiments using entangled light particles, photons. This and other experiments confirm that quantum mechanics is correct and pave the way for quantum computers, quantum networks and quantum encrypted communication."

The second edition includes descriptions of the *Schrödinger equation* and *quantum superposition* on pages 303-6. It also includes a paper [Experiments testing macroscopic quantum superpositions must be slow. (September, 2015). Mari, A., De Palma, D. & Giovannetti, V.], which considered a thought experiment where the preparation of a macroscopically massive or charged particle in a quantum superposition and the associated dynamics of a distant test particle apparently allowed for superluminal communication and concluded that macroscopic quantum substitutions must be slow.

I would like to acknowledge Wikipedia, in particular, which provided much of this material, as well as other referenced sources.

Trevor G. Underwood
18 SE 10th Ave
Fort Lauderdale, FL33301.

August 6, 2025.

PART I Electron spin.

Spin was first discovered in the context of the emission spectrum of alkali metals. Starting around 1910, many experiments on different atoms produced a collection of relationships involving quantum numbers for atomic energy levels partially summarized in Bohr's model for the atom. Transitions between levels obeyed selection rules and the rules were known to be correlated with even or odd atomic number. Additional information was known from changes to atomic spectra observed in strong magnetic fields, known as the *Zeeman effect*. In 1925, Wolfgang Pauli used this large collection of empirical observations to propose a new degree of freedom, introducing what he called a "two-valuedness not describable classically" associated with the electron in the outermost shell.

The physical interpretation of Pauli's "degree of freedom" was initially unknown. Ralph Kronig, one of Alfred Landé's assistants, suggested in early 1925 that it was produced by the self-rotation of the electron. When Pauli heard about the idea, he criticized it severely, noting that the electron's hypothetical surface would have to be moving faster than the speed of light in order for it to rotate quickly enough to produce the necessary angular momentum. *This would violate the theory of relativity.* Largely due to Pauli's criticism, Kronig decided not to publish his idea.

In the autumn of 1925, the same thought came to Dutch physicists George Uhlenbeck and Samuel Goudsmit at Leiden University. Under the advice of Paul Ehrenfest, they published their results. [Uhlenbeck, G. E. & Goudsmit, S. (November, 1925). Ersetzung der Hypothese vom unmechanischen Zwang durch eine Forderung bezuglich des inneren Verhaltens jedes einzelnen Elektrons. (Replacement of the hypothesis of unmechanical coercion by a requirement regarding the internal behavior of each individual electron.) See below.] The young physicists immediately regretted the publication: Hendrik Lorentz and Werner Heisenberg both pointed out problems with the concept of a spinning electron.

Pauli was especially unconvinced and continued to pursue his two-valued degree of freedom. This allowed him to formulate the *Pauli exclusion principle*, stating that no two electrons can have the same quantum state in the same quantum system.

Fortunately, by February 1926 Llewellyn Thomas managed to resolve a factor-of-two discrepancy between experimental results for the fine structure in the hydrogen spectrum and calculations based on Uhlenbeck and Goudsmit's (and Kronig's unpublished) model. This discrepancy was due to a *relativistic effect*, the difference between the electron's rotating rest frame and the nuclear rest frame; the effect is now known as Thomas precession. Thomas' result convinced Pauli that *electron spin* was the correct interpretation of his two-valued degree of freedom, while he continued to insist that the classical rotating charge model is invalid.

In 1927 Pauli formalized the theory of *spin* using the theory of quantum mechanics invented by Erwin Schrödinger and Werner Heisenberg. He pioneered the use of Pauli matrices as a representation of the spin operators and introduced a two-component spinor wave-function. *Pauli's theory of spin was non-relativistic.*

In 1928, Paul Dirac published his *relativistic* electron equation, using a four-component spinor (known as a "*Dirac spinor*") for the electron wave-function. *Relativistic spin* explained gyromagnetic anomaly. In 1940, Pauli provided a relativistic proof of the *spin–statistics theorem*, which states that fermions have half-integer spin, and bosons have integer spin. [Pauli, W. (1940). The Connection Between Spin and Statistics. See below.]

The first direct experimental evidence of the *electron spin* was the Stern–Gerlach experiment of 1922. However, the correct explanation of this experiment was only given in 1927. The original interpretation assumed the two spots observed in the experiment were due to *quantized orbital angular momentum*. However, in 1927 Ronald Fraser showed that Sodium atoms are isotropic with no *orbital angular momentum* and suggested that the observed magnetic properties were due to *electron spin*. In same year, Phipps and Taylor applied the *Stern-Gerlach technique* to hydrogen atoms; the ground state of hydrogen has zero *angular momentum* but the measurements again showed two peaks. Once the quantum theory became established, it became clear that *the original interpretation could not have been correct*: the possible values of *orbital angular momentum* along one axis is always an odd number, unlike the observations. Hydrogen atoms have a single electron with *two spin states* giving the two spots observed; silver atoms have closed shells which do not contribute to the *magnetic moment* and only the unmatched outer electron's *spin* responds to the field.

Spin is an intrinsic form of angular momentum carried by elementary particles, and thus by composite particles such as hadrons, atomic nuclei, and atoms. Spin is quantized, and accurate models for the interaction with spin require *relativistic* quantum mechanics or quantum field theory. [???]

The existence of electron spin angular momentum is inferred from experiments, such as the Stern–Gerlach experiment, in which silver atoms were observed to possess two possible discrete angular momenta despite having no orbital angular momentum.

Spin is described mathematically as a *vector* for some particles such as *photons*, and as *spinors* and *bispinors* for other particles such as *electrons*. *Spinors* and *bispinors* behave similarly to *vectors*: they have definite magnitudes and change under rotations; however, they use an unconventional "direction". *All elementary particles of a given kind have the same magnitude of spin angular momentum*, though its direction may change. These are indicated by assigning the particle a *spin quantum number*.

The SI units of *spin* are the same as *classical angular momentum* (i.e., N·m·s, J·s, or $kg \cdot m^2 \cdot s^{-1}$). *In quantum mechanics, angular momentum and spin angular momentum take discrete values proportional to the Planck constant*. In practice, *spin is usually given as a dimensionless spin quantum number by dividing the spin angular momentum by the reduced Planck constant ħ.* Often, the "*spin quantum number*" is simply called "*spin*".

Rotating charged mass

The earliest models for *electron spin* imagined *a rotating charged mass*, but *this model fails* when examined in detail: the required space distribution does not match limits on the electron radius: the required rotation speed exceeds the speed of light. *In the Standard Model, the fundamental particles are all considered "point-like"; they have their effects through the field that surrounds them.* Any model for *spin* based on mass rotation would need to be consistent with that model.

The study of the behavior of such "spin models" is a thriving area of research in condensed matter physics. For instance, the *Ising model* describes spins (dipoles) that have only two possible states, up and down, whereas in the Heisenberg model the spin vector is allowed to point in any direction. These models have many interesting properties, which have led to interesting results in the *theory of phase transitions*.

Pauli's "classically non-describable two-valuedness"

Wolfgang Pauli, a central figure in the history of *quantum spin*, initially rejected any idea that the "degree of freedom" he introduced to explain experimental observations was related to rotation. He called it "*classically non-describable two-valuedness*". Later he allowed that it is related to *angular momentum*, but insisted on considering *spin* an abstract property. *This approach allowed Pauli to develop a proof of his fundamental Pauli exclusion principle*, a proof now called the *spin-statistics theorem*. In retrospect this insistence and the style of his proof initiated the modern particle physics era, *where abstract quantum properties derived from symmetry properties dominate*. Concrete interpretation became secondary and optional.

Circulation of classical fields

The first classical model for spin proposed a small rigid particle rotating about an axis, as ordinary use of the word may suggest. Angular momentum can be computed from a classical field as well. By applying Frederik Belinfante's approach to calculating the angular momentum of a field, Hans C. Ohanian showed that "*spin is essentially a wave property...generated by a circulating flow of charge in the wave field of the electron*". This same concept of spin can be applied to gravity waves in water: "spin is generated by subwavelength circular motion of water particles".

Unlike classical wavefield circulation which allows continuous values of angular momentum, *quantum wavefields allow only discrete values*. Consequently, *energy transfer to or from spins states always occurs in fixed quantum steps*. Only a few steps are allowed: for many qualitative purposes the complexity of the spin quantum wavefields can be ignored and the system properties can be discussed in terms of "*integer*" or "*half-integer*" *spin* models.

Relation to orbital angular momentum

As the name suggests, spin was originally conceived as the rotation of a particle around some axis. Historically *orbital angular momentum* related to particle *orbits*. While the names based on mechanical models have survived, the physical explanation has not. *Quantization fundamentally alters the character of both spin and orbital angular momentum*.

Since elementary particles are point-like, self-rotation is not well-defined for them. However, *spin* implies that the phase of the particle depends on the angle as $e^{iS\theta}$, for rotation of angle θ around the axis parallel to the spin S. This is equivalent to the quantum-mechanical interpretation of *momentum* as *phase dependence in the position*, and of *orbital angular momentum* as *phase dependence in the angular position*.

For *fermions*, the picture is less clear. *Angular velocity* is equal by the *Ehrenfest theorem* to the *derivative of the Hamiltonian to its conjugate momentum*, which is the *total angular momentum operator* J = L + S. Therefore, if the *Hamiltonian* H is dependent upon the *spin* S, dH/dS is non-zero, and *the spin causes angular velocity, and hence actual rotation*, i.e. *a change in the phase-angle relation over time*. However, whether this holds for a free electron is ambiguous, since for an electron, S^2 is constant, and therefore *it is a matter of interpretation whether the Hamiltonian includes such a term*. Nevertheless, *spin* appears in the Dirac equation, and thus the *relativistic* Hamiltonian of the *electron*, treated as a Dirac field, can be interpreted as including a dependence in the spin S.

Quantum number

Spin obeys the mathematical laws of angular momentum quantization. The specific properties of *spin angular momenta* include:

- *Spin quantum numbers may take either half-integer or integer values.*
- Although the *direction* of its *spin* can be changed, *the magnitude of the spin of an elementary particle cannot be changed.*
- *The spin of a charged particle is associated with a magnetic dipole moment with a g-factor that differs from 1.* (In the classical context, this would imply the internal charge and mass distributions differing for a rotating object.)

The conventional definition of the *spin quantum number* is s = n/2, where n can be any non-negative integer. Hence the allowed values of s are 0, 1/2, 1, 3/2, 2, etc. The value of s for an *elementary particle* depends only on the type of particle and cannot be altered in any known way (in contrast to the spin direction described below). The *spin angular momentum* S of any physical system is quantized. The allowed values of S are

$$S = \hbar\sqrt{s(s+1)} = h/2\pi \sqrt{n/2\ (n+2)/2} = h/4\pi \sqrt{n(n+2)},$$

where h is the Planck constant, and $\hbar = h/2\pi$ is the reduced Planck constant. In contrast, *orbital angular momentum* can only take on integer values of s; i.e., even-numbered values of n.

Fermions and bosons

Those particles with half-integer spins, such as 1/2, 3/2, 5/2, are known as fermions, while those particles with integer spins, such as 0, 1, 2, are known as bosons. The two families of particles obey different rules and broadly have different roles in the world around us. A key distinction between the two families is that *fermions obey the Pauli exclusion principle*: that is, *there cannot be two identical fermions simultaneously having the same quantum numbers* (meaning, roughly, having the same position, velocity and spin direction). *Fermions obey the rules of Fermi–Dirac statistics.* In contrast, *bosons obey the rules of Bose–Einstein statistics* and have *no such restriction*, so they may "bunch together" in identical states. Also, *composite particles can have spins different from their component particles.* For example, a helium-4 atom in the ground state has spin 0 and behaves like a boson, even though the quarks and electrons which make it up are all fermions.

This has some profound consequences:

- *Quarks and leptons (including electrons and neutrinos), which make up what is classically known as matter, are all fermions with spin 1/2.* The common idea that "matter takes up space" actually comes from the *Pauli exclusion principle* acting on these particles to prevent the fermions from being in the same quantum state. Further compaction would require electrons to occupy the same energy states, and therefore a kind of pressure (sometimes known as degeneracy pressure of electrons) acts to resist the fermions being overly close. Elementary fermions with other spins (3/2, 5/2, etc.) are not known to exist.

- *Elementary particles which are thought of as carrying forces are all bosons with spin 1*. They include the *photon*, which carries the *electromagnetic force*, the *gluon* (*strong force*), and the *W and Z bosons* (*weak force*). The ability of bosons to occupy the same quantum state is used in the *laser*, which aligns many photons having the same quantum number (the same direction and frequency), *superfluid liquid helium* resulting from helium-4 atoms being bosons, and *superconductivity*, where pairs of electrons (which individually are fermions) act as single composite bosons.
- *Elementary bosons with other spins (0, 2, 3, etc.) were not historically known to exist*, although they have received considerable theoretical treatment and are well established within their respective mainstream theories. In particular, theoreticians have proposed the *graviton* (predicted to exist by some quantum gravity theories) with *spin* 2, and the *Higgs boson* (explaining *electroweak symmetry* breaking) with *spin* 0. Since 2013, the Higgs boson with spin 0 has been considered proven to exist. *It is the first scalar elementary particle (spin 0) known to exist in nature.*
- *Atomic nuclei* have *nuclear spin* which may be either half-integer or integer, so that the nuclei may be either fermions or bosons.

Pauli's spin–statistics theorem

Pauli's *spin–statistics theorem* splits particles into two groups: *bosons* and *fermions*, where *bosons obey Bose–Einstein statistics*, and *fermions obey Fermi–Dirac statistics* (and therefore the *Pauli exclusion principle*). Specifically, *the theorem requires that particles with half-integer spins obey the Pauli exclusion principle while particles with integer spin do not*. As an example, electrons have *half-integer spin* and are *fermions* that obey the Pauli exclusion principle, while *photons have integer spi*n and do not. The theorem which was derived by Wolfgang Pauli in 1940 relies on both quantum mechanics and the *theory of special relativity*. [Pauli, W. (1940). The Connection Between Spin and Statistics. *Phys. Rev.*, 58, 8, 716–22; doi:10.1103/PhysRev.58.716.]

Magnetic moments

Particles with spin can possess a magnetic dipole moment, just like a rotating electrically charged body in classical electrodynamics. These *magnetic moments* can be experimentally observed in several ways, e.g. by the deflection of particles by inhomogeneous magnetic fields in a *Stern–Gerlach experiment*, or by measuring the *magnetic fields* generated by the particles themselves.

The *intrinsic magnetic moment* μ of a spin-1/2 particle with *charge* q, *mass* m, and *spin angular momentum* S, is

$$\mu = g_s q/2m\ S,$$

where the dimensionless quantity g_s is called the *spin g-factor*. For exclusively orbital rotations it would be 1 (assuming that the mass and the charge occupy spheres of equal radius).

However, based on experiments on rotation by magnetization (the Einstein-Richardson-de-Haas effect) as well as on the converse magnetization by rotation (Barnett effect) and the anomalous Zeeman effect, *the ratio of spin magnetic moment M_s to spin angular momentum P_s* has twice the classical value – e/2mc for *the ratio of orbital magnetic moment to orbital angular momentum*, so that

$$M_s/P_s = -\ e/mc.$$

The *electron*, being a charged elementary particle, possesses a nonzero *magnetic moment*. One of the triumphs of the *theory of quantum electrodynamics* is its accurate prediction of the *electron g-factor*, which has been experimentally determined to have the value −2.00231930436092(36), with the digits in parentheses denoting measurement uncertainty in the last two digits at one standard deviation. *The value of –2 arises from the Dirac equation, a fundamental equation connecting the electron's spin with its electromagnetic properties*, and *the deviation from −2 arises from the electron's interaction with the surrounding electromagnetic field, including its own field.*

Composite particles also possess *magnetic moments* associated with their *spin*. In particular, *the neutron possesses a non-zero magnetic moment despite being electrically neutral*. This fact was an early indication that *the neutron is not an elementary particle. In fact, it is made up of quarks, which are electrically charged particles. The magnetic moment of the neutron comes from the spins of the individual quarks and their orbital motions.*

Neutrinos are both elementary and electrically neutral. The *minimally extended Standard Model* that takes into account non-zero neutrino masses *predicts neutrino magnetic moments* of:

$$\mu_\nu \approx 3 \times 10^{-19}\ \mu_B\ m_\nu c^2/\text{eV},$$

where the μ_ν are the *neutrino magnetic moments*, m_ν are the *neutrino masses*, and μ_B is the *Bohr magneton*. New physics above the *electroweak scale* could, however, lead to significantly higher *neutrino magnetic moments*. It can be shown in a model-independent way that *neutrino magnetic moments* larger than about 10^{-14} μ_B are "unnatural" because they would also lead to large radiative contributions to the *neutrino mass*. Since the *neutrino masses* are known to be at most about 1 eV/c^2, fine-tuning would be necessary in order to prevent large contributions to the *neutrino mass* via radiative corrections. *The measurement of neutrino magnetic moments is an active area of research*. Experimental results have put the neutrino *magnetic moment* at less than 1.2×10^{-10} times the electron's *magnetic moment*.

On the other hand, *elementary particles with spin but without electric charge, such as a photon* or a Z boson, *do not have a magnetic moment*.

Curie temperature and loss of alignment

In ordinary materials, the magnetic dipole moments of individual atoms produce magnetic fields that cancel one another, because each dipole points in a random direction, with the overall average being very near zero. *Ferromagnetic materials* below their Curie temperature [named for Pierre Curie], however, *exhibit magnetic domains in which the atomic dipole moments spontaneously align locally, producing a macroscopic, non-zero magnetic field* from the domain. These are the ordinary "magnets" with which we are all familiar.

In paramagnetic materials, the magnetic dipole moments of individual atoms will partially align with an externally applied magnetic field. In diamagnetic materials, on the other hand, the magnetic dipole moments of individual atoms align oppositely to any externally applied magnetic field, even if it requires energy to do so.

Spin projection quantum number and multiplicity

In classical mechanics, the angular momentum of a particle possesses not only a magnitude (how fast the body is rotating), but also a direction (either up or down on the axis of rotation of the particle). Quantum-mechanical *spin* also contains information about direction, but in a more subtle form. *Quantum mechanics states that the component of angular momentum for a spin-s particle measured along any direction can only take on the values*

$$S_i = \hbar s_i, \qquad s_i \in \{-s, -(s-1), \ldots, s-1, s\},$$

where S_i is the *spin component* along the *i*-th axis (either x, y, or z), s_i is the *spin projection quantum number* along the *i*-th axis, and s is the *principal spin quantum number*.

Conventionally the direction chosen is the z axis:

$$S_z = \hbar s_z, \qquad s_z \in \{-s, -(s-1), \ldots, s-1, s\},$$

where S_z is the *spin component* along the z axis, s_z is the *spin projection quantum number* along the z axis.

One can see that there are 2s + 1 possible values of s_z. The number "2s + 1" is the *multiplicity of the spin system*. For example, there are only two possible values for a *spin-1/2 particle*: $s_z = +\frac{1}{2}$ and $s_z = -\frac{1}{2}$. *These correspond to quantum states in which the spin component is pointing in the +z or −z directions respectively, and are often referred to as "spin up" and "spin down"*. For a spin-3/2 particle, like a delta baryon, the possible values are +3/2, +1/2, −1/2, −3/2.

Spin vector

For a given *quantum state*, one could think of a *spin vector* $\langle S \rangle$ whose components are the expectation values of the *spin components* along each axis, i.e., $\langle S \rangle = [\langle S_x \rangle, \langle S_y \rangle, \langle S_z \rangle]$. This vector then would describe the "*direction*" in which the spin is pointing, corresponding to the classical concept of the axis of rotation. *It turns out that the spin vector is not very useful in actual quantum-mechanical calculations, because it cannot be measured directly*: s_x, s_y and s_z cannot possess simultaneous definite values, *because of a quantum uncertainty relation between them*. However, for statistically large collections of particles that have been placed in the same pure *quantum state*, such as through the use of a *Stern–Gerlach apparatus*, the *spin vector* does have a well-defined experimental meaning: It specifies the direction in ordinary space in which a subsequent detector must be oriented in order to achieve the maximum possible probability (100%) of detecting every particle in the collection. For spin-1/2 particles, this probability drops off smoothly as the angle between the *spin vector* and the detector increases, until at an angle of 180°—that is, for detectors oriented in the opposite direction to the *spin vector*—the expectation of detecting particles from the collection reaches a minimum of 0%.

As a qualitative concept, the *spin vector* is often handy because it is easy to picture classically. For instance, *quantum-mechanical spin* can exhibit phenomena analogous to classical gyroscopic effects. For example, one can exert a kind of "torque" on an *electron* by putting it in a *magnetic field* (the field acts upon the electron's *intrinsic magnetic dipole moment*). The result is that *the spin vector undergoes precession, just like a classical gyroscope*. This phenomenon is known as *electron spin resonance* (ESR). The equivalent behavior of protons in atomic nuclei is used in *nuclear magnetic resonance* (NMR) spectroscopy and imaging.

Mathematically, *quantum-mechanical spin states* are described by vector-like objects known as *spinors*. There are subtle differences between the behavior of spinors and vectors under coordinate rotations. For example, *rotating a spin-1/2 particle by 360° does not bring it back to the same quantum state, but to the state with the opposite quantum phase*; this is detectable, in principle, with interference experiments. To return the particle to its exact original state, one needs a 720° rotation. *A spin-zero particle can only have a single quantum state, even after torque is applied. Rotating a spin-2 particle 180° can bring it back to the same quantum state, and a spin-4 particle should be rotated 90° to bring it back to the same quantum state.* The spin-2 particle can be analogous to a straight stick that looks the same even after it is rotated 180°, and a spin-0 particle can be imagined as sphere, which looks the same after whatever angle it is turned through.

Mathematical formulation

Spin obeys commutation relations analogous to those of the orbital angular momentum:

$$[S^\wedge_j, S^\wedge_k] = i\hbar\varepsilon_{jkl} S^\wedge_l,$$

where ε_{jkl} is the *Levi-Civita symbol*. It follows (as with *angular momentum*) that the eigenvectors of $S^{\wedge 2}$ and $S^\wedge_z$ (expressed as kets in the total S basis) are

$$S^{\wedge 2} \mid s, m_s \rangle = \hbar^2 s(s + 1) \mid s, m_s \rangle,$$
$$S^\wedge z \mid s, m_s \rangle = \hbar m_s \mid s, m_s \rangle.$$

The *spin raising and lowering operators* acting on these *eigenvectors* give

$$S^\wedge_\pm \mid s, m_s \rangle = \hbar \sqrt{\{s(s + 1) - m_s(m_s \pm 1)} \mid s, m_s \pm 1 \rangle,$$

where
$$S^\wedge_\pm = S^\wedge_x \pm i S^\wedge_y.$$

But unlike *orbital angular momentum*, the eigenvectors are not spherical harmonics. They are not functions of θ and φ. There is also no reason to exclude half-integer values of s and m_s.

All quantum-mechanical particles possess an intrinsic spin s (though this value may be equal to zero). The projection of the spin s on any axis is quantized in units of the reduced Planck constant, such that the *state function* of the particle is, say, not $\psi = \psi(\boldsymbol{r})$, but $\psi = \psi(r, s_z)$, where s_z can take only the values of the following discrete set:

$$s_z \in \{- s\hbar, - (s - 1)\hbar, ..., + (s - 1)\hbar, + s\hbar\}.$$

One distinguishes bosons (*integer spin*) and fermions (*half-integer spin*). The *total angular momentum* conserved in interaction processes is then the sum of the *orbital angular momentum* and the *spin*.

Pauli matrices

The quantum-mechanical *operators* associated with spin-1/2 *observables* are

$$S^{\wedge} = \hbar/2\ \sigma,$$

where in Cartesian components

$$S_x = \hbar/2\ \sigma_x, \quad S_y = \hbar/2\ \sigma_y, \quad S_z = \hbar/2\ \sigma_z.$$

For the special case of *spin-1/2 particles*, σ_x, σ_y and σ_z are the three Pauli matrices:

$$\sigma_x = \begin{pmatrix} 0 & 1 \\ 1 & 0 \end{pmatrix}, \quad \sigma_y = \begin{pmatrix} 0 & -i \\ i & 0 \end{pmatrix}, \quad \sigma_z = \begin{pmatrix} 1 & 0 \\ 0 & -1 \end{pmatrix}.$$

Pauli exclusion principle

The Pauli exclusion principle states that the *wavefunction* $\psi(\boldsymbol{r}_1, \sigma_1, \ldots, \boldsymbol{r}_N, \sigma_N)$ for a system of N identical particles having *spin* s must change upon interchanges of any two of the N particles as

$$\psi(\ldots, \boldsymbol{r}_i, \sigma_i, \ldots, \boldsymbol{r}_j, \sigma_j, \ldots) = (-1)^{2s}\, \psi(\ldots, \boldsymbol{r}_j, \sigma_j, \ldots, \boldsymbol{r}_i, \sigma_i, \ldots).$$

Thus, for *bosons* the prefactor $(-1)^{2s}$ will reduce to +1, for *fermions* to −1. This permutation postulate for N-particle *state functions* has most important consequences in daily life, e.g. the *periodic table* of the *chemical elements*.

Rotations

As described above, quantum mechanics states that *components* of *angular momentum* measured along any direction can only take a number of discrete values. *The most convenient quantum-mechanical description of particle's spin is therefore with a set of complex numbers corresponding to amplitudes of finding a given value of projection of its intrinsic angular momentum on a given axis*. For instance, for a spin-1/2 particle, we would need two numbers $a \pm 1/2$, giving amplitudes of finding it with projection of angular momentum equal to + ħ/2 and − ħ/2, satisfying the requirement

$$|\, a_{+1/2}\,|^2 + |\, a_{-1/2}\,|^2 = 1.$$

For a generic particle with *spin* s, we would need 2s + 1 such parameters. Since these numbers depend on the choice of the axis, they transform into each other non-trivially when this axis is rotated. It is clear that the transformation law must be linear, so we can represent it by associating a matrix with each rotation, and the product of two transformation matrices corresponding to rotations A and B must be equal (up to phase) to the matrix representing rotation AB. Further, rotations preserve the quantum-mechanical inner product, and so should transformation matrices.

Mathematically speaking, these matrices furnish a unitary projective representation of the rotation group SO(3). Each such representation corresponds to a representation of the covering group of SO(3), which is SU(2). There is one n-dimensional irreducible representation of SU(2) for each dimension, though this representation is n-dimensional real for odd n and n-dimensional complex for even n (hence of real dimension 2n).

Recalling that a *generic spin state* can be written as a ***superposition*** *of states* with definite m, we see that if s is an integer, the values of m are all integers, and this matrix corresponds to the identity operator.

> [***Quantum superposition*** is a fundamental principle of quantum mechanics that ***states that linear combinations of solutions to the Schrödinger equation are also solutions of the Schrödinger equation.*** This follows from the fact that the Schrödinger equation is a linear differential equation in time and position. More precisely, the state of a system is given by a linear combination of all the eigenfunctions of the Schrödinger equation governing that system.]

However, if s is a half-integer, the values of m are also all half-integers, giving $(-1)^{2m} = -1$ for all m, and hence upon rotation by 2π the state picks up a minus sign. *This fact is a crucial element of the proof of the spin–statistics theorem.*

Lorentz transformations

We could try the same approach to determine the behavior of spin under general Lorentz transformations, but we would immediately discover a major obstacle. Unlike SO(3), the group of Lorentz transformations SO(3,1) is *non-compact* and therefore does not have any faithful, unitary, finite-dimensional representations.

In case of spin-1/2 particles, it is possible to find a construction that includes both a finite-dimensional representation and a scalar product that is preserved by this representation. We associate a 4-component *Dirac spinor* ψ with each particle. These *spinors* transform under Lorentz transformations according to the law

$$\psi' = \exp(1/8\omega_{\mu\nu}[\gamma_\mu, \gamma_\nu])\psi,$$

where γ_ν are *gamma matrices*, and $\omega_{\mu\nu}$ is an *antisymmetric* 4×4 matrix parametrizing the transformation. It can be shown that the *scalar product*

$$\langle\psi|\phi\rangle = \psi^{-}\phi = \psi^{\dagger}\gamma_0\phi$$

is preserved. *It is not, however, positive-definite, so the representation is not unitary.*

Measurement of spin along the x, y, or z axes

Each of the *(Hermitian) Pauli matrices* of spin-1/2 particles has two *eigenvalues*, +1 and −1. By the postulates of quantum mechanics, an experiment designed to measure the *electron spin* on the x, y, or z axis can only yield an *eigenvalue* of the corresponding *spin operator* (S_x, S_y or S_z) on that axis, i.e. $\hbar/2$ or $-\hbar/2$. The *quantum state* of a particle (with respect to spin), can be represented by a two-component *spinor*:

$$\psi = \begin{pmatrix} a + bi \\ c + di \end{pmatrix}.$$

When the *spin* of this particle is measured with respect to a given axis (in this example, the x axis), the probability that its *spin* will be measured as $\hbar/2$ is just $|\langle\psi_{x+} \mid \psi\rangle|^2$. Correspondingly, the probability that its *spin* will be measured as $-\hbar/2$ is just $|\langle\psi_{x-} \mid \psi\rangle|^2$. *Following the measurement, the spin state of the particle collapses into the corresponding eigenstate*. As a result, *if the particle's spin along a given axis has been measured to have a given eigenvalue, all measurements will yield the same eigenvalue* (since $|\langle\psi_{x+} \mid \psi_{x+}\rangle|^2 = 1$, etc.), *provided that no measurements of the spin are made along other axes*.

Compatibility of spin measurements

Since the Pauli matrices do not commute, measurements of spin along the different axes are incompatible. This means that if, for example, we know the *spin* along the x axis, and we then measure the *spin* along the y axis, we have invalidated our previous knowledge of the x axis *spin*. This can be seen from the property of the *eigenvectors* (i.e. *eigenstates*) of the *Pauli matrices* that

$$|\langle\psi_{x\pm} \mid \psi_{y\pm}\rangle|^2 = |\langle\psi_{x\pm} \mid \psi_{z\pm}\rangle|^2 = |\langle\psi_{y\pm} \mid \psi_{z\pm}\rangle|^2 = \tfrac{1}{2}.$$

So *when physicists measure the spin of a particle along the x axis as, for example, $\hbar/2$, the particle's spin state collapses into the eigenstate* $|\psi_{x+}\rangle$. When we then subsequently measure the particle's *spin* along the y axis, the *spin state* will now collapse into either

$|\psi_{y+}\rangle$ or $|\psi_{y-}\rangle$, each with probability ½. Let us say, in our example, that we measure $-\hbar/2$. When we now return to measure the particle's *spin* along the x axis again, the probabilities that we will measure $\hbar/2$ or $-\hbar/2$ are each ½ (i.e. they are $|\langle\psi_{x+}|\psi_{y-}\rangle|^2$ and $|\langle\psi_{x-}|\psi_{y-}\rangle|^2$ respectively). *This implies that the original measurement of the spin along the x axis is no longer valid*, since the *spin* along the x axis will now be measured to have either *eigenvalue* with equal probability.

Parity

In tables of the *spin quantum number* s for *nuclei* or *particles*, the *spin* is often followed by a "+" or "−". This refers to the *parity* with "+" for *even parity* (wave function unchanged by spatial inversion) and "−" for *odd parity* (wave function negated by spatial inversion). For example, see the isotopes of bismuth, in which the list of isotopes includes the column *nuclear spin* and *parity*. For Bi-209, the longest-lived isotope, the entry 9/2− means that the *nuclear spin* is 9/2 and the *parity* is odd.

Measuring spin

The *nuclear spin of atoms* can be determined by sophisticated improvements to the original Stern-Gerlach experiment. *A single-energy (monochromatic) molecular beam of atoms in an inhomogeneous magnetic field will split into beams representing each possible spin quantum state*. For an atom with *electronic spin* S and *nuclear spin* I, there are $(2S + 1)(2I + 1)$ *spin states*. For example, neutral Na atoms, which have $S = 1/2$, were passed through a series of inhomogeneous magnetic fields that selected one of the two *electronic spin states* and separated the *nuclear spin states*, from which four beams were observed. Thus, the *nuclear spin* for ^{23}Na atoms was found to be $I = 3/2$.

Applications

Spin has important theoretical implications and practical applications. Well-established direct applications of *spin* include:

- *Nuclear magnetic resonance* (NMR) *spectroscopy* in chemistry;
- *Electron spin resonance* (ESR or EPR) *spectroscopy* in chemistry and physics;
- *Magnetic resonance imaging* (MRI) in medicine, a type of applied NMR, which relies on proton spin density;
- *Giant magnetoresistive* (GMR) *drive-head technology* in modern hard disks.

Electron spin plays an important role in magnetism, with applications for instance in computer memories. The manipulation of *nuclear spin* by radio-frequency waves (*nuclear magnetic resonance*) is important in chemical spectroscopy and medical imaging.

Spin–orbit coupling leads to the fine structure of atomic spectra, which is used in atomic clocks and in the modern definition of the second. Precise measurements of the *g-factor* of the *electron* have played an important role in the development and verification of *quantum electrodynamics*. *Photon spin* is associated with the *polarization of light* (photon polarization).

An emerging application of *spin* is as a binary information carrier in *spin transistors*. The original concept, proposed in 1990, is known as Datta–Das *spin transistor*. Electronics based on *spin transistors* are referred to as *spintronics*. The manipulation of *spin* in dilute magnetic semiconductor materials, such as metal-doped ZnO or TiO_2 imparts a further degree of freedom and has the potential to facilitate the fabrication of more efficient electronics.

There are many indirect applications and manifestations of spin and the associated *Pauli exclusion principle*, starting with the *periodic table of chemistry*.

Arthur Holly Compton (September 10, 1892–March 15, 1962).

Compton was an American physicist who won the Nobel Prize in Physics in 1927 for his 1923 discovery of the *Compton effect*, which demonstrated the particle nature of electromagnetic radiation. It was a sensational discovery at the time: the wave nature of light had been well-demonstrated, but the idea that light had both wave and particle properties was not easily accepted. He is also known for his leadership over the Metallurgical Laboratory at the University of Chicago during the Manhattan Project, and served as chancellor of Washington University in St. Louis from 1945 to 1953.

Compton was born on September 10, 1892, in Wooster, Ohio, the son of Elias and Otelia Catherine (née Augspurger) Compton, who was named American Mother of the Year in 1939. They were an academic family. Elias was dean of the University of Wooster (later the College of Wooster), which Arthur also attended. Arthur's eldest brother, Karl, who also attended Wooster, earned a Doctor of Philosophy (PhD) degree in physics from Princeton University in 1912, and was president of the Massachusetts Institute of Technology from 1930 to 1948. His second brother Wilson likewise attended Wooster, earned his PhD in economics from Princeton in 1916 and was president of the State College of Washington, later Washington State University from 1944 to 1951.

Compton was initially interested in astronomy, and took a photograph of Halley's Comet in 1910. Around 1913, he described an experiment where an examination of the motion of water in a circular tube demonstrated the rotation of the earth, a device now known as the Compton generator. That year, he graduated from Wooster with a Bachelor of Science degree and entered Princeton, where he received his Master of Arts degree in 1914. Compton then studied for his PhD in physics under the supervision of Hereward L. Cooke, writing his dissertation on *The Intensity of X-Ray Reflection, and the Distribution of the Electrons in Atoms*.

When Compton earned his PhD in 1916, he, Karl and Wilson became the first group of three brothers to earn PhDs from Princeton. Later, they would become the first such trio to simultaneously head American colleges. Their sister Mary married a missionary, C. Herbert Rice, who became the principal of Forman Christian College in Lahore. In June 1916, Compton married Betty Charity McCloskey, a Wooster classmate and fellow graduate. They had two sons, Arthur Alan Compton and John Joseph Compton.

Compton spent a year as a physics instructor at the University of Minnesota in 1916–17, then two years as a research engineer with the Westinghouse Lamp Company in Pittsburgh, where he worked on the development of the sodium-vapor lamp. During World War I he developed aircraft instrumentation for the Signal Corps.

In 1919, Compton was awarded one of the first two National Research Council Fellowships that allowed students to study abroad. *He chose to go to the University of Cambridge's Cavendish Laboratory in England.* Working with George Paget Thomson, the son of J. J. Thomson, *Compton studied the scattering and absorption of gamma rays.* He observed that the scattered rays were more easily absorbed than the original source.

Compton was greatly impressed by the Cavendish scientists, especially Ernest Rutherford, Charles Galton Darwin and Arthur Eddington, and he ultimately named his second son after J. J. Thomson.

Returning to the United States, Compton was appointed Wayman Crow Professor of Physics, and head of the Department of Physics at Washington University in St. Louis in 1920. *In 1922, he found that X-ray quanta scattered by free electrons had longer wavelengths and, in accordance with Planck's relation, less energy than the incoming X-rays, the surplus energy having been transferred to the electrons. This discovery, known as the "Compton effect" or "Compton scattering", demonstrated the particle concept of electromagnetic radiation.*

In 1923, Compton published a paper in the *Physical Review* that explained the X-ray shift by attributing particle-like momentum to photons, something Einstein had invoked for his 1905 Nobel Prize–winning explanation of the photo-electric effect. First postulated by Max Planck in 1900, these were conceptualized as elements of light "quantized" by containing a specific amount of energy depending only on the frequency of the light. In his paper, Compton derived the mathematical relationship between the shift in wavelength and the scattering angle of the X-rays *by assuming that each scattered X-ray photon interacted with only one electron*. His paper concludes by reporting on experiments that verified his derived relation:

$$\lambda_\theta - \lambda_0 = h/mc\,(1 - \cos\theta)$$

where

λ_0 is the initial wavelength,
λ_θ is the wavelength after scattering,
h is the Planck constant,
m is the electron rest mass,
c is the speed of light, and
θ is the scattering angle.

[This formula is not included in his 1923 paper but can be derived from the formula that were. See paper below.]

The quantity h/mc is known as the Compton wavelength of the electron; it is equal to 2.43×10^{-12} m. The wavelength shift $\lambda_\theta - \lambda_0$ lies between zero (for $\theta = 0°$) and twice the Compton wavelength of the electron (for $\theta = 180°$). He found that some X-rays experienced no wavelength shift despite being scattered through large angles; in each of these cases the photon failed to eject an electron. In these cases, the magnitude of the shift is related not to the Compton wavelength of the electron, but to the Compton wavelength of the entire atom, which can be upwards of 10,000 times smaller.

"When I presented my results at a meeting of the American Physical Society in 1923", Compton later recalled, "it initiated the most hotly contested scientific controversy that I have ever known." The wave nature of light had been well demonstrated, and the idea that it could have a dual nature was not easily accepted. It was particularly telling that diffraction in a crystal lattice could only be explained with reference to its wave nature. It earned Compton the Nobel Prize in Physics in 1927. Compton and Alfred W. Simon developed the method for observing at the same instant individual scattered X-ray photons and the recoil electrons. In Germany, Walther Bothe and Hans Geiger independently developed a similar method.

In 1923, Compton moved to the University of Chicago as professor of physics, a position he would occupy for the next 22 years. In 1925, he demonstrated that the scattering of 130,000-volt X-rays from the first sixteen elements in the periodic table (hydrogen through sulfur) were polarized, a result predicted by J. J. Thomson. He used X-rays to investigate ferromagnetism, concluding that it was a result of the alignment of electron spins.

Compton's first book, *X-Rays and Electrons*, was published in 1926. In it he showed how to calculate the densities of diffracting materials from their X-ray diffraction patterns. He revised his book with the help of Samuel K. Allison to produce *X-Rays in Theory and Experiment* (1935). This work remained a standard reference for the next three decades.

In 1926, he became a consultant for the Lamp Department at General Electric. In 1934, he returned to England as Eastman visiting professor at Oxford University. While there, General Electric asked him to report on activities at General Electric Company plc's research laboratory at Wembley. Compton was intrigued by the possibilities of the research there into fluorescent lamps. His report prompted a research program in America that developed it.

By the early 1930s, Compton had become interested in cosmic rays. At the time, their existence was known but their origin and nature remained speculative. Their presence could be detected using a spherical "bomb" containing compressed air or argon gas and measuring its electrical conductivity. Trips to Europe, India, Mexico, Peru and Australia gave Compton the opportunity to measure cosmic rays at different altitudes and latitudes. Along with other groups who made observations around the globe, they found that *cosmic rays were 15% more intense at the poles than at the equator*. Compton attributed this to the effect of cosmic rays being made up principally of charged particles, rather than photons as Robert Millikan had suggested, with the latitude effect being due to Earth's magnetic field.

During World War II, Compton was a key figure in the Manhattan Project that developed the first nuclear weapons. His reports were important in launching the project. In April 1941, Vannevar Bush, head of the wartime National Defense Research Committee (NDRC), created a special committee headed by Compton to report on the NDRC uranium program. Compton's report, which was submitted in May 1941, foresaw the prospects of developing radiological weapons, nuclear propulsion for ships, and nuclear weapons using uranium-235 or the recently discovered plutonium. In October he wrote another report on the practicality of an atomic bomb. For this report, he worked with Enrico Fermi on calculations of the critical mass of uranium-235, conservatively estimating it to be between 20 kilograms (44 lb) and 2 tonnes (2.0 long tons; 2.2 short tons). He also discussed the prospects for uranium enrichment with Harold Urey, spoke with Eugene Wigner about how plutonium might be produced in a nuclear reactor, and with Robert Serber about how the plutonium produced in a reactor might be separated from uranium. His report, submitted in November, stated that a bomb was feasible, although he was more conservative about its destructive power than Mark Oliphant and his British colleagues.

The final draft of Compton's November report made no mention of using plutonium, but after discussing the latest research with Ernest Lawrence, Compton became convinced that a plutonium bomb was also feasible. In December, Compton was placed in charge of the plutonium project. He hoped to achieve a controlled chain reaction by January 1943, and to have a bomb by January 1945. In 1942, he had the research groups working on plutonium and nuclear reactor design at Columbia University, Princeton University and the University of California, Berkeley, concentrated together as the Metallurgical Laboratory in Chicago. Its objectives were to produce reactors to convert uranium to plutonium, to find ways to chemically separate the plutonium from the uranium, and to design and build an atomic bomb.

In June 1942, the United States Army Corps of Engineers assumed control of the nuclear weapons program and Compton's Metallurgical Laboratory became part of the Manhattan Project. That month, Compton gave Robert Oppenheimer responsibility for bomb design. It fell to Compton to decide which of the different types of reactor designs that the Metallurgical Laboratory scientists had devised should be pursued, even though a successful reactor had not yet been built. When labor disputes delayed construction of the Metallurgical Laboratory's new home in the Red Gate Woods, Compton decided to build Chicago Pile-1, the first nuclear reactor, under the stands at Stagg Field. Under Fermi's direction, it went critical on December 2, 1942. The Metallurgical Laboratory was also responsible for the design and operation of the X-10 Graphite Reactor at Oak Ridge, Tennessee.

A major crisis for the plutonium program occurred in July 1943, when Emilio Segrè's group confirmed that plutonium created in the X-10 Graphite Reactor at Oak Ridge contained high levels of plutonium-240. Its spontaneous fission ruled out the use of plutonium in a gun-type nuclear weapon. Oppenheimer's Los Alamos Laboratory met the challenge by designing and building an implosion-type nuclear weapon.

Compton was at the Hanford site in September 1944 to watch the first reactor being brought online. The first batch of uranium slugs was fed into Reactor B at Hanford in November 1944, and shipments of plutonium to Los Alamos began in February 1945. Throughout the war, Compton would remain a prominent scientific adviser and administrator. In 1945, he served, along with Lawrence, Oppenheimer, and Fermi, on the Scientific Panel that recommended military use of the atomic bomb against Japan. He was awarded the Medal for Merit for his services to the Manhattan Project.

After the war ended, Compton resigned his chair as Charles H. Swift Distinguished Service Professor of Physics at the University of Chicago and returned to Washington University in St. Louis, where he was inaugurated as the university's ninth chancellor in 1946. Compton retired as chancellor in 1954, but remained on the faculty as Distinguished Service Professor of Natural Philosophy until his retirement from the full-time faculty in 1961. In retirement he wrote *Atomic Quest*, a personal account of his role in the Manhattan Project, which was published in 1956.

Before his death, he was professor-at-large at the University of California, Berkeley for spring 1962. Compton died in Berkeley, California, from a cerebral hemorrhage on March 15, 1962.

Compton, A. H. (August, 1921). The Magnetic Electron*.

Journ. Frankl. Inst., 192, 2, 145-55; https://www.semanticscholar.org/paper/The-magnetic-electron-Compton/f602176e15e52ed703a67865f4c87eeb7a83048e.

* Based on a paper read before Section B of the American Association for the Advancement of Science, December 27, 1920.

Washington University, St. Louis.

Compton's paper on investigations of ferromagnetic substances with X-rays was the first to introduce the idea of *electron spin*. Compton hypothesized that the electron's *magnetic moment* was intrinsically connected to the electron's *spin* and pointed out the possible bearing of this idea on the origin of the natural unit of magnetism.

The evidence brought forward by the speakers who have preceded me has shown that many magnetic phenomena find a satisfactory explanation on the hypothesis that *matter contains a large number of minute elementary magnets*. The theories of *para-* and *ferro-magnetism* as developed by Langevin, Weiss and others, though based upon the hypothesis of such ultimate magnetic particles, make no assumptions concerning their nature. The explanation of *diamagnetism*, on the other hand, is based upon the view that this effect owes its origin to the circulation of electricity in resistanceless paths. The success of these theories in explaining the principal characteristics of magnetism gives us confidence in the real existence of these magnetic particles. Let us see, therefore, if it is possible to identify these elementary magnets with any of the fundamental divisions of matter.

The original investigations of *ferromagnetism* which led to the hypothesis of an elementary magnetic particle credited molecules with the properties of small permanent magnets. This view finds some support in the profound effect of heating, mechanical jarring, etc., on the ease of magnetization of iron. The dependence of magnetic permeability upon the chemical condition of a substance suggests the same view. But perhaps the strongest argument that has been brought forward in support of the idea of molecular magnets has been the discovery of the Heusler alloys, in which by melting together elements which are only slightly magnetic an alloy with ferromagnetic properties is produced. It is, however, difficult to imagine what mechanism could reasonably give to a group of atoms, such as the chemical molecule, the properties of a single magnetic particle. Moreover, if on magnetization such a group of atoms should actually turn around within a crystal, as the elementary magnets are supposed to do, the resulting change in the positions of the atoms composing the molecule should produce a change in the crystal form; since, as we know, the form of the crystal is dependent upon the arrangement. of its component atoms. It is, however, a matter of common observation that a magnetic field effects no such change in the form of a magnetic crystal.

Perhaps the most natural, and certainly the most generally accepted view of the nature of the elementary magnet, is that the revolution of electrons in orbits within the atom give to the atom as a whole the properties of a tiny permanent magnet. Support of this view is found in the quantitative explanation which it affords of the *Zeeman effect*. It seems but a step from the explanation of this effect to Langevin's explanation of *diamagnetism* as another result of the induced electronic currents within the atom. On Langevin s view the electronic orbits act as resistanceless circuits in which an external magnetic field induces changes of current. By Lenz's law these induced currents will always be in the direction to give the electronic orbit a magnetic polarity opposite to the applied field, thus accounting for the atom's *diamagnetic* properties. This theory offers a satisfactory qualitative explanation of *diamagnetism*, and accounts for the fact that *diamagnetism* is independent of temperature. But quantitatively it is inadequate. For, in order to explain the magnitude of the observed *diamagnetic* susceptibility on this view, one must suppose either that the atom possesses a number of electrons equal to several times its atomic number, or the distance between the electrons in the atom must be several times as great as is estimated by more direct methods. Moreover, the experiments of Barnett[1] and J. Q. Stewart[2] show that the *ratio of charge to mass* of the elementary magnet, though of the same order of magnitude, *is appreciably greater than one would expect if the magnetic moment is due solely to electrons revolving in orbits*.

[1] Barnett, S. J. (1915). *Phys. Rev.*, 6, 240.
[2] Stewart, J. Q. (1918). *Phys. Rev.*, 11, 100.

But perhaps a more serious difficulty with the usual electron theory of *diamagnetism* is that the induced change in magnetic moment of the electronic orbit involves also a change in its angular momentum. It is obvious, *according to the classical electrical theory, that any electron revolving in an orbit will soon radiate its energy*. Any angular momentum induced by an applied magnetic field will, on this theory, therefore, rapidly disappear so that *diamagnetism* should be merely a transient effect. *Let us then assume with Bohr that if each electron has some definite angular momentum such as $h/2\pi$, no radiation occurs*. On this view the electrons in the normal atom will all possess the requisite angular momentum, and when an external magnetic field is applied the induced change in angular momentum will put the electrons in an unstable condition. *On this view also, therefore, the additional rotational energy induced by an applied magnetic field will not be permanent, but will soon be dissipated*. In fact, the theory of atomic structure has yet to be proposed according to which *diamagnetism*, accounted for by the induced magnetic moment of electrons revolving in orbits, can be more than a transient phenomenon.

Besides the molecule and the atom, we have the other two fundamental divisions of matter, the atomic nucleus and the electron. The sign of the Richardson-Barnett effect indicates that *it is negative electricity which is chiefly responsible for magnetic effects*, which makes the view that the positive nucleus is the elementary magnet difficult to defend. On the other hand, many of the magnetic properties of matter receive a satisfactory explanation on Parson's hypothesis[3], that the electron is a continuous ring of negative electricity spinning rapidly about an axis perpendicular to its plane, and therefore possessing a magnetic moment as well as an electric charge.

[3] Parson, A. L. (1915). *Smithsonian Misc. Collections.*

Thus, for example, the fact that such a ring can rotate without radiating enables this hypothesis to account for *diamagnetism* as a permanent instead of a transient effect. While retaining Parson's view of a magnetic electron of comparatively large size, we may suppose with Nicholson that instead of being a ring of electricity, the electron has a more nearly isotropic form with a strong concentration of electric charge near the center and a diminution of electric density as the radius increases. It is natural to suppose that the mass of such an electron is concentrated principally near its center and that the ratio of the charge to the mass of its external portions will be greater than that for the electron as whole. While the explanation of the inertia of such a charge of electricity is perhaps not obvious, it is at least consistent with our usual conceptions and it has the advantage of offering an explanation for the large value of e/m observed in Barnett and Stewart's experiments. It also makes possible an explanation of the relatively large induced currents required to account for *diamagnetism* without introducing the assumption of a prohibitively large radius for the electric charge. A series of experiments has recently been performed, designed to determine which of these fundamental divisions of matter is identical with the elementary magnet in ferromagnetic substances. The first of these, due to K. T. Compton and E. A. Trousdale[4], had for its object the detection of any displacement of the atoms of a substance on magnetization.

[4] Compton, K. T. & Trousdale, E. A. (1915). *Phys. Rev.*, 5. 315.

If the elementary magnet consists of a group of atoms such as the chemical molecule, the rotation of this elementary magnet into alignment with an applied external field will cause a displacement of the individual atoms. It is known, however, that the position of the spots on a Lane photograph depends upon the arrangement of the atoms within the crystal employed. If then, such a photograph is taken with a magnetic crystal, the character of the diffraction pattern should change when the direction of magnetization of the crystal is altered. In these experiments, however, no effect of this character was found. The obvious conclusion is that the ultimate magnetic particle does not consist of any group of atoms such as the chemical molecule. The second of these experiments, performed by Mr. Rognley and myself[5],

[5] Compton, A. H. & Rognley, 0. (1920). *Phys. Rev.*, 16, 464.

was based upon the fact that the intensity of reflection of X-rays from the surface of a crystal depends not only upon the arrangement of the atoms within the crystal, but also upon the distribution of the electrons within the atoms. Let us suppose that the atom acts as a tiny magnet due to the orbital motion of its component electrons. Magnetization of the crystal will orient these atomic magnets and in so doing will change the planes of revolution of the electrons. This change in the electronic distribution should, therefore, affect the intensity of reflection of a beam of X-rays from the crystal's surface. An attempt was made to detect such a change in the intensity of X-ray reflection from a crystal of magnetite when strongly magnetized. Apparatus sufficiently sensitive to detect a change in intensity of less than one per cent. was employed, but magnetization of the crystal failed to produce any measurable effect. ... The following table shows in the first column the order of the X-ray spectrum line which was being studied; in the second column the calculated ratio of intensity from the magnetized to that from the unmagnetized crystal, supposing the atom to have the Rutherford form; and the third and fourth columns represent the similar ratios as estimated from a cubic form of atom. ... According to experiment the value of these ratios was always unity, at least within one per cent. It is clear that none of the types of atoms considered could be oriented by a magnetic field without producing a noticeable effect. In fact, it is difficult to imagine any form of magnetic atom which would be so nearly isotropic that it would have given no effect in our experiment. *It is, therefore, difficult to avoid the conclusion that the elementary magnet is not the atom as a whole.*

Since neither the molecule nor the atom gives a satisfactory explanation of these experiments, the view suggests itself that it is something within the atom, presumably the electron, which is the ultimate magnetic particle. Let us see then if we can find any positive evidence for the existence of an electron with a magnetic moment.

On the basis of the classical dynamics we should expect the electron, whatever its form, to possess thermal energy of rotational motion, equal on the average to that of a molecule or atom at the same temperature. On Planck's more recent quantum hypothesis, however, which is perhaps the more reasonable view, at the absolute zero of temperature each particle of matter - including the electron - should retain an average amount of energy

½ hv for each degree of freedom for motion. For a rotating system this corresponds to, an angular momentum of $h/2\pi$. Thus, whatever view we adopt, the thermal motions of the electron will give to it an appreciable magnetic moment. For a particle of the small moment of inertia of the electron, the frequency of rotation corresponding to an angular momentum $h/2\pi$ will be exceedingly high, and the corresponding energy ½ hv will be large compared with the additional energy which it may acquire due to an increase in temperature. Thus, the angular momentum, and hence also the magnetic moment of the electron, will be nearly the same at different temperatures - a property characteristic of the elementary magnets. *It is interesting to notice, also, that the magnitude of the magnetic moment of an electron spinning with an angular momentum $h/2\pi$ is of the proper order to account for ferromagnetic properties, being about one-third the magnetic moment of the iron atom.*

If an electron with such an angular momentum is to have a peripheral velocity which does not approach that of light, it is necessary that the radius of gyration of the electron shall be greater than 10^{-11} cm. While such an electron is much larger than the spherical electron of Lorentz, recent experiments on the scattering of X-rays and gamma rays indicate the electron's diameter may be even greater than the minimum value thus required to explain magnetic properties. Experiment shows that the scattering of very high frequency radiation is considerably less than theory demands if the electron is supposed to have negligible dimensions. In the case of hard gamma rays, indeed, I have found the scattering at certain angles to fall below 1/1000, the intensity predicted on the usual theory[6].

[6] Compton, A. H. *Phil. Mag.*, (in printer's hands).

The only adequate explanation of these experiments seems to be that interference occurs between the rays scattered from the different parts of the same electron. Such an explanation clearly implies that the diameter of the electron is comparable with the wave-length of the radiation employed, which means that the effective radius of the electron is of the order of 10^{-10} cm. *Considerations of the size of the electron, therefore, support rather than oppose the view that the electron may have an appreciable magnetic moment.*

Further evidence that the electron possesses properties other than those of an electric charge of negligible dimensions is afforded by a study of the white X-radiation emitted at the target of an X-ray tube. It was noticed by Kaye that the X-rays emitted in the direction of the cathode ray beam are harder and more intense than those traveling in the opposite direction. The difference in both hardness and intensity of the radiation at different angles is in good accord with the view proposed by D. L. Webster that the particles emitting the radiation are moving in the direction of the cathode-ray beam, giving rise to a Doppler effect. Indeed, it is very difficult to give any other explanation of the difference in wave-length of the radiation in different directions. But, on this view, in order to account for the difference in hardness observed in the case of gamma rays, *the radiating particles must have a velocity of about one-half the speed of light*. Since the highest known speeds at which atoms travel is only about one-tenth the velocity of light, as observed in the case of alpha particles, the swiftly moving radiators giving rise to this high-frequency X-radiation must therefore be free electrons. If this view is correct, it follows, as Webster has pointed out, that *the electron must be a system capable of emitting radiation*, and is therefore, not a mere charge of electricity of negligible dimensions. *On the present view we may well suppose that the electron is spinning like a gyroscope and on traversing matter is set into mutational oscillations*, resulting in the observed radiation.

Strong evidence that *the electron possesses a magnetic moment* is afforded by H. S. Allen's recent explanation of the rotation of the plane of polarization by optically active substances[7].

[7] Allen, H. S. (1920). *Phil. Mag*., 40, 426.

You will remember in Drude's classical work it is found that optical rotation may be explained if the electrons, when made to oscillate by a passing electric wave, do not move exactly in the plane of the electric vector. *He supposes rather that there is a component of motion at right angles to the electric vector and finds that such a motion will account for the observed rotation*. Allen shows that the motion perpendicular to the electric vector which Drude assumes is *a natural consequence of the view that the electron is magnetic and has an appreciable diameter*. It would take us too far afield to discuss the details of this work, but the significance of the result is obvious, since it has heretofore been difficult to give a reasonable account of the type of motion postulated by Drude.

Finally, I wish to discuss a phenomenon, first noticed by C. T. R. Wilson and brought to my attention by Mr. Shimizu, which, if its obvious explanation is correct, gives *direct evidence that free electrons possess magnetic polarity*. Suppose that a magnetic electron is placed in a homogeneous *paramagnetic* medium. Every part of the medium will be slightly magnetized in the direction of the lines of force, and the magnetic field at the electron due to the magnetic moment of each portion of the medium will have a positive component in the direction of the electron's magnetic axis. Thus, the magnetization induced in the surrounding medium will give rise to a magnetic force at the electron in the direction of its own magnetic axis. The case is exactly analogous to placing a bar magnet in a field of iron filings. The iron filings will be magnetized by induction in the direction of the *lines of force* and if the bar magnet is removed, there still exists a magnetic field where the magnet was because of the magnetization of the surrounding iron filings. If now the electron is in motion, this induced magnetic field will produce the same effect as would an externally applied field of the same intensity. That is, the force due to the magnetic field from the surrounding medium acting on the moving electric charge will make it follow a curved instead of a straight path. If, because of its gyroscopic action, the axis of the electron does not change its direction, the induced magnetic field will always be in the same direction, and the electron will describe a helical orbit. In any actual medium, composed of discreet particles and therefore not homogeneous on an electronic scale, this spiral motion will be superposed upon an irregular motion due to collisions, and the axis of the electron will not remain fixed in direction. Thus, any spiral motion that may appear should be rather broken. A rough calculation, assuming an electron to be projected into air with a speed corresponding to a drop through 10,000 volts, which is about that of the secondary cathode rays produced by ordinary X-rays, and having a magnetic moment corresponding to the angular momentum $h/2\pi$, indicates that the induced magnetic field at the electron should be of the order of 3000 gauss, if the permeability of the medium is that of ordinary air. This field is strong enough to produce a very decided curvature in the electron's path, so in spite of the irregularities in the electron's motion we might hope to observe experimentally the predicted helical tracks.

Below are a few of C. T. R. Wilson's photographs of the tracks of secondary cathode rays and beta particles. ...

...

Let us then review the different lines of evidence that have given us information concerning the nature of the elementary magnet. In the first place, the Richardson-Barnett effect shows that *magnetism is due chiefly to the circulation of negative electricity whose ratio of charge to mass is not greatly different from that of the electron*. In the second place, *experiments on the diffraction of X-rays by magnetic crystals* indicate that the elementary magnet is not any group of atoms, such as the chemical molecule, nor even the atom itself; but *lead rather to the view that it is the electron rotating about its own axis which is responsible for the ferro-magnetism*. And finally, *positive evidence in favor of the hypothesis of some form of magnetic electron is supplied by a consideration of the curvature of the tracks of beta rays through air*. May I then conclude that *the electron itself, spinning like a tiny gyroscope, is probably the ultimate magnetic particle*.

Wolfgang Ernst Pauli (April 25, 1900 – December 15, 1958)

Pauli was an Austrian theoretical physicist and one of the pioneers of quantum physics. Pauli made many important contributions as a physicist, primarily in the field of quantum mechanics. He seldom published papers, preferring lengthy correspondences with colleagues such as Niels Bohr from the University of Copenhagen in Denmark and Werner Heisenberg, with whom he had close friendships. Many of his ideas and results were never published and appeared only in his letters, which were often copied and circulated by their recipients.

In 1945, after having been nominated by Albert Einstein, Pauli received the Nobel Prize in Physics for his "decisive contribution through his discovery of a new law of Nature, the exclusion principle or Pauli principle". "In Niels Bohr's model of the atom, electrons move in fixed orbits around a nucleus. As this model developed, electrons were assigned certain quantum numbers corresponding to distinct states of energy and movement. In 1925, Wolfgang Pauli introduced two new numbers and formulated the Pauli principle, which proposed that no two electrons in an atom could have identical sets of quantum numbers. It was later discovered that protons and neutrons in nuclei could also be assigned quantum numbers and that Pauli's principle applied here too." [Wolfgang Pauli – Facts. NobelPrize.org. https://www.nobelprize.org/prizes/physics/1945/pauli/facts/.]

Pauli was born in Vienna to a chemist, Wolfgang Joseph Pauli (né Wolf Pascheles, 1869–1955), and his wife, Bertha Camilla Schütz; his sister was Hertha Pauli, a writer and actress. Pauli's middle name was given in honor of his godfather, physicist Ernst Mach. Pauli's paternal grandparents were from prominent Jewish families of Prague; his great-grandfather was the Jewish publisher, Wolf Pascheles. Pauli's mother, Bertha Schütz, was raised in her mother's Roman Catholic religion; her father was Jewish writer Friedrich Schütz. Pauli was raised as a Roman Catholic, although eventually he and his parents left the Church.

Pauli attended the Döblinger-Gymnasium in Vienna, graduating with distinction in 1918. Two months later, he published his first paper, on Albert Einstein's theory of general relativity. He attended the Ludwig-Maximilians University in Munich, working under Arnold Sommerfeld, where he received his PhD in July 1921 for his thesis on the quantum theory of ionized diatomic hydrogen (H_2^+).

Sommerfeld asked Pauli to review the theory of relativity for the Encyklopädie der mathematischen Wissenschaften (Encyclopedia of Mathematical Sciences). Two months after receiving his doctorate, Pauli completed the article, which came to 237 pages. Einstein praised it; published as a monograph, it remains a standard reference on the subject.

Pauli spent a year at the University of Göttingen as the assistant to Max Born, and the next year at the Institute for Theoretical Physics in Copenhagen (later the Niels Bohr Institute).

In 1921 Pauli worked with Bohr to create the Aufbau Principle, which described building up electrons in shells based on the German word for building up, as Bohr was also fluent in German.

From 1923 to 1928, he was a professor at the University of Hamburg. In 1924, Pauli proposed a new quantum degree of freedom (or quantum number) with two possible values, to resolve inconsistencies between observed molecular spectra and the developing theory of quantum mechanics. The *Pauli exclusion principle*, perhaps his most important work, which stated that no two electrons could exist in the same quantum state, identified by four quantum numbers including his new two-valued degree of freedom. A year later, George Uhlenbeck and Samuel Goudsmit identified Pauli's new degree of freedom as electron spin, in which Pauli for a very long time wrongly refused to believe.

In 1926, shortly after Heisenberg published the matrix theory of modern quantum mechanics, Pauli used it to derive the observed spectrum of the hydrogen atom. This result was important in securing credibility for Heisenberg's theory.

Pauli introduced the 2×2 Pauli matrices as a basis of spin operators, thus solving the *nonrelativistic* theory of spin. Dirac invented similar but larger (4 x 4) spin matrices for use in his *relativistic* treatment of fermionic spin.

Soon after reading the manuscript of Dirac's *quantum electrodynamics* paper [Dirac, P. A. M. (March, 1927). The quantum theory of the emission and absorption of radiation], Pauli embarked, in collaboration with Heisenberg, on a program to construct his own version of quantum electrodynamics.

In 1928, Pauli was appointed Professor of Theoretical Physics at ETH Zurich in Switzerland.

In 1929, Heisenberg and Pauli succeeded in formulating a general gauge-invariant *relativistic quantum field theory* by treating particles and fields as separate entities interacting through the intermediaries of field quanta. [Heisenberg, W. & Pauli, W. (July, 1929). Zur Quantendynamik der Wellenfelder. (On the quantum dynamics of wave fields.) *Zeit. Phys.*, 56, 1-61; (January, 1930). Zur Quantendynamik der Wellenfelder II. (On the quantum dynamics of wave fields II.) *Zeit. Phys.*, 59, 168-190.]

In 1930, Pauli considered the problem of beta decay. In a letter of December 4 to Lise Meitner et al., beginning, "Dear radioactive ladies and gentlemen", he proposed the existence of a hitherto unobserved neutral particle with a small mass, no greater than 1% the mass of a proton, to explain the continuous spectrum of beta decay. In 1934, Enrico

Fermi incorporated the particle, which he called a neutrino, "little neutral one" in Fermi's native Italian, into his theory of beta decay. The neutrino was first confirmed experimentally in 1956 by Frederick Reines and Clyde Cowan, two and a half years before Pauli's death. On receiving the news, he replied by telegram: "Thanks for message. Everything comes to him who knows how to wait. Pauli."

In 1929, Pauli married Käthe Margarethe Deppner, a cabaret dancer. The marriage was unhappy, ending in divorce after less than a year. At the end of 1930, shortly after his postulation of the neutrino and immediately after his divorce and his mother's suicide, Pauli experienced a personal crisis. In January 1932 he consulted psychiatrist and psychotherapist Carl Jung, who also lived near Zurich. Jung immediately began interpreting Pauli's deeply archetypal dreams based on the I Ching, and Pauli became one of Jung's best students. He married again in 1934 to Franziska Bertram (1901–1987). They had no children.

Pauli held visiting professorships at the University of Michigan in 1931 and the Institute for Advanced Study in Princeton in 1935. In 1933 Pauli published the second part of his book on Physics, *Handbuch der Physik*, which was considered the definitive book on the new field of quantum physics. Oppenheimer called it "the only adult introduction to quantum mechanics."

The German annexation of Austria in 1938 made Pauli a German citizen, which became a problem for him in 1939 after World War II broke out. In 1940, he tried in vain to obtain Swiss citizenship, which would have allowed him to remain at the ETH.

In 1940, Pauli moved to the United States, where he was employed as a professor of theoretical physics at the Institute for Advanced Study. In the same year, he re-derived the *spin-statistics theorem*, a critical result of quantum field theory that states that particles with half-integer spin are fermions, while particles with integer spin are bosons. [Pauli, W. (1940). The Connection Between Spin and Statistics. *Phys. Rev.*, 58, 8, 716–22. See below.]

In 1946, after the war, he became a naturalized U.S. citizen and returned to Zurich, where he mostly remained for the rest of his life. In 1949, he was granted Swiss citizenship. Pauli was elected a Foreign Member of the Royal Society (ForMemRS) in 1953. In 1958 he became a foreign member of the Royal Netherlands Academy of Arts and Sciences; and he was also awarded the Max Planck medal. The same year, he fell ill with pancreatic cancer. When his last assistant, Charles Enz, visited him at the Rotkreuz hospital in Zurich, Pauli asked him, "Did you see the room number?" It was 137. Throughout his life, Pauli had been preoccupied with the question of why the fine-structure constant, a dimensionless fundamental constant, has a value nearly equal to 1/137. Pauli died in that room on 15 December 1958, at age 58.

Pauli's Exclusion Principle.

In the early 20th century, it became evident that *atoms and molecules with even numbers of electrons are more chemically stable than those with odd numbers of electrons*. In the 1916 article "*The Atom and the Molecule*" by Gilbert N. Lewis, for example, the third of his six postulates of chemical behavior states that the atom tends to hold an even number of electrons in any given shell, *and especially to hold eight electrons*, which he assumed to be typically arranged symmetrically at the eight corners of a cube. In 1919 chemist Irving Langmuir suggested that the periodic table could be explained if the electrons in an atom were connected or clustered in some manner. Groups of electrons were thought to occupy a set of electron shells around the nucleus. In 1922, Niels Bohr updated his model of the atom by assuming that certain numbers of electrons (for example 2, 8 and 18) corresponded to stable "closed shells". [Bohr, N. (1922). *The Theory of Spectra and Atomic Constitution; three essays*. Cambridge: Cambridge University Press. The third essay "*The structure of the atom and the physical and chemical properties of the elements*" was based on a Danish address, given before a joint meeting of the Physical and Chemical Societies of Copenhagen on the 18th of October 1921, and printed in *Fysisk Tidsskrift*, xix. p. 153, 1921.]

> Essay III, pp. 74-75: "The importance of spatial electronic configurations has, in addition, been pointed out by Lewis and Langmuir in connection with their atomic models. Thus Lewis, who in several respects independently came to the same conclusions as Kossel, suggested that *the number 8 characterizing the first groups of the periodic system might indicate a constitution of the outer atomic groups where the electrons within each group formed a configuration like the corners of a cube*. He emphasized how a configuration of this kind leads to instructive models of the molecular structure of chemical combinations. It is to be remarked, however, that such a "static" model of electronic configuration will not be possible if we assume the forces within the atom to be due exclusively to the electric charges of the particles. Langmuir, who has attempted to develop Lewis' conceptions still further and to account not only for the occurrence of the first octaves, but also for the longer periods of the periodic system, supposes therefore the structure of the atoms to be governed by forces whose nature is unknown to us. He conceives the atom to possess a "cellular structure," so that each electron is in advance assigned a place in a cell and these cells are arranged in shells in such a manner, that the various shells from the nucleus of the atom outward contain exactly the same number of places as the periods in the periodic system proceeding in the direction of increasing atomic number. Langmuir's work has attracted much attention among chemists, since it has to some extent thrown light on the conceptions with which empirical chemical science is concerned. On his theory the explanation of the

properties of the various elements is based on a number of postulates about the structure of the atoms formulated for that purpose. Such a descriptive theory is sharply differentiated from one where an attempt is made to explain the specific properties of the elements with the aid of general laws applying to the interaction between the particles in each atom. *The principal task of this lecture will consist in an attempt to show that an advance along these lines appears by no means hopeless, but on the contrary that with the aid of a consistent application of the postulates of the quantum theory it actually appears possible to obtain an insight into the structure and stability of the atom.*"

Pauli looked for an explanation for these numbers, which were at first only empirical. At the same time, he was trying to explain experimental results of the *Zeeman effect* in atomic spectroscopy and in *ferromagnetism*. He found an essential clue in a 1924 paper by Edmund C. Stoner, which pointed out that, for a given value of the principal quantum number (n), the number of energy levels of a single electron in the alkali metal spectra in an external magnetic field, where all degenerate energy levels are separated, is equal to the number of electrons in the closed shell of the noble gases for the same value of n. This led Pauli to realize that *the complicated numbers of electrons in closed shells can be reduced to the simple rule of one electron per state if the electron states are defined using four quantum numbers*. For this purpose, he introduced a new two-valued quantum number, identified by Samuel Goudsmit and George Uhlenbeck as *electron spin*. [Pauli, W. (1925). Über den Zusammenhang des Abschlusses der Elektronengruppen im Atom mit der Komplexstruktur der Spektren. *Zeit. Phys.*, 31, 1, 765–83; doi:10.1007/BF02980631.]

Pauli's Exclusion Principle states that two or more identical particles with half-integer spins (i.e. fermions) cannot simultaneously occupy the same quantum state within a system that obeys the laws of quantum mechanics. This principle was formulated by Pauli in 1925 *for electrons*, and later extended to all *fermions* with his *spin–statistics theorem* of 1940.

In the case of electrons in atoms, the exclusion principle can be stated as follows: in a poly-electron atom *it is impossible for any two electrons to have the same two values of all four of their quantum numbers*, which are: n, the *principal quantum number*; ℓ, the *azimuthal quantum number*; m_ℓ, the *magnetic quantum number*; and m_s, the *spin quantum number*. For example, *if two electrons reside in the same orbital, then their values of n, ℓ, and m_ℓ are equal. In that case, the two values of m_s (spin) pair must be different*. Since the only two possible values for the *spin* projection m_s are +1/2 and −1/2, it follows that one electron must have m_s = +1/2 and one m_s = -1/2.

Particles with an *integer spin* (*bosons*) are not subject to the Pauli exclusion principle. *Any number of identical bosons can occupy the same quantum state*, such as *photons* produced by a laser, or atoms found in a *Bose–Einstein condensate*.

A more rigorous statement is: *under the exchange of two identical particles, the total (many-particle) wave function is antisymmetric for fermions and symmetric for bosons*. This means that *if the space and spin coordinates of two identical particles are interchanged, then the total wave function changes sign for fermions, but does not change sign for bosons*.

So, if hypothetically two *fermions* were in the same state—for example, in the same atom in the same *orbital* with the same *spin*—then interchanging them would change nothing and the *total wave function* would be unchanged. However, the only way a total wave function can both change sign (required for fermions), and also remain unchanged is that such a function must be zero everywhere, which means such a state cannot exist. This reasoning does not apply to bosons because the sign does not change.

Overview

The *Pauli Exclusion Principle* describes the behavior of all *fermions* (particles with *half-integer spin*), while *bosons* (particles with *integer spin*) are subject to other principles. *Fermions* include elementary particles such as quarks, electrons and neutrinos. Additionally, *baryons such as protons and neutrons (subatomic particles composed from three quarks) and some atoms (such as helium-3) are fermions*, and are therefore described by the Pauli exclusion principle as well. *Atoms can have different overall spin, which determines whether they are fermions or bosons*: for example, helium-3 has spin 1/2 and is therefore a fermion, whereas helium-4 has spin 0 and is a boson. The *Pauli Exclusion Principle* underpins many properties of everyday matter, from its large-scale stability to the chemical behavior of atoms.

Half-integer spin means that the *intrinsic angular momentum* value of *fermions* is $\hbar = h/2\pi$ (reduced Planck's constant) times a half-integer (1/2, 3/2, 5/2, etc.). In the theory of quantum mechanics, *fermions are described by antisymmetric states*. In contrast, *particles with integer spin (bosons) have symmetric wave functions and may share the same quantum states*. *Bosons* include the *photon*, the Cooper pairs which are responsible for superconductivity, and the W and Z bosons. *Fermions* take their name from the *Fermi–Dirac statistical distribution*, which they obey, and *bosons* take theirs from the *Bose–Einstein distribution*.

Connection to quantum state symmetry

In his Nobel lecture, Pauli clarified the importance of quantum state symmetry to the exclusion principle: ["Wolfgang Pauli, Nobel lecture (December 13, 1946)" https://www.nobelprize.org/uploads/2018/06/pauli-lecture.pdf.]

> "Among the different classes of symmetry, the most important ones (which moreover for two particles are the only ones) are the symmetrical class, in which

> the wave function does not change its value when the space and spin coordinates of two particles are permuted, and the antisymmetrical class, in which for such a permutation the wave function changes its sign...[The antisymmetrical class is] the correct and general wave mechanical formulation of the exclusion principle."

The Pauli exclusion principle with a single-valued many-particle wavefunction is equivalent to requiring the wavefunction to be antisymmetric with respect to exchange.

Advanced quantum theory

According to the spin–statistics theorem, particles with integer spin occupy symmetric quantum states, and particles with half-integer spin occupy antisymmetric states; furthermore, only integer or half-integer values of spin are allowed by the principles of quantum mechanics. [In *relativistic* quantum field theory, the Pauli principle follows from applying a rotation operator in imaginary time to particles of half-integer spin.]

Applications

Atoms

The *Pauli Exclusion Principle* helps explain a wide variety of physical phenomena. *One particularly important consequence of the principle is the elaborate electron shell structure of atoms and the way atoms share electrons, explaining the variety of chemical elements and their chemical combinations*. An electrically neutral atom contains bound electrons equal in number to the protons in the nucleus. Electrons, being *fermions*, cannot occupy the same quantum state as other electrons, so electrons have to "stack" within an atom, i.e. have different *spins* while at the same electron *orbital* as described below.

An example is the *neutral helium atom* (He), which has two bound electrons, both of which can occupy the lowest-energy (1s) states by acquiring opposite spin; as spin is part of the quantum state of the electron, the two electrons are in different quantum states and do not violate the Pauli principle. *However, the spin can take only two different values (eigenvalues)*. In a *lithium atom* (Li), with three bound electrons, the third electron cannot reside in a 1s state and must occupy a higher-energy state instead. The lowest available state is 2s, so that the ground state of Li is $1s^2 2s$. Similarly, successively larger elements must have shells of successively higher energy. *The chemical properties of an element largely depend on the number of electrons in the outermost shell*; atoms with different numbers of occupied electron shells but the same number of electrons in the outermost shell have similar properties, which gives rise to the *periodic table* of the *elements*.

Solid state properties

In conductors and semiconductors, there are very large numbers of molecular orbitals which effectively form a continuous band structure of energy levels. In strong conductors (metals) electrons are so degenerate that they cannot even contribute much to the thermal capacity of a metal. Many mechanical, electrical, magnetic, optical and chemical properties of solids are the direct consequence of Pauli exclusion.

Stability of matter

The stability of each electron state in an atom is described by the quantum theory of the atom, which shows that close approach of an electron to the nucleus necessarily increases the electron's kinetic energy, an application of the *Uncertainty Principle* of Heisenberg. However, stability of large systems with many electrons and many nucleons is a different question, and requires the *Pauli Exclusion Principle*.

It has been shown that the Pauli Exclusion Principle is responsible for the fact that ordinary bulk matter is stable and occupies volume. This suggestion was first made in 1931 by Paul Ehrenfest, who pointed out that the electrons of each atom cannot all fall into the lowest-energy orbital and must occupy successively larger shells. Atoms, therefore, occupy a volume and cannot be squeezed too closely together.

The first rigorous proof was provided in 1967 by Freeman Dyson and Andrew Lenard (de), who considered the balance of attractive (electron–nuclear) and repulsive (electron–electron and nuclear–nuclear) forces and showed that ordinary matter would collapse and occupy a much smaller volume without the Pauli principle. A much simpler proof was found later by Elliott H. Lieb and Walter Thirring in 1975. They provided a lower bound on the quantum energy in terms of the Thomas-Fermi model, which is stable due to a theorem of Teller. The proof used a lower bound on the kinetic energy which is now called the Lieb-Thirring inequality.

The consequence of the Pauli principle here is that electrons of the same spin are kept apart by a repulsive exchange interaction, which is a short-range effect, acting simultaneously with the long-range electrostatic or Coulombic force. This effect is partly responsible for the everyday observation in the macroscopic world that two solid objects cannot be in the same place at the same time.

Astrophysics

Dyson and Lenard did not consider the extreme magnetic or gravitational forces that occur in some astronomical objects. In 1995 Elliott Lieb and coworkers showed that the Pauli principle still leads to stability in intense magnetic fields such as in neutron stars, although at a much higher density than in ordinary matter.

Astronomy provides a spectacular demonstration of the effect of the Pauli principle, in the form of white dwarf and neutron stars. In both bodies, the atomic structure is disrupted by extreme pressure, but the stars are held in hydrostatic equilibrium by degeneracy pressure, also known as Fermi pressure. This exotic form of matter is known as degenerate matter. The immense gravitational force of a star's mass is normally held in equilibrium by thermal pressure caused by heat produced in thermonuclear fusion in the star's core. In white dwarfs, which do not undergo nuclear fusion, an opposing force to gravity is provided by *electron degeneracy pressure*. In *neutron stars*, subject to even stronger gravitational forces, electrons have merged with protons to form neutrons. Neutrons are capable of producing an even higher degeneracy pressure, *neutron degeneracy pressure*, albeit over a shorter range. This can stabilize neutron stars from further collapse, but at a smaller size and higher density than a white dwarf. *Neutron stars* are the most "rigid" objects known; their Young modulus (or more accurately, bulk modulus) is 20 orders of magnitude larger than that of diamond. However, even this enormous rigidity can be overcome by the gravitational field of a *neutron star* mass exceeding the Tolman–Oppenheimer–Volkoff limit, leading to the formation of a black hole.

Pauli, W. (February, 1925). Über den Zusammenhang des Abschlusses der Elektronengruppen im Atom mit der Komplexstruktur der Spektren. (On the connection between the completion of electron groups in an atom and the complex structure of spectra.)

Zeit. Phys., 31, 1, 765–83; https://doi.org/10.1007/BF02980631; translation at http://www.fisicafundamental.net/relicario/doc/Pauli_1925.pdf.

Institut für theoretische Physik, Hamburg.

Received January 16, 1925.

Pauli first reviewed the established theories for the energy differences *of the triplet levels of the alkaline earths*, based respectively, on *the anomaly of the relativity correction* of the *optically active electron*, and *the dependence of the interaction between the electron and the atom core on the relative orientation of these two systems*. He noted a serious difficulty with the former is the connection of these ideas with the *correspondence principle*, which was well known to be a necessary means to explain the selection rules for the *quantum numbers* k_1, j, and m and the polarization of the Zeeman components, in particular, that *it was necessary that the totality of the stationary states of an atom corresponded to a collection (class) of orbits with a definite type of periodicity properties*. The dynamic explanation of this kind of motion of the *optically active electron*, which was based upon the assumption of deviations of the forces between the *atom* core and the *electron* from central symmetry, *seemed to be incompatible with the possibility to represent the alkali doublet (and thus also the magnitude of the corresponding precession frequency) by relativistic formulae*. Consequently, Pauli, decided to pursue instead the alternative *non-relativistic* theory to the problem of *completion of electron groups in an atom*, in order to draw conclusions only about the *number of possible stationary states* of an *atom* when several equivalent *electrons* are present. But this did not address the position and relative order of the term values. On the basis of these results, Pauli obtained a general classification of every *electron* in the *atom* by the principal quantum number n and two auxiliary quantum numbers k_1 and k_2 to which he added a further quantum number m_1 in the presence of an external field, in agreement with experiments. In particular, his rule explained Stoner's result in a natural way and with it the period lengths 2, 8 18, 32.

Especially in connection with Millikan and Landé's observation that the alkali doublet can be represented by relativistic formulae and with results obtained in an earlier paper, it is suggested that this doublet and its *anomalous Zeeman effect* expresses a classically non-describable *two-valuedness* of the quantum theoretical properties of the optically active electron [German: *Leuchtelektron*], without any participation of the closed rare gas configuration of the atom core in the form of a core *angular momentum* or as the seat of the magneto-mechanical anomaly of the *atom*. We then attempt to pursue this point of view, taken as a temporary working hypothesis, as far as possible in its consequences also for *atoms* other than the alkali *atoms*, notwithstanding its difficulties from the point of view of principle. First of all it turns out that it is possible, in contrast to the usual ideas, to assign for the case of a *strong external magnetic field*, which is so strong that we can neglect the coupling between the atomic core and the optically active electrons, to those two systems, as far as the number of their *stationary states*, the values of their *quantum numbers*, and their *magnetic energy* is concerned, no other properties than those of the free atomic core of the *optically active electron* of the *alkalis*. *On the basis of these results, one is also led to a general classification of every electron in the atom by the principal quantum number n and two auxiliary quantum numbers* k_1 *and* k_2 *to which is added a further quantum number* m_1 *in the presence of an external field.* In conjunction with a recent paper by E. C. Stoner this classification leads to a *general quantum theoretical formulation of the completion of electron groups in atoms*.

1. The Permanence of Quantum Numbers (Principle of Gradual Construction [German: *Aufbauprinzip*]) in Complex Structures and the Zeeman Effect.

In a previous paper [Pauli, W. (1925). *Zeit. Phys*., 31, 373] it was emphasized that the usual ideas, according to which *the inner, completed electron shells of an atom play an essential part in the complex structure of optical spectra and their anomalous Zeeman effect in the shape of core angular momenta and as the real seat of the magneto-mechanical anomaly*, are subject to several serious difficulties. It seems therefore plausible to set against these ideas that especially the doublet structure of the alkali spectra and their *anomalous Zeeman effect* are caused by a classically indescribable *two-valuedness* of the quantum theoretical properties of the *optically active electron*.

> [The *Normal Zeeman Effect* is the splitting of spectral lines of an atomic spectrum due to the interaction between an external magnetic field and *the orbital magnetic momentum*. This effect can be observed *in the absence of electron spins*. The normal Zeeman effect can be observed as a *triplet in the observed spectrum* instead of a single spectral line in the expected spectrum. There the single spectral line has been split into three lines with equal spaces between them.
>
> The *Anomalous Zeeman Effect*, which was discovered by Thomas Preston in Dublin, Ireland, is the splitting of spectral lines of an atomic spectrum caused by the interaction between an external magnetic field and *the combined orbital and intrinsic magnetic moment*. This effect can be observed as a complex splitting of spectral lines. Sometimes the spaces between the spectral lines are wider than

expected. This happens due to the effects of *electron spin.* Since the *spin* of electrons contributes to the *angular momentum*, splitting becomes more complicated.

See Zeeman, P. (March, 1897). On the influence of magnetism on the nature of the light emitted by a substance. *Phil. Mag.*, 5, 43, 226-39; republished in *The Astrophysical Journal*, 5, 332-47; https://doi.org/_10.1086/140355; and Preston, T. (January, 1899). Radiation Phenomena in the Magnetic Field. *Nature*, 59, 1523, 224-9; https://doi.org/10.1038/059224c0; first published as Preston, T. (April, 1898). Radiation Phenomena in a Strong Magnetic Field. *Scientific Transactions of the Royal Dublin Society*, 6, 385-389; (1898). Radiation phenomena in the magnetic field. *Phil. Mag.*, 45, 275, 325-39; https://doi.org/10.1080/14786449808621140; also, both in Underwood, T. G. (2023). *Quantum Electrodynamics - annotated sources, Volume I.*]

This idea is particularly based upon the results of Millikan and Landé that the optical doublets of the alkalis are similar to the *relativity* doublets in X-ray spectra and that their magnitude is determined by a *relativistic formula.* If we now pursue this point of view, we shall assign -- as was done by Bohr and Coster for the X-ray spectra -- to the *stationary states* of the *optically active electron* involved in the emission of the alkali spectra two auxiliary *quantum numbers* k_1 and k_2 as well as the *principal quantum number* n. The first *quantum number* k_1 (usually simply denoted by k) has the values 1, 2, 3, ... for the s, p, d, . . . terms and changes by unity in the allowed transition processes; *it determines the magnitude of the central force interaction forces of the valence electron with the atom core*. The second *quantum number* k_2 is for the two terms of a doublet (e.g., p_1 and p_2) equal to $k_1 - 1$ and k_1, in the transition processes it changes by Â ± 1 or 0 and determines the magnitude of the *relativity correction* (which is modified according to Landé to take into account *the penetration of the optically active electron in the atom core*). If we follow Sommerfeld to define the *total angular momentum quantum number* j of an atom in general as the maximum value of the *quantum number* m_1 (usually simply denoted by m) which determines *the component of the angular momentum along an external field*, we must put $j = k_2 - 1$ for the alkalis. The number of *stationary states* in a *magnetic field* for given k_1 and k_2 is $2j + 1 = 2k_2$, and the number of these states for both doublet terms with given k_1 is altogether $2(2k_1 - 1)$.

If we now consider the case of *strong field* (*Paschen-Back effect*), we can introduce apart from k_1 and the just mentioned *quantum number* m_1, instead of k_2 also a *magnetic quantum number* m_2 which *determines directly the energy of the atom in the magnetic field*, that is, the component of the *magnetic moment* of the *valence electron* parallel to the field. For the two terms of the doublet it has, respectively, the values $m_1 + 1/2$ and $m_1 - 1/2$. Just as in the doublet structure of the alkali spectra the "anomaly of the *relativity* correction" is expressed (the magnitude of which is mainly determined by another *quantum number*, as is the magnitude of the central force *interaction energy* of the *optically active electron* and the *atom* core), so appears in the deviations of Zeeman structure from the normal Lorentz triplet the "*magnetomechanical anomaly*" which is similar to the other anomaly (the magnitude of the *magnetic moment* of the *optically active electron* is mainly determined by another *quantum number*, as is the *angular momentum*). Clearly, *the appearance of half-odd-integral (effective) quantum numbers and the thereby formally caused value* $g = 2$ *of the splitting factor of the s-term of the alkalis is closely connected with the twofoldness of the energy level*. We shall here, however, not attempt a more detailed theoretical analysis of this state of affairs and use the following considerations of the Zeeman effect of the *alkalis* as empirical data.

Without worrying about the difficulties encountered by our point of view, which we shall mention presently, *we now try to extend this formal classification of the optically active electron by four quantum numbers* n, k_1, k_2, m_1 *to atoms, more complex than the alkalis*. It now turns out that we can retain completely on the basis of this classification the *principle of permanence of quantum numbers* (*Aufbauprinzip*) also for the complex structure of the spectra and the *anomalous Zeeman effect* in contrast to the usual ideas. This principle, due to Bohr, states that *when a further electron is added to a -- possibly charged -- atom, the quantum numbers of the electrons which are already bound to the atom retain the same values as correspond to the appropriate state of the free atom core.*

Let us first of all consider the *alkaline earths*. The spectrum consists in this case of a singlet and a triplet system. The *quantum states* with a well-defined value of the quantum number k_1 of the *optically active electron* correspond then for the first system to altogether $1(2k_1 - 1)$ and in the last system to $3(2k_1 - 1)$ *stationary states* in an *external magnetic field*. Up to now this was interpreted as meaning that in strong fields the optically active electron in each case could take up $2k_1 - 1$ positions, while the *atom* core was able to take up in the first case one, and in the last case three positions. *The number of these positions is clearly different from the number 2 of the positions of the free atom core (alkali-like s-term) in a field.* Bohr [Bohr, N. (1923). Ann. Phys., 71, 228; especially p. 276] called this state of affairs a "*constraint*" [German: *Zwang*] which is *not analogous to the action of external fields of force*. Now, however, we can simply interpret the total $4(2k_1 - 1)$ states of the *atom* as meaning that the *atom* core always has two positions in a field, and the *optically active electron* as for the alkalis $2(2k_1 - 1)$ states.

More generally, a branching rule formulated by Heisenberg and Landé states that a stationary state of the *atom* core with N *states* in a field leads *through the addition of one more electron* to two systems of terms, corresponding to altogether $(N + 1)(2k_1 - 1)$ and

$(N-1)(2k_1-1)$ *states* in a field, respectively, for a given value of the *quantum number* k_1 of the last *electron*. According to our interpretation, the $2N(2k_1-1)$ states of the complete atom in a *strong field* come about through N *states* of the atom core and $2(2k_1-1)$ *states* of the *optically active electron*. In the present quantum theoretical classification of the electrons the term *multiplicity* required by the branching rule is simply a consequence of the "*Aufbauprinzip*". According to the ideas presented here Bohr's *constraint* expresses itself not in a violation of the permanence of quantum numbers when the series electron is coupled to the atom core, but only in the peculiar *two-valuedness* of the quantum theoretical properties of each electron in the *stationary states* of an atom.

We can, however, from this point of view use the "Aufbauprinzip" to calculate not only the number of *stationary states*, but also *the energies in the case of strong fields* (at least that part which is proportional to the field) *additively from those of the free atom core and of the optically active electron*, where the latter can be taken from the alkali spectra. Because, in this case, both the total component $\mathbf{m}^-_1$ of the *angular momentum* of the *atom* along the field (in units $[\textit{h-bar}]$) as well as the component $\mathbf{m}^-_2$ of the *magnetic moment* of the *atom* in the same direction (in Bohr magnetons) are equal to the sum of the quantum numbers m_1 and m_2 of the single *electrons*:

$$\mathbf{m}^-_1 = Sm_1, \qquad\qquad \mathbf{m}^-_2 = Sm_2. \qquad\qquad (1)$$

The latter can independently run through all values corresponding to the values of the *angular momentum quantum numbers* k_1 and k_2 of the *electrons* in the *stationary state* of the *atom* considered. ($\mathbf{m}^-_2oh$ is here thus the part of the *energy* of the *atom* proportional to the *field strength*; o = *Larmor frequency*.)

Let us consider as an example the two *s-terms* (singlet- and triplet S-term) of the *alkaline earths*. To begin with it is sufficient to consider only the two *valence electrons*, as the contribution of the other electrons to the sums in (1) vanish when taken together. According to our general assumption we must for each of the two *valence electrons* take (independently of the other electron) the values $m_1 = -1/2$, $m_2 = -1$ and $m_1 = 1/2$, $m_2 = 1$ of the s-terms of the alkalis. According to (1) we then get the following values for the quantum numbers $\mathbf{m}^-_1$ and $\mathbf{m}^-_2$ of the *total atom*:

$$\mathbf{m}^-_1 = -1/2 - 1/2, \quad -1/2 + 1/2, \quad +1/2 - 1/2, \quad +1/2 + 1/2$$
$$\mathbf{m}^-_2 = \quad -1 - 1, \qquad -1 + 1, \qquad 1 - 1, \qquad 1 + 1,$$

or

$$\mathbf{m}^-_1 = -1 \qquad 0 \qquad 1$$
$$\mathbf{m}^-_2 = -2 \qquad 0,0 \qquad 0\ 2$$

[Corresponding to one term with $j = 0$ and one with $j = 1$ in *weak fields*.][a]

[a] One notes that one must assign to the two cases $m_l = -1/2$ for the first and $m_2 = 1/2$ for the second *electron*, or $m_l = +1/2$ for the first and $m_2 = -1/2$ for the second *electron* two different terms (as far as the part of the energy independent of the field is concerned). This is perhaps a blemish of the classification given here. It will later on, however, turn out that if the inner and the outer *valence electron* are equivalent, these two terms are in fact identical.

To obtain the p-, d-, . . . terms of the *alkaline earths*, one must combine in (1) the unchanged contribution of the first *valence electron* (S-term) in an appropriate manner with the m_l- and m_l-values of the p-, d-, . . . terms of the alkalis for the second *electron*.

The rule (1) leads in general exactly to the procedure for calculating the energy values in strong field proposed recently by Landé which has been shown by this author to give correct results also in complicated cases. According to Landé this procedure leads, for instance, to the correct Zeeman terms of neon (at least in the case of *strong fields*) if one assumes[b] that in the *atom* core there is one active *electron* in a p-term (instead of in an s-term as above) and if one lets the *optically active electron* go through s-, p-, d-, f-, . . . terms.

[b] The replacement here of a seven-shell (*atom* core of neon) by one *electron* will be given a theoretical basis in the next section.

This result now suggests that we characterize in general each electron in an atom not only by a principal quantum number n, but also by the two auxiliary quantum numbers k_1 and k_2, even when several equivalent electrons or completed electron groups are present. Moreover, we shall allow (also in the just-mentioned cases) in our thoughts *such a strong magnetic field that we can assign to each electron, independently of the other electrons not only the quantum numbers n and k_1, but also the two quantum numbers m_1 and m_2* (where the last one determines the contribution of the *electron* to the *magnetic energy* of the *atom*). The connection between k_2 and m_2 for given k_1 and m_1 must be taken from the alkali spectra.

Before we apply in the next section this *quantum theoretical classification* of the *electrons* in an *atom* to the problem of the completion of the electron groups, we must discuss in more detail the difficulties encountered by the here - proposed ideas of the complex structure and the *anomalous Zeeman effect* and the limitations of the meaning of our ideas.

First of all, these ideas do not pay proper regard to the, in many respects independent, separate appearance of the different term systems (e.g., the singlet and the triplet systems of the *alkaline earths*), which also play a role in the position of the terms of these systems and in the *Landé interval rule*. Certainly, *one cannot assume two different causes for the energy differences of the triplet levels of the alkaline earths*, both the anomaly of the *relativity correction* of the optically active electron and the dependence of the interaction between the *electron* and the *atom* core on the relative orientation of these two systems.

Incompatibility with Einstein's *theory of special relativity.*

A more serious difficulty, raising a matter of principle, is however the connection of these ideas with the *correspondence principle* which is well known to be a necessary means to explain the selection rules for the *quantum numbers* k_1, j, and m and the polarization of the Zeeman components. It is, to be sure, not necessary according to this principle to assign in a definite *stationary state* to each *electron* an orbit uniquely determined in the sense of usual kinematics; however, *it is necessary that the totality of the stationary states of an atom corresponds to a collection (class) of orbits with a definite type of periodicity properties*. In our case, for instance, the above-mentioned selection and polarization rules require according to the *correspondence principle* a kind of motion *corresponding to a central force orbit on which is superposed a precession of the orbital plane around a definite axis of the atom to which is added in weak external magnetic fields also a precession around an axis through the nucleus in the direction of the field.* The dynamic explanation of this kind of motion of the *optically active electron*, which was based upon the assumption of deviations of the forces between the *atom* core and the *electron* from central symmetry, *seems to be incompatible with the possibility to represent the alkali doublet (and thus also the magnitude of the corresponding precession frequency) by relativistic formulae*. The situation with respect to the kind of motion in the case of *strong fields* is similar.

The difficult problem thus arises how to interpret the appearance of the kind of motion of the *optically active electron* which is required by the *correspondence principle* independently of its special dynamic interpretation which has been accepted up to now *but which can hardly be retained.* There also seems to be a close connection between this problem and the question of the magnitude of the term values of the Zeeman effect (especially of the alkali spectra).

As long as this problem remains unsolved, the ideas about the complex structure and the *anomalous Zeeman effect* suggested here can certainly not be considered to be a sufficient physical basis for the explanation of these phenomena, especially as they were in many respects better reproduced in the usually accepted point of view. *It is not impossible that in the future one will succeed in merging these two points of view*. In the present state of the problem, it seemed of interest to us to pursue as far as possible also the first point of view to see what its consequences are. This is the sense in which one must consider our discussions in the next section of the application of the tentative point of view, presented here to the problem of the completion of *electron* groups in an *atom*, notwithstanding the objections which can be made against it. *We shall here draw conclusions only about the number of possible stationary states of an atom when several equivalent electrons are present, but not about the position and relative order of the term values.*

2. *On a General Quantum Theoretical Rule for the Possibility of the Occurrence of Equivalent Electrons in an Atom.*

It is well known that the appearance of several equivalent electrons, that is, electrons which are fully equivalent both with respect to their quantum numbers and with respect to their binding energies, in an atom is possible only under special circumstances which are closely connected with the regularities of the complex structure of spectra. For instance, the ground state of the *alkaline earths* in which the two *valence electrons* are equivalent corresponds to a singlet S-term, while in those *stationary states* of the *atom* which belong to the triplet system the *valence electrons* are never bound equivalently, as the lowest triplet s-term has a *principal quantum number* exceeding that of the *ground state* by unity. Let us now as second example consider the *neon spectrum*. This consists of two groups of terms with different series limits, corresponding to different states of the *atom* core. The first group, belonging to the removal of an *electron* with the quantum numbers $k_1 = 2$, $k_2 = 1$ from the *atom* core can be considered to be composed of a singlet and a triplet system, while the second group, belonging to the removal of an electron with $k_1 = k_2 = 2$ from the *atom* core, can be said to be a triplet and quintet system. The ultraviolet resonance lines of neon have not yet been observed, but there can hardly be any doubt that the *ground state* of a Ne-atom must be considered to be a p-term as far as its combination with the known *excited states* of the *atom* is concerned; in accordance with the unique definiteness and the diamagnetic behavior of the inert gas configuration there can be only one such term, namely with the value $j = 0$.[c]

[c] As already indicated, the value of j is defined here and henceforth as the maximum value of the quantum number m_1.

As the only p-terms with $j = 0$ are the (lowest) triplet terms of the two groups, we can thus conclude that for Ne for the value 2 of the *principal quantum number* only those two triplet terms exist and moreover are identical for both groups of terms.

In general, we can thus expect *that for those values of the quantum numbers n and k_1 for which already some electrons are present in the atom, certain multiplet terms of spectra are absent or coincide*. The question arises what quantum theoretical rules decide this behavior of the terms.

As is already clear from the example of the neon spectrum, this question is closely connected with the problem of the completion of *electron groups* in an *atom*, which determines the lengths 2, 8, 18, 32, . . . of the periods in the periodic table of the elements. This completion consists in that an n-quantum *electron group* neither through emission or absorption of radiation nor through other external influences is able to accept more than $2n^2$ electrons.

It is well known that Bohr in his *theory of the periodic table*, which contains a unified summary of spectroscopic and chemical data and especially a quantum theoretical basis for the occurrence of chemically similar elements such as the platinum and iron metals and the rare earths in the later period of the table, has introduced a subdivision of these *electron groups* into subgroups. By characterizing each *electron* in the *stationary states* of the *atom* by analogy with the stationary states of a central force motion by a symbol n_k with k [less than or equal to] n, he obtained in general for an *electron group* with a value n of the *principal quantum number* n subgroups. In this way Bohr was led to the scheme of the structure of the inert gases given in Table 1. He has, however, emphasized himself that the equality, assumed here, of *the number of electrons in the different subgroups of a maingroup is highly hypothetical* and that for the time being no complete and satisfying theoretical explanation of the completion of the electron groups in the atom, and especially of the period lengths 2, 8, 18, 32, . . . in the *periodic table* could be given.

. . .

Recently essential progress was made in the problem of the completion of the electron groups in an atom by the considerations of E. C. Stoner[6].

[6] Stoner, E. C. (1924). *Phil. Mag.*, 48, 719.

This author suggests first of all a scheme for the atomic structure of the inert gases in which in contrast to Bohr no opening of a completed subgroup is allowed by letting other electrons of the same main group be added to it, so that the number of electrons in a closed subgroup depends only on the value of k, but not on the value of n, that is, on the existence of other subgroups in the same main group. …

. . .

We can now make this idea of Stoner's more precise and more general, if we apply the ideas about the complex structure of the spectra and the *anomalous Zeeman effect*, discussed in the previous section, to the case where equivalent *electrons* are present in an *atom*. In that case *we arrived, on the basis of an attempt to retain the permanence of quantum numbers, at a characterization of each electron in an atom by both the principal quantum number n and the two auxiliary quantum numbers k_1 and k_2.* In *strong magnetic* *fields* also an *angular momentum quantum number* m_1 was added to this for each *electron* and, furthermore, one can use apart from k_1 and m_1 also a *magnetic moment quantum number m_2, instead of k_2*. First of all, we see that the use of the two quantum numbers k_1 and k_2 for each *electron* is in excellent agreement of Stoner's subdivision of Bohr's subgroup. Secondly, by considering the case of *strong magnetic fields* we can reduce Stoner's result, that *the number of electrons in a completed subgroup is the same as the number of the corresponding terms of the Zeeman effect of the alkali spectra*, to the following more general rule about the occurrence of equivalent electrons in an atom:

> *There can never be two or more equivalent electrons in an atom for which in strong fields the values of all quantum numbers n, k_1, k_2, m_1 (or, equivalently, n, k_1, m_1, m_1) are the same. If an electron is present in the atom for which these quantum numbers (in an external field) have definite values, this state is "occupied".*

We must bear in mind that the *principal quantum number* occurs in an essential way in this rule; of course, several (not equivalent) *electrons* may occur in an *atom* which have the same values of the *quantum numbers* k_1, k_2, m_1, but have different values of the *principal quantum number* n.

We cannot give a further justification for this rule, but it seems to be a very plausible one. It refers, as mentioned, first of all to the case of *strong fields*. However, from thermodynamic arguments (invariance of statistic weights under adiabatic transformations of the system) it follows that the number of *stationary states* of an *atom* must be the same in strong and weak fields for given values of the numbers k_1 and k_2 of the separate electrons and a value of $\mathbf{m}^{-}_{1} = Sm_1$ (see (1)) for the whole *atom*. We can therefore also in the latter case make definite statements about the number of *stationary states* and the corresponding values of j (for a given number of equivalent electrons belonging to different values of k_1 and k_2). We can thus find the number of possibilities of realizing various *incomplete electron shells* and give an unambiguous answer to the question posed at the beginning of this section about the absence or coincidence of certain *multiplet terms in spectra* for values of the *principal quantum number* for which several equivalent *electrons* are present in an *atom*. We can, however, only say something about the number of terms and the values of their quantum numbers, but not about their magnitude and about interval relations.

We must now show that the consequences of our rule agree with experiment in the simplest cases. We must wait and see whether it will also prove itself in comparison with experiment in more complicated cases or whether it will need modifications in that case; this will become clear when complicated spectra are sorted out.

First of all, we see that Stoner's result and with it the period lengths 2, 8, 18, 32, . . . are immediately included in a natural way in our rule. Clearly, for given k_1 and k_2 there cannot be more equivalent *electrons* in an *atom* than the appropriate value of m_l, (that is, $2k_2$) and in the completed group there corresponds exactly one *electron* to each of these values of m_l. Secondly, it turns out that our rule has an immediate consequence that the triplet s-term with the same *principal quantum number* as the *ground state* is absent for the *alkaline earths*. If we investigate the possibilities for the equivalent binding of two *electrons* in s-terms (in that case we have thus $k_1 = 1$ and k_2 can also only have the value 1), according to our rule the cases are excluded in strong fields where both *electrons* have $m_l = 1/2$ or both have $m_l = -1/2$; rather, we can only have $m_l = 1/2$ for the first *electron* and $m_l = -1/2$ for the second *electron*, or the other way round[f] so that the *quantum number* $\mathbf{m}^{-}_{\mathbf{l}} = Sm_l$ for the *total atom* can only have the value 0.

[f] The second case corresponds to an interchange of the two equivalent *electrons* and gives us therefore here no new *stationary state* (compare the footnote lettered a). However, in this two-fold realizability of the *quantum state* considered is contained the fact that its statistical weight with respect to the exchangeability of the two *electrons* must be multiplied by two (compare also the discussion of statistical weights by Stoner)[6].

Therefore, also in *weak fields* (or when there is no field) only the value $j = 0$ is possible (singlet S-term).

We now investigate the case that one *electron* is removed from a closed shell, as will occur in X-ray spectra. Clearly when an *electron* is missing from one of Stoner's part-subgroups, the case is always possible that no *electron* is present with the value m_l; we call this the "*hole-value*" of m_l. The other *electrons* are then uniquely divided over the other values of m_l so that for each of those values we have one *electron*. The sum of these other values of m_l and thus the *quantum number* of the *total atom* is clearly in each case equal to the opposite of the *hole-value* of m_l. If we let it go through all possible values and take into account that an *electron* can be removed from every part-subgroup, we see that in *strong fields* the multiplicity of the *hole-values* of m_l and thus also that of the values of is the same as that of the m_l value of a single *electron*. Due to the invariance of statistical weights, it follows thus also for *weak fields* that the numbers of *stationary states* and of j-values of single ionized closed *electron shells* (X-ray spectra) are the same as in the alkali spectra, in accordance with experiment.

This is a special case of a general *reciprocity law*: *For each arrangement of electrons there exists a conjugate arrangement in which the hole-values of m_l and the occupied values of m_l are interchanged.* This interchange may refer to a single part-subgroup while the other part-subgroups are unchanged, or to a *Bohr subgroup*, or to the whole of a main group, since the different part-subgroups are completely independent of one another as far as possible arrangements are concerned. *The electron numbers of the two conjugate arrangements add up to the number of electrons in the completed state of the group (or subgroup) considered, while the j-values of the two arrangements are the same*. The latter follows from the fact that the sum of the *hole-values* of m_1 of an arrangement always is the opposite of the sum of the occupied m_1-values. Therefore, the *quantum numbers* $\mathbf{\bar{m}_1}$ of the *whole atom* are the opposite of one another for conjugate arrangements. As the j-values are defined as the upper limit of the set of $\mathbf{\bar{m}_1}$ -values, and as this set is symmetric around zero, it follows that the j-values are the same (compare the examples discussed below). Because of this *periodicity law* to some extent the relations at the end of a *period* of the *periodic table* reflect those at the beginning of a *period*. We must emphasize, however, that this for the time being refers only to the number of *stationary states* of the *shell* in question and the values of their *quantum numbers*, whereas *we can say nothing about the magnitude of their energies or about interval relations*[g].

[g] However, because of the equality of the number of m_2-values for conjugate arrangements it follows that also in weak fields the "g-sums" (taken over terms with the same j) of the appropriate terms are the same.

As an application of our rule, *we shall discuss now the special case of the gradual formation of the eight-shell (where of the principal quantum number considered no electrons with k = 2 are present in the ground state)*; this gives us at the same time another example of the just-derived *reciprocity rule*. The binding of the first two electrons in this shell has already been discussed and in what follows we shall assume for the sake of simplicity that no electron is missing from the $k_1 = 1$ subgroup so that it is closed (…). According to Stoner, for the following elements until the completion of the *eight-shell* (e.g., from B to Ne), the *ground state* will always be a *p-term*, in agreement with all experimental data up to now. Especially follows the alkali-like spectrum, corresponding to the binding of the third electron of the *eight-shell*, with the well-known absence of the *s-term* with the same *principal quantum number* as the *ground state*.

We can thus immediately go over to the binding of the fourth *electron* of the *eight-shell*, which appears in the not-yet analyzed arc spectrum of *carbon* and the partially already unraffled arc spectrum of *lead*. According to the *Landé-Heisenberg branching rule* (see previous section) the corresponding spectrum should have in general the same structure as the *neon* spectrum, that is, consist of a singlet-triplet group and a triplet-quintet group with different series limits, corresponding to the *$2p_1$ - and the $2p_2$ - doublet term* of the ion considered. We shall show, however, that according to our rules these spectra must differ essentially, as far as the number and *f-values* of the *p-terms* of the maximum *principal quantum number* (n = 2 for C, n = 6 for Pb) is concerned, from the Ne-spectrum (where, as we mentioned at the beginning of this section, apart from the ground state with $j = 0$ no further *p-term* exists with *principal quantum number* 2); this is in contrast to the structure of the excited states which we expect to be similar.

We must distinguish three cases, according to the number of electrons in the two part-subgroups with $k_1 = 2$, $k_2 = 1$ and with $k_1 = 2$, $k_2 = 2$ over which we must distribute two *electrons* (we have already assumed that the first two *electrons* are bound in *s-terms*, $k_1 = k_2 = 1$).

> (a) Two equivalent n_{21}-*electrons*: Corresponding to the p_1-*term* of the alkalis m_1 can for this part-subgroup only take on the two values $m_1 = \hat{A} \pm 1/2$. It is thus closed in this case with $\mathbf{m}^-_1 = 0$ and $j = 0$.
>
> (b) One n_{21}- and one n_{22}-*electron*: For the second part-subgroup m_1 can, corresponding to the p_2-*term* of the alkalis take on the four values $\hat{A} \pm 1/2$, $\hat{A} \pm 3/2$ and these can be combined in all possible ways with the above-mentioned values $m_1 = +1/2$ of the first *electron*, since the two *electrons* are in different part-subgroups and are thus not equivalent[h].

> > [h] Because of this we must count the case $m_1 = +1/2$ for the first and $m_1 = -1/2$ for the second *electron* different from the case $m_1 = -1/2$ for the first and $m_1 = +1/2$ for the second *electron*. Compare the footnote lettered a.

We have thus $\mathbf{m}^-_1 = (-3/2, -1/2, 1/2, 3/2) + (-1/2, 1/2)$
$$= \hat{A} \pm (3/2 + 1/2), \hat{A} \pm (3/2 - 1/2), \hat{A} \pm (1/2 + 1/2), \hat{A} \pm (1/2 - 1/2)$$
$$= \hat{A} \pm 2, \hat{A} \pm 1, \hat{A} \pm 1, 0, 0.$$

> From this we see immediately that the terms split in two series with [absolute value] $\mathbf{m}^-_1$ [less than or equal to] 2 and with $\mathbf{m}^-_1$ [less than or equal to] 1. In the field free case these correspond clearly to two terms: one with $j = 2$, and one with $j = 1$.
>
> (c) Two equivalent n_{22}-*electrons*: According to our rule the m_1-values of the two *electrons* must be different and we find for the possible values of $\mathbf{m}^-_1$:
> $$\mathbf{m}^-_1 = \hat{A} \pm (3/2 + 1/2), \hat{A} \pm (3/2 - 1/2), (3/2 - 3/2), (1/2 - 1/2)$$
> $$= \hat{A} \pm 2, \hat{A} \pm 1, 0, 0.$$

If there is no magnetic field, we find thus one term with $j = 2$ and one with $j = 0$.

Altogether we find thus for the four-shell five different p-terms with maximum principal quantum number, of which two have $j = 2$, one $j = 1$, and two $j = 0$.

We can say nothing about the energies or the interval relations of this group of terms. However, we can make definite statements about the Zeeman splittings of these terms to be expected.

By substituting the m_2-values (taken from the Zeeman terms of the *alkalis* in *strong fields*) for the separate *electrons* corresponding to the given m_l-values, we find from rule (1) the Zeeman splittings for the five *p-terms* of the *four-shell* in strong fields:

$\bar{m}_1$	–2	–1	0	1	2
$\bar{m}_2$	–3, –2	–2, –1, 0	0, 0, 0, 0, 0	1, 1, 2	2, 3

Using the same rule applied by Landé to higher-order multiplets, one obtains from this for the determination of the sum of the *g-values* for the two j = 2 terms (denoted by Sg_2) and for the *g-value* for the j = 1 term (denoted by g_1) the equations

$$2Sg_2 = 2 + 3 = 5, \qquad Sg_2 + g_1 = 1 + 1 + 2 = 4,$$

or

$$Sg_2 = 5/2, \qquad g_1 = 3/2.$$

The earliest test of this theoretical result for the *four-shell* is possible for *lead*. Observations certainly show four *p-terms*, while the existence of a fifth *p-term* is doubtful. So far unpublished measurements by E. Back of a few *lead* lines make it, moreover, very likely that the first four *p-terms* have *j-values* 2, 2, 1, 0, and that the *g-values* of these terms also agree with the theoretically expected ones.

Let us now return to the discussion of the gradual construction of the *eight-shell*. By means of the *reciprocity rule*, applied to the whole of the *Bohr subgroup* with k = 2, which contains in its closed state six *electrons*, we can immediately apply the results obtained for the *four-shell* to the number of possibilities to realize the *six-shell* (from electrons with $k_1 = 2$), which occurs, for instance, for O. The following cases of the *six-shell* are clearly conjugate to the cases (a), (b), and (c):

(a) Four equivalent n_{22}-*electrons* (two empty spaces in the n_{21}-group). This part-subgroup is closed; hence as before sub (a) one term with j = 0.

(b) One n_{21}-, three equivalent n_{22}-*electrons* (one empty space in the n_{21}-, and one empty space in the n_{22}-group). As before: one term with j = 2 and one term with j = 1.

(c) Two equivalent n_{21}- and two equivalent n_{22}-*electrons* (two empty spaces in the n_{22}-group). The first part-subgroup is closed. As before: one term with j = 2, one term with j = 0.

We must thus also here, for instance for *oxygen*, expect five *p-terms* with the smallest *principal quantum number*. So far only three such terms have been observed for O and S, with *j-values* of 2, 1, 0. We must wait and see whether two more *p-terms* of the same *principal quantum number* can be found from the observations, or whether our rule must be modified in this case.

As yet there are no observations about the *five-shell* (3 electrons with $k_1 = 2$) and we shall therefore give only the result of the discussion; according to our rule this shell gives rise to five *p-terms*, one term with $j = 5/2$, three terms with $j = 3/2$, and one term with $j = 1/2$. For the *seven-shell*, realised in x-ray spectra we get -- as we mentioned before -- terms similar to the *alkalis*.

We shall not discuss here further special cases, before experimental data are available, but it should be clear from the examples given that in each case our rule is able to give a unique answer to the question about the possibilities of realizing the different *shells* for a given number of equivalent *electrons*. To be sure, only in the simplest cases was it possible to verify that the results obtained in this way are in agreement with experiment.

In general, we may note that the discussions given here are in principle based, as far as the transition from strong to weak or vanishing fields is concerned, upon the invariance of the statistical weights of quantum states. However, on the basis of the results obtained there seem to be no reasons for a connection between the problem of the completion of *electron* groups in an *atom* and the *correspondence principle*, as Bohr suspected to be the case. It is probably necessary to improve the basic principles of quantum theory before we can successfully discuss the problem of a better foundation of the general rules, suggested here, for the occurrence of equivalent *electrons* in an *atom*.

George Eugene Uhlenbeck (December 6, 1900–October 31, 1988).

Uhlenbeck was a Dutch-American theoretical physicist.

He was the son of Eugenius and Anne Beeger Uhlenbeck. He attended the Hogere Burgerschool (High School) in The Hague, from which he graduated in 1918. He subsequently entered Delft University of Technology as a student in chemical engineering. During the next year, he transferred to Leiden University, to study physics and mathematics, and he earned his bachelor's degree in 1920 (Dutch: Kandidaatsexamen).

Uhlenbeck was then admitted by Ehrenfest (a student of Boltzmann's) to the Wednesday evening physics colloquium in Leiden. Ehrenfest became the most important scientific influence in his life. From 1922 to 1925 Uhlenbeck was the tutor of the younger son of the Dutch ambassador in Rome. While there, he attended lectures by Tullio Levi-Civita and Vito Volterra and met his longtime friend, Enrico Fermi. In 1923, Uhlenbeck received his master's degree from Leiden (Dutch: Doctoraalexamen).

He returned to Leiden in 1925 to become Ehrenfest's assistant. Ehrenfest assigned him to work with his graduate student, Samuel Goudsmit for a quick update on "what was currently happening in physics". In mid-September 1925, Uhlenbeck and Goudsmit introduced *electron spin*, which posits *intrinsic angular momentum* for the *electron.* [Uhlenbeck, G. E. & Goudsmit, S. (November, 1925). Ersetzung der Hypothese vom unmechanischen Zwang durch eine Forderung bezuglich des inneren Verhaltens jedes einzelnen Elektrons. (Replacement of the hypothesis of unmechanical coercion by a requirement regarding the internal behavior of each individual electron.) *Naturw.*, 13, 47, 953-4.]

In 1927 Uhlenbeck earned his Ph.D. degree under Ehrenfest with his thesis titled: "*Over Statistische Methoden in de Theorie der Quanta*" ("On Statistical Methods in the Quantum Theory").

Uhlenbeck married Else Ophorst in Arnhem, Netherlands in August 1927. He received a doctorate from the Leiden University in the same year.

In 1927, Uhlenbeck took a position as an instructor in physics at the University of Michigan in Ann Arbor, Michigan. He stayed there until 1935, when he succeeded H. A. Kramers as a professor of theoretical physics in Utrecht. During his eight years in Ann Arbor, Uhlenbeck organized the noted "Summerschool" in theoretical physics.

In 1938, Uhlenbeck spent half a year as visiting professor at Columbia University in New York City, and then he returned to Ann Arbor as a professor of theoretical physics during the next year. Because of the rise of Nazism in Europe, he and Else decided to leave his position in the Netherlands and return to America.

During part of World War II, from 1943 through 1945, Uhlenbeck led a theory group at the Radiation Laboratory in Cambridge, Massachusetts, which was doing radar research. In 1945, he returned to Ann Arbor, where he was named the Henry Smith Carhart Professor of Physics in 1954. He remained in Ann Arbor until 1960, when he joined the Rockefeller Institute for Medical Research (now the Rockefeller University) in New York City as a professor and member of the institute.

Uhlenbeck retired in 1971, but remained scientifically active until the early 1980s. He died on October 31, 1988, in Boulder, Colorado, at the age of 87.

Samuel Abraham Goudsmit (July 11, 1902–December 4, 1978).

Goudsmit was a Dutch-American physicist famous for jointly proposing the concept of electron spin with George Eugene Uhlenbeck in 1925. [Uhlenbeck, G. E. & Goudsmit, S. (November, 1925). Ersetzung der Hypothese vom unmechanischen Zwang durch eine Forderung bezuglich des inneren Verhaltens jedes einzelnen Elektrons. (Replacement of the hypothesis of unmechanical coercion by a requirement regarding the internal behavior of each individual electron.) *Naturw.*, 13, 47, 953-4.]

Goudsmit was born in The Hague, Netherlands, of Dutch Jewish descent. He was the son of Isaac Goudsmit, a manufacturer of water-closets, and Marianne Goudsmit-Gompers, who ran a millinery shop. In 1943, his parents were deported to a concentration camp by the German occupiers of the Netherlands and were murdered there.

Goudsmit studied physics at the University of Leiden under Paul Ehrenfest, where he obtained his PhD in 1927. Goudsmit married Jaantje Logher, in 1927. Their daughter, Esther Marianne Goudsmit was born in 1933 in Ann Arbor, Michigan.

After receiving his PhD, Goudsmit served as a professor at the University of Michigan between 1927 and 1946. In 1930 he co-authored a text with Linus Pauling titled *The Structure of Line Spectra*.

During World War II he worked at the Massachusetts Institute of Technology. As scientific head of the Alsos Mission, he successfully reached a German group of nuclear physicists around Werner Heisenberg and Otto Hahn at Hechingen (then French zone) in advance of French physicist Yves Rocard, who had previously succeeded in recruiting German scientists to come to France.

Alsos, part of the Manhattan Project, was designed to assess the progress of the Nazi atomic bomb project. In the book *Alsos*, published in 1947, Goudsmit concludes that the Germans did not get close to creating a weapon. He attributed this to the inability of science to function under a totalitarian state and to Nazi scientists' lack of understanding of how to engineer an atomic bomb. Both of these conclusions have been disputed by later historians (see Heisenberg) and contradicted by the fact that the totalitarian Soviet state produced the bomb shortly after the book's release. However, that statement overlooks the actions of physicist Klaus Fuchs who sent "many intelligence reports directly from Los Alamos".

After the war he was briefly a professor at Northwestern University, and from 1948 to 1970 was a senior scientist at the Brookhaven National Laboratory, chairing the Physics Department 1952–1960. He meanwhile became well known as editor-in-chief of the leading physics journal *Physical Review*, published by the American Physical Society. In

July 1958 he started the journal *Physical Review Letters*, which offers short notes with attendant brief delays.

Goudsmit and Jaantje Logher divorced in 1960, and in the same year Goudsmit married Irene Bejach. Like Goudsmit's parents, Irene's father, a German medical doctor and Berlin public health official, Curt Dietrich Bejach, had been murdered by the Nazis. He perished at the Auschwitz concentration camp.

On his retirement as editor in 1974, Goudsmit moved to the faculty of the University of Nevada, Reno, where he remained until his death four years later on December 4, 1978, at the age of 76.

Uhlenbeck, G. E. & Goudsmit, S. (November, 1925). Ersetzung der Hypothese vom unmechanischen Zwang durch eine Forderung bezuglich des inneren Verhaltens jedes einzelnen Elektrons. (Replacement of the hypothesis of unmechanical coercion by a requirement regarding the internal behavior of each individual electron.)

Naturw., 13, 47, 953-4; https://doi.org/10.1007/BF01558878; translation by T. G. Underwood; also in Underwood, T. G. (2023). *Quantum Electrodynamics - annotated sources*, Volume I, pp. 282-6.

October 17, 1925.

Instituut voor Theoretische Natuurkunde, Leiden.

The idea of a *quantized spinning of the electron* was put forward for the first time by Compton in August 1921. Without being aware of Compton's suggestion Uhlenbeck and Goudsmit noted doublets in the alkali spectra that did not conform to current models of the atom. They proposed applying the model of the *spinning electron* to interpret a number of features of the quantum theory of the *anomalous Zeeman effect*, and applied the classical formula for spherical rotating electron with finite radius and surface charge.

---

§ 1. As is well known, the structure and magnetic behavior of the spectra can be described in detail with the help of *Landé's vector model* R, K, J and m[1].

[1] See Back, E. & Landé, A. (1925). *Zeemaneffekt und Multiplettstruktur der Spektrallinien.* (Zeeman effect and Multiplet structure of the spectral lines.) Berlin: Verlag von Julius Springer).

Here, R denotes the *momentum* moment of the atomic remnant ~ i.e. of the atom without the luminous electron - K the *momentum* moment of the luminous electrons, J their resultant and m the projection of J on the direction of an *external magnetic field*, all expressed in the branch quantum units:

(a) that for the rest of the atom the behavior of the *magnetic moment* to the *mechanical* is *twice as large as you would expect classically.*

(b) that in the formulae, where R^2, K^2, J^2 occurs, you can do this by using these expressions $R^2 - ¼$, $K^2 - ¼$, $J^2 - ¼$. [The Heisenberg Averaging[2])].

[2] Heisenberg, W. (1924). Über eine Änderung der formalen Regeln der Quantentheorie in einem Problem anomaler Zeeman-Effekte. (On an alteration to the formal rules of quantum theory in a problem of anomalous Zeeman effect.) *Zeit. Phys.*, 26, 291-307.

This model has shown itself to be very robust and has, among other things, fought to unravel the most complicated spectra.

§ 2. However, one starts to encounter difficulties as soon as one tries to connect *Landé's vector model* to our ideas about the formation of the atom from electrons. E.g.:

a) Pauli[3] has already shown that in the case of the alkali atoms, the atomic radical must be magnetically ineffective, otherwise the influence of the *relativity correction* would cause a dependency of the Zeeman effect on the nuclear charge, which is not perceived in these spectra.

> [3] Pauli Jr., W. (1925). Über den Einfluss der Geschwindigkeitsabhängigkeit der Elektronenmasse auf den Zeemaneffekt. (On the influence of the velocity dependence of the electron mass on the Zeeman effect.) *Zeit. Phys*., 31, 373.

b) In *Lande's model*, one must not identify the *momentum moment* of the atomic radical with that of the positive ions, as one would expect it according to the definition of the atomic radical. [Landé-Heisenberg branching theorem[4] — unmechanical coercion].

> [4] See Back, E. & Landé, A. (1925). *Zeemaneffekt und Multiplettstruktur der Spektrallinien*. *Loc. cit*., pages 55ff.

c) For some spectra recently analyzed with the help of Lande's scheme (e.g. vanadium, titanium), the K of the basic term did not correspond at all with the values expected from the Bohr-Stone periodic system.

§ 3. The above-mentioned difficulties point all in the same direction, namely that *the meaning of which is attributed to Lande's vectors is probably not correct*. Pauli[5] has already embarked on a new path, which is particularly difficult.

> [5] Pauli Jr., W. (1925). Über den Zusammenhang des Abschlusses der Elektronengruppen im Atom mit der Komplexstruktur der Spektren. (On the relationship between the completion of the electron groups in the atom and the complex structure of the spectra.) *Zeit. Phys*., 31, 765.

From this he concluded that in the case of alkali spectra, all quantum numbers must be written to the luminous *electron* alone. According to Pauli, each *electron* in the *magnetic field* then gets 4 independent quantum numbers. With the help of Bohr's construction principle and a few general sentences, he was then able to achieve the same results as Landé in a simple way[6].

> [6] Compare: Goudsmit, S. (December, 1925). Über die Komplexstruktur der Spektren. *Zeit. Phys*., 32, 1, 794-98; https://doi.org/10.1007/BF01331715; Heisenberg, W. (1925). Quantentheorie der multiplen Struktur und des abnormalen Zeeman-Effekts. (Quantum theory of multiple structure and the abnormal Zeeman effect.) *Zeit. Phys*., 32, 841-60;

Hund, F. (1925). Zur Deutung verwickelter Spektren, insbesondere der Elemente Scandium bis Nickel. (On the interpretation of entangled spectra, in particular the elements scandium to nickel.) *Zeit. Phys*., 33, 345-71; http://dx.doi.org/ 10.1007/BF01328319.

The difficulties mentioned in § 2 disappear completely in the Pauli procedure. The connection to the Bohr-Stoner periodic system is achieved, and new aspects are still opened[7].

[7] See those in 5) below.

§ 4. In both cases, however, the appearance of the so-called *relativistic doublet* in the rontgen and alkali spectra remains an enigma. To explain this fact, one has recently come to the assumption of a classically indescribable ambiguity in the quantum theoretical properties of the *electron*[1].

[1] Heisenberg, W. (1925). Quantentheorie der multiplen Struktur und des abnormalen Zeeman-Effekts. (Quantum theory of multiple structure and the abnormal Zeeman effect.) *Zeit. Phys*., 32, 841-60; *loc. cit.*

§ 5 There seems to us to be another way open. Pauli does not bind himself to a model idea. The 4 quantum numbers assigned to each electron have lost their original Landé meaning. It is now obvious to give to each electron with its 4 quantum numbers 4 degrees of freedom. One can then give the quantum numbers, for example, the following meaning: n and k remain as before the main and azimuthal quantum number of the electron in its orbit. *R, however, will be assigned its own rotation of the electron*[2].

[2] Note that the quantum numbers of the *electron* occurring here must be taken from the alkali spectra. R therefore has only the value 1 for each *electron* (in Landé standardization).

The other quantum numbers retain their old meaning. Through our imagination, the conceptions of Landé and Pauli with all their advantages have formally merged with each other[3].

[3] For example, the meaning of the Heisenberg's Scheme III is now becoming more understandable, in which one has to assemble both the R and the K of the *electrons* for an entire atom.

The *electron* must now take over the still misunderstood property (referred to in § I under a), which Landé attributed to the atomic remnant.

The closer quantitative implementation of this idea will probably depend heavily on the choice of the *electron* model. In order to come into line with the facts, the following demands must therefore be made of this model:

a) *The ratio of the magnetic moment of the electron to the mechanical one must be twice as large for the self-rotation as for the orbital motion*[4].

[4] For example, for a spherical rotating electron with surface charge, the Abraham formulas can be used [Abraham, M. (1903). Prinzipien der Dynamik des Elektrons. (Principles of electron dynamics.) *Ann. Phys.*, 315, 105-79] read:

Rotational energy $1/9\ e^2 a/c^2\ \dot{\varphi}^2$ (a = electron radius),

also: $p_\varphi = 2/9\ e^2 a/c^2\ \dot{\varphi}$

Magnetic moment: $\Phi = 1/3\ e a^2/c\ \dot{\varphi}$

Mass: $m = 2/3\ e^2/c^2 a$

Also: $\Phi/p_\varphi = 3/2\ ac/e = 2 \times e/2mc$,

in fact, twice as much as in the orbital motion. Note, however, that when quantizing this rotational motion, *the peripheral speed of the electron is far from the speed of light*.

b) The different orientations from the R to the orbital plane (or K) of the *electron* must be able to provide the explanation of *relativity-doublets*, perhaps in connection with a Heisenberg -Wentzel averaging rule[5].

[5] Heisenberg, W. *loc. cit.* Wentzel, G. (1925). *Ann. Phys.*, 76, 803.

Uhlenbeck, G. E. & Goudsmit, S. (February 20, 1926). Spinning Electrons and the Structure of Spectra.

Nature, 117, 264-5; https://doi.org/10.1038/117264a0.

Abstract.

So far as we know, the idea of a quantized spinning of the electron was put forward for the first time by A. K. Compton [(August, 1921). The magnetic electron. *Journ. Frankl. Inst.*, 192, 145-55], who pointed out the possible bearing of this idea on the origin of the natural unit of magnetism. Without being aware of Compton's suggestion, we have directed attention in a recent note [Uhlenbeck, G. E. & Goudsmit, S. (November, 1925). Ersetzung der Hypothese vom unmechanischen Zwang durch eine Forderung bezuglich des inneren Verhaltens jedes einzelnen Elektrons. *Naturw.*, 13, 47, 953-4] to the possibility of applying the spinning electron to interpret a number of features of the quantum theory of the Zeeman effect, which were brought to light by the work especially of van Lohuizen, Sommerfeld, Landé and Pauli, and also of the analysis of complex spectra in general. In this letter we shall try to show how our hypothesis enables us to overcome certain fundamental difficulties which have hitherto hindered the interpretation of the results arrived at by those authors.

Paul Adrien Maurice Dirac (August 8, 1902–October 20, 1984).

Dirac was an English theoretical physicist who is regarded as one of the most significant physicists of the 20th century. Dirac shared the 1933 Nobel Prize in Physics with Erwin Schrödinger "for the discovery of new productive forms of atomic theory". "During the intense period of 1925-26 quantum theories were proposed that accurately described the energy levels of electrons in atoms. These equations needed to be adapted to Einstein's theory of *relativity*, however. In 1928 Paul Dirac formulated a fully *relativistic* quantum theory. The equation gave solutions that he interpreted as being caused by a particle equivalent to the electron, but with a positive charge. This particle, the positron, was later confirmed through experiments." [Paul A. M. Dirac – Facts. NobelPrize.org. https://www.nobelprize.org/prizes/physics/1933/dirac/facts/.]

Dirac made fundamental contributions to the early development of both *quantum mechanics* and *quantum electrodynamics*. Among other discoveries, he formulated the *Dirac equation* which describes the behavior of fermions and predicted the existence of *antimatter*. The notion of an antiparticle to each fermion particle – e.g. the positron as antiparticle to the electron – stems from his equation. He was the first to develop *quantum field theory*, which underlies all theoretical work on sub-atomic or "elementary" particles today. He also made significant contributions to the reconciliation of *general relativity* with quantum mechanics. He proposed and investigated the concept of a magnetic monopole, an object not yet known empirically, as a means of bringing even greater symmetry to *Maxwell's Equations* of electromagnetism.

Dirac was born at his parents' home in Bristol, England, on August 8, 1902, and grew up in the Bishopston area of the city. His father, Charles, a Swiss national from Saint-Maurice, Switzerland, immigrated to London in 1890, where he worked as a teacher of French. In 1896 he moved to Bristol, where he was appointed Head of Modern Languages at the Merchant Venturers' School, where he supplemented his income with private language classes. His mother, Florence, née Holten, was the daughter of a ship's captain. Charles met her shortly after his arrival, when she was working as a librarian at the Bristol Central Library. Paul had a younger sister, Béatrice, known as Betty, and an older brother, Reginald, known as Felix, who died by suicide in March 1925.

Charles and the children were officially Swiss nationals until they became naturalized in 1919. Dirac's father was strict and authoritarian, although he disapproved of corporal punishment. Dirac had a strained relationship with his father. Charles forced Dirac to speak to him only in French so that he might learn the language. When Dirac found that he could not express what he wanted to say in French, he chose to remain silent. He grew to dislike eating, largely on account of his parents' insistence that he eat every morsel of food on his plate.

Dirac was educated first at Bishop Road Primary School, which was just around the corner from his home. Although initially he only just made the top third of his class, he steadily improved so that by the age of 10 he was consistently near the top of his class. At home he pursued his extra-curricular hobby of astronomy. The school did not teach science but gave classes in technical drawing, that provided Dirac with his unique way of thinking about science. Like many parents, Charles entered all his children for scholarship exams. Although Felix and Betty each failed one, Dirac passed every one, so was educated at minimal expense to his parents.

Dirac started his secondary education at the all-boys Merchant Venturers' School (later Cotham School), where his father worked, shortly after the outbreak of the 1st World War on August 4, 1914. For Charles it was a fifteen-minute cycle to the school, but he made his sons to walk there and back twice a day, as they had lunch at home, rather than taking the tram. The school was an institution attached to the University of Bristol, which shared grounds and staff. It emphasized technical subjects like bricklaying, shoemaking and metal work, technical drawing, and modern languages. This was unusual at a time when secondary education in Britain was still dedicated largely to the classics. It took only weeks for Dirac to establish himself as a stellar pupil. Except for history and German, he shone at every academic subject, and was usually ranked as the top student of his class. He excelled at science, including chemistry, where he learned about atoms; and he began mulling over the nature of space and time. In particular, it further advanced Dirac's ability to visualize objects and their movements in three dimensions.

Dirac's teacher, Arthur Pickering, gave up on teaching Dirac with the other boys, instead sending him to the library with a reading list. He suggested that he look beyond simple geometry to the theories of the German mathematician Bernhard Riemann.

Dirac was a workaholic, very quiet, and had no interest in sports. As the gap between the abilities of Felix and Dirac widened, their relationship deteriorated until they were no longer on speaking terms. In 1918, shortly before the end of the 1st World War, although Dirac could have taken his pick from dozens of science courses, and considered taking a degree in mathematics, he decided to follow his brother by studying engineering on a City of Bristol University Scholarship at the University of Bristol's engineering faculty, which was co-located with the Merchant Venturers' School.

On November 7, 1919, the London *Times* published its famous article about the "Revolution in Science", reporting the verification of Einstein's *Theory of General Relativity*, by Arthur Eddington's claim that they had verified the predicted bending of light by the Sun during the recent eclipse. *Relativity became Dirac's new passion*, but it was not easy to find an accessible technical account of the theory. It would also several decades before Einstein's *Theory of Special Relativity*, which applied to observers moving relative to each other at uniform speed in a straight line, could be convincingly demonstrated. In the meantime, Einstein's reasoning made it possible to amend the description of everything given by Newton's theory and produce a "special relativistic" version. Dirac began transcribing every bit of physics expressed in non-relativistic form to make it fit with special relativity. *This appears to be the origin of Dirac's unquestioning obsession with introducing special relativity into quantum theory.*

Shortly before he completed his degree in 1921, he sat for the entrance examination for St. John's College, Cambridge. He passed and was awarded a £70 scholarship, but this fell short of the amount of money required to live and study at Cambridge. Despite his having graduated with a first-class honors Bachelor of Science degree in engineering, the economic climate of the post-war depression was such that he was unable to find work as an engineer. Instead, he took up an offer to study for a Bachelor of Arts degree in mathematics at the University of Bristol free of charge. He was permitted to skip the first year of the course owing to his engineering degree.

In 1923, Dirac graduated, once again with first-class honors, and received a £140 scholarship from the Department of Scientific and Industrial Research. Along with his £70 scholarship from St John's College, this was enough to live at Cambridge. There, *Dirac pursued his interests in the theory of general relativity*, an interest he had gained earlier as a student in Bristol, and in the nascent field of *quantum physics*, under the supervision of Ralph Fowler. From 1925 to 1928 he held an 1851 Research Fellowship from the Royal Commission for the Exhibition of 1851.

Dirac's first step into a new *quantum theory* was taken late in September 1925. Fowler had received a proof copy of Heisenberg's paper [Heisenberg, W. (July, 1925). Über quantentheoretische Umdeutung kinematischer und mechanischer Beziehungen. (On the quantum-theoretical re-interpretation of kinematic and mechanical relations.) *Zeit. Phys.*, 33, 879-93], which Fowler sent on to Dirac, who was on vacation in Bristol, asking him to look into this paper carefully.

Dirac's attention was drawn to a mysterious mathematical relationship, at first sight unintelligible, that Heisenberg had established, between *non-commuting variables*. Several weeks later, back in Cambridge, Dirac suddenly recognized that this mathematical form had the same structure as the Poisson brackets that occur in the classical dynamics of particle motion. From this thought he restated Heisenberg's quantum theory in terms of *non-commuting dynamical variables* represented by Poisson brackets, and demonstrated mathematically some of the assumptions that Heisenberg had made by appealing to the *Correspondence Principle*. This led him at the age of 25 to a formulation of quantum mechanics that allowed him to obtain the *quantization rules* in a novel and illuminating manner. Dirac described the quantization of the *electromagnetic field* as an ensemble of harmonic oscillators with the introduction of the concept of creation and annihilation operators of particles. For this work, [Dirac, P. A. M. (December, 1925). The Fundamental Equations of Quantum Mechanics. *Roy. Soc. Proc., A*, 109, 752, 642-53; received November 7, 1925.] published in June 1926, the first thesis on quantum mechanics to be submitted anywhere, Dirac received a PhD from Cambridge.

Dirac was regarded by his friends and colleagues as unusual in character. In a 1926 letter to Paul Ehrenfest, Albert Einstein wrote of Dirac, "I have trouble with Dirac. This balancing on the dizzying path between genius and madness is awful." In another letter concerning the Compton effect he wrote, "I don't understand Dirac at all."

He wrote a series of papers, published mainly in the Proceedings of the Royal Society, leading up to his *relativistic* theory of the electron (1928) and the theory of *holes* (1930). This latter theory required the existence of a positive particle having the same mass and charge as the known (negative) electron. This, the positron was discovered experimentally at a later date (1932) by C. D. Anderson, while its existence was likewise proved by Blackett and Occhialini (1933) in the phenomena of "pair production" and "annihilation".

In 1928, building on 2×2 spin matrices, which Dirac purported to have discovered independently of Wolfgang Pauli's work on *non-relativistic* spin systems, he proposed the *Dirac equation* as a *relativistic equation of motion* for the *wave function* of the electron. This work led Dirac to predict the existence of the positron, the electron's antiparticle, which he interpreted in terms of what came to be called the *Dirac sea*. The *Dirac equation* also contributed to explaining the origin of *quantum spin* as a relativistic phenomenon. *However, introduction of special relativity into the wave equation resulted in a second class of solutions of the wave equation in which the energy of a free electron was negative.* [Dirac, P. A. M. (February, 1928). The Quantum Theory of the Electron. *Roy. Soc. Proc., A*, 117, 778, 610–24]; introduces vectors with *4 components* resulting in a *relativistic equation of motion* for the wave function of the electron referred to as the *Dirac equation* that describes all spin-½ particles with mass.

In the spring of 1929, he was a visiting professor at the University of Wisconsin–Madison. An anecdote recounted in a review of the 2009 biography [Pais, A. (2009). *Paul Dirac: The Man and His Work.* Cambridge University Press.] tells of Heisenberg and Dirac sailing on an ocean liner to a conference in Japan in August 1929:

> "Both still in their twenties, and unmarried, they made an odd couple. Heisenberg was a ladies' man who constantly flirted and danced, while Dirac—'an Edwardian geek', as biographer Graham Farmelo puts it—suffered agonies if forced into any kind of socializing or small talk. 'Why do you dance?' Dirac asked his companion. 'When there are nice girls, it is a pleasure,' Heisenberg replied. Dirac pondered this

> notion, then blurted out: 'But, Heisenberg, how do you know beforehand that the girls are nice?'"

Dirac's *The Principles of Quantum Mechanics*, published in 1930, is a landmark in the history of science. It quickly became one of the standard textbooks on the subject and is still used today. In that book, Dirac incorporated the previous work of Heisenberg on matrix mechanics and of Schrödinger on wave mechanics into a single mathematical formalism that associates measurable quantities to operators acting on the Hilbert space of vectors that describe the state of a physical system. The book also introduced the *Dirac delta function*. Following his 1939 article, he also included the *bra–ket notation* in the third edition of his book, thereby contributing to its universal use nowadays.

Whilst Dirac was relaxing on the Crimean coast, during one of his visits to the Soviet Union, in July 1932, Carl Anderson, working on the effects of cosmic rays in his cloud chamber at Caltech, was the first to detect the positive electron (positron) predicted by Dirac. By the autumn of 1932 this was confirmed by Patrick Blackett and an Italian visitor, Guiseppe Occhialini at the Cavendish, Cambridge University.

In the autumn of 1932, Dirac returned to considering how quantum mechanics can be developed by analogy with classical mechanics, finding another way of doing this other than by using Newton's laws, by generalizing the property of classical physics that enables the path of any object to be calculated, using the Lagrangian, where the Lagrangian is the difference between an object's kinetic and potential energy. The path taken between two points in any specified time interval is the *path of least action*, where the *action* associated with the object's path is obtained by adding the values of the Lagrangian along the path. When he generalized the idea to quantum mechanics, he found that a quantum particle has an infinite number of paths centered around the path predicted by classical mechanics. He found a way of taking into account all of the available paths by calculating their probability. [Dirac, P. A. M. (1933). The Lagrangian in Quantum Mechanics. *Phys. Zeit. Sowjet.*, 3, 1, 64-72]; alternative formulation of quantum mechanics in terms of Lagrangian in place of Hamiltonian, "*many-time*" theory.

> [***Lagrangian mechanics*** is a formulation of classical mechanics founded on the d'Alembert principle of virtual work. It was introduced by the Italian-French mathematician and astronomer Joseph-Louis Lagrange in his presentation to the Turin Academy of Science in 1760 culminating in his 1788 grand opus, *Mécanique analytique*. Lagrangian mechanics describes a mechanical system as a pair (M, L) consisting of a configuration space M and a smooth function L within that space called a ***Lagrangian***. For many systems, L = T − V, where T and V are the ***kinetic*** and ***potential energy*** of the system, respectively. The stationary ***action principle*** requires that the action functional of the system derived from L must remain at a

stationary point (specifically, a maximum, minimum, or saddle point) throughout the time evolution of the system. This constraint allows the calculation of the equations of motion of the system using Lagrange's equations.

Newton's laws and the concept of forces are the usual starting point for teaching about mechanical systems. This method works well for many problems, but for others the approach is nightmarishly complicated. Lagrangian mechanics adopts **energy** rather than **force** as its basic ingredient, leading to more abstract equations capable of tackling more complex problems. ***Lagrange's approach was to set up independent generalized coordinates for the position and speed of every object, which allows the writing down of a general form of Lagrangian (total kinetic energy minus potential energy of the system) and summing this over all possible paths of motion of the particles yielded a formula for the 'action', which he minimized to give a generalized set of equations***. This summed quantity is minimized along the path that the particle actually takes. This choice eliminates the need for the constraint force to enter into the resultant generalized system of equations. There are fewer equations since one is not directly calculating the influence of the constraint on the particle at a given moment.]

Normally, he would submit a paper like this to a British journal but this time he chose to demonstrate his support for Soviet physics by sending the paper to a new Soviet journal that was about to publish his collaborative paper on field theory. [Dirac, P. A. M., Fock, V. A., Podolsky, B. (1932). On quantum electrodynamics. *Phys. Zeit. Sowjet.*, 2, 468]; *relativistic* model in which a fixed number of electrons interact through a second-quantized electromagnetic field, applies Dirac's *interaction representation* formulation of quantum field theory to full electrodynamics. Dirac was quietly pleased with his "little paper". It was not until almost a decade later that a few theoreticians in the next generation recognized the significance of the paper. [Farmelo, G. (2009). *The Strangest Man. The hidden life of Paul Dirac*. Basic Books, New York.]

In 1933 Dirac was awarded the Nobel Prize in Physics.

In 1934, he published a paper showing how expressions for the *electric* and *current densities* can be separated into two parts, where one contains the singularities that result in an infinite number of negative-energy electrons with infinite energies, and other describes the densities physically present. [Dirac, P. A. M. (March, 1934). Discussion of the infinite distribution of electrons in the theory of the positron. *Proc. Camb. Phil. Soc.*, 30, 2, 150-63]; attempts to addresses problem of electrons with negative energy with relativistic 'hole' theory.

However, further studies by Felix Bloch with Arnold Nordsieck, and Victor Weisskopf, in 1937 and 1939, revealed that such computations were reliable only at a first order of perturbation theory, a problem already pointed out by Robert Oppenheimer. At higher orders in the series infinities emerged, making such computations meaningless and casting serious doubts on the internal consistency of the theory itself. With no solution for this problem known at the time, *it appeared that a fundamental incompatibility existed between special relativity and quantum mechanics.*

Meanwhile, after the Gamows fled the Soviet Union following the Solvay conference in 1933 and arrived in Cambridge in 1934, Dirac had a dalliance with George Gamow's wife Rho, a strikingly attractive brunette, who taught him Russian, Dirac's fourth language. Then, after the Gamows left for Copenhagen, he had another with the wife of Fellow of St. Johns, a Russian émigré poet, Lydia Jackson, who continued with his Russian tuition.

On the day after Dirac arrived in Princeton at the end of September, as a visitor at the Institute for Advanced Studies, he ran into one of his new colleagues, Eugene Wigner, having lunch with his sister, Margit, known as Manci, who was visiting from their native Hungary, Dirac, the "lonely-looking man at the next table" was invited to join them. In 1937, Dirac married Margit and adopted Margit's two children, Judith and Gabriel. Paul and Margit Dirac had two children together, both daughters, Mary Elizabeth and Florence Monica.

Einstein was at Princeton at the time of Dirac's visit, having arrived with his wife in in October 1933, who was fifty-four but looked older. The two men respected each other but there was no special warmth between them. Einstein admired the success of quantum theory but mistrusted it. During 1935, Einstein completed his collaboration with his younger research associates, Boris Podolsky and Nathan Rosen, on a paper that cast serious doubts on the conventional interpretation of the theory. [Einstein, A., Podolsky, B. & Rosen, N. (May, 1935). Can Quantum-Mechanical Description of Physical Reality Be Considered Complete? *Phys. Rev.*, 47, 777-80.]

In 1942, Dirac gave his Bakerian Lecture, which was well received. [Dirac, P. A. M. (March, 1942.) Bakerian Lecture - The physical interpretation of quantum mechanics. *Roy. Soc. Proc., A*, 180, 980, 1-40.] And, in 1945 and 1948 made two more important contributions. [Dirac, P. A. M. (April, 1945). On the Analogy Between Classical and Quantum Mechanics. *Rev. Mod. Phys.*, 17, 195]; Dirac's proposal of the *path integral formulation* of quantum mechanics, extensively developed in Feynman, R. P. (1948). Space-Time Approach to Non-Relativistic Quantum Mechanics; and [Dirac, P. A. M. (May, 1948). Quantum Theory of Localizable Dynamical Systems. *Phys. Rev.*, 73, 9, 1092-103]; *relativistic* quantum theory in terms of variables on a space-like surface in space-time, referenced in Schwinger (1948). Quantum Electrodynamics. I. A Covariant Formulation.

A possible way out of the difficulties facing quantum theory, was given by Hans Bethe in 1947, who made the first *non-relativistic* computation of the shift of the lines of the hydrogen atom as measured by Lamb and Rutherford. Despite the limitations of the computation, agreement was excellent. The idea was simply to attach infinities to corrections of *mass* and *charge* that were actually fixed to a finite value by experiments. In this way, the infinities get absorbed in those constants and yield a finite result in good agreement with experiments. This procedure was named *renormalization.*

Even though *renormalization* works very well in practice, Dirac never accepted it, Dirac commented in 1975: "I must say that I am very dissatisfied with the situation because this so-called 'good theory' does involve neglecting infinities which appear in its equations, neglecting them in an arbitrary way. This is just not sensible mathematics. Sensible mathematics involves neglecting a quantity when it is small – not neglecting it just because it is infinitely great and you do not want it!" [Kragh, Helge (1990). Dirac: A Scientific Biography. Cambridge: Cambridge University Press.]. His final judgment on quantum field theory in his last paper was that "These rules of *renormalization* give surprisingly, excessively good agreement with experiments. Most physicists say that these working rules are, therefore, correct. I feel that is not an adequate reason. Just because the results happen to be in agreement with observation does not prove that one's theory is correct." [Dirac, P. A. M. (1987). The inadequacies of quantum field theory. In *Paul Adrien Maurice Dirac*, page 194. B. N. Kursunoglu and E. P. Wigner, eds., Cambridge University Press.] Nor was Feynman entirely comfortable with its mathematical validity, even referring to *renormalization* as a "shell game" and "hocus pocus" [Feynman, Richard (1985). *QED: The Strange Theory of Light and Matter*, page 128. Princeton University Press.]

Dirac was the Lucasian Professor of Mathematics at the University of Cambridge, was a member of the Center for Theoretical Studies, University of Miami, and spent the last decade of his life at Florida State University. Dirac was also awarded the Royal Medal in 1939 and both the Copley Medal and the Max Planck Medal in 1952. He was elected a Fellow of the Royal Society in 1930, an Honorary Fellow of the American Physical Society in 1948, and an Honorary Fellow of the Institute of Physics, London in 1971. He received the inaugural J. Robert Oppenheimer Memorial Prize in 1969. Dirac became a member of the Order of Merit in 1973, having previously turned down a knighthood as he did not want to be addressed by his first name. In 1984, Dirac died in Tallahassee, Florida, and was buried at Tallahassee's Roselawn Cemetery.

Dirac, P. A. M. (February, 1928). The Quantum Theory of the Electron.

Roy. Soc. Proc., A, 117, 778, 610–24; https://doi.org/10.1098/rspa.1928.0023.

Communicated by R. H. Fowler, F.R.S.

Received January 2, 1928.

St. John's College, Cambridge.

Dirac noted that the new quantum mechanics applied to the problem of the structure of the atom with *point-charge electrons* did not give results in agreement with experiment. The discrepancies consisted of "duplexity" phenomena; the observed number of stationary states for an electron in an atom being twice the number given by the theory. Goudsmit and Uhlenbeck introduced the idea of an electron with a *spin*. Previous *relativity* treatments by Gordon and Klein obtained the operator of the wave equation by the same procedure as in the *non-relativity* theory; they substituted classical *quantum differential operators* for the *momentum vector* in the amended *relativistic Hamiltonian equation* and applied the resulting differential operator to the *wave function* to obtain the *Klein-Gordon equation*. Dirac noted that Gordon and Klein's treatments gave rise to two difficulties. The *first difficulty* was in the physical interpretation of wave-mechanical expressions for the *charge* and the *current*. This was satisfactory for emission and absorption of radiation, but only provided the probability of any dynamical variable at any specific time having a value between specified limits if they referred to the *position* of the electron, but, unlike the *non-relativity* theory, *not if they refer to its momentum or any other dynamical variable*. The *second difficulty* was that the conjugate imaginary of the wave equation was the same as that for an electron with charge – e and negative energy. *This paper only addressed the removal of the first of difficulties*. The resulting theory was only an approximation but appeared sufficient to address the duplexity problems without further assumptions. Dirac applied the method of *q-numbers* and using non-commutative algebra exhibited the properties of a free electron and of an electron in a central field of electric force. He showed that simplest Hamiltonian for a *point charge electron satisfying requirements of both relativity and the general transformation theory* of quantum mechanics led to an explanation of all duplexity phenomena of number of stationary states being twice the observed value *without further assumption about spin*. In contrast to the Schrödinger equation which described wave functions of only one complex value, Dirac introduced *vectors of four complex numbers* (known as bispinors). This resulted in a *relativistic equation of motion* for the *wave function of the electron* referred to as the *Dirac equation*, $\{p_0 + \rho_1 (\boldsymbol{\sigma}, \mathbf{p}) + \rho_3 mc\}\ \psi = 0$, where **p** is the *momentum* vector, and $\boldsymbol{\sigma}$ denotes the vector $(\sigma_1, \sigma_2, \sigma_3)$. This included a term equal to the spin correction given by Darwin and Pauli. It described all spin-½ particles with mass, but did not address the second class of solutions of the wave equation in which *charge of the electron is positive* and *energy of a free electron is negative*.

[This work led Dirac to predict the existence of the *positron*, the electron's antiparticle, which he interpreted in terms of what came to be called the *Dirac sea*. The *positron* was observed by Carl Anderson in 1932.]

The new quantum mechanics, when applied to the problem of the *structure of the atom with point-charge electrons*, does not give results in agreement with experiment. The discrepancies consist of "duplexity" phenomena, the observed number of stationary states for an electron in an atom being twice the number given by the theory. To meet the difficulty, Goudsmit and Uhlenbeck have *introduced the idea of an electron with a spin angular momentum of half a quantum* and a *magnetic moment* of one Bohr magneton[1].

[1] Uhlenbeck, G.E. & Goudsmit, S.A. (1925). Ersetzung der Hypothese vom unmechanischen Zwang durch eine Forderung bezüglich des inneren Verhaltens jedes einzelnen Elektrons. (Replacement of the hypothesis of unmechanical coercion by a requirement regarding the internal behavior of each individual electron.) *Naturwiss.*, 13, 953-54; http://dx.doi.org/10.1007/BF01558878.

This model for the electron has been fitted into the new mechanics by Pauli*, and Darwin[#], working with an equivalent theory, has shown that it gives results in agreement with experiment for hydrogen-like spectra to the first order of accuracy.

* Pauli, W. (September, 1927). Zur Quantenmechanik des magnetischen Elektrons. (On the quantum mechanics of magnetic electrons.) *Zeit. Phys.*, 43, 601-23[; shows how the *non-relativistic* formulation by Dirac [Dirac (January, 1927). The Physical Interpretation of the Quantum Dynamics] and Jordan using the general canonical transformations of the Schrödinger functions enables a quantum-mechanical representation of electrons by the method of *eigenfunctions*, the differential equations for the *eigenfunctions* of the magnetic electron that are given in the present paper can be regarded as only provisional and approximate, like the Heisenberg-Jordan matrix formulation they *are not written down in a relativistically-invariant way*, for the hydrogen atom they are valid only in the approximation in which the dynamical behavior of the proper moment can be considered to be a secular perturbation].

[#] Darwin, C. G. (September, 1927). The Electron as a Vector Wave. *Roy. Soc. Proc.*, A, 116, 773, 227-53[; difficulties in interpretation of the spinning electron in terms of wave theory, *wave functions with 2 components*, should be interpreted in terms of a *vector*, but vector found to be in some degree arbitrary, when *relativity* transformation is applied to identify the "doublet effect" with the Zeeman effect gives value for the doublet separation twice as great as it should be, not at present possible to see what form the Thomas correction should take in the wave theory, the trouble is no doubt connected with the fact that the hydrogen spectrum has only been verified to a first approximation and goes wrong in the second—a difficulty at present shared by all theories].

The question remains as to why Nature should have chosen this particular model for the electron instead of being satisfied with the *point-charge*. One would like to find some incompleteness in the previous methods of applying quantum mechanics to the *point-charge electron* such that, when removed, the whole of the duplexity phenomena follow without arbitrary assumptions. In the present paper it is shown that this is the case, *the incompleteness of the previous theories lying in their disagreement with relativity*, or, alternatively, *with the general transformation theory of quantum mechanics*. It appears that *the simplest Hamiltonian for a point-charge electron satisfying the requirements of both relativity and the general transformation theory leads to an explanation of all duplexity phenomena without further assumption*.

All the same there is a great deal of truth in the spinning electron model, at least as a first approximation. *The most important failure of the model seems to be that the magnitude of the resultant orbital angular momentum of an electron moving in an orbit in a central field of force is not a constant*, as the model leads one to expect.

§ 1. *Previous Relativity Treatments*.

The *relativity Hamiltonian* according to the classical theory *for a point electron moving in an arbitrary electro-magnetic field with scalar potential A_0 and vector potential A is*

$$F = (W/c + e/c\,A_0)^2 + (\mathbf{p} + e/c\,\mathbf{A})^2 + m^2c^2,$$

where **p** is the *momentum* vector. It has been suggested by Gordon*

* Gordon, W. (January, 1927). Der Comptoneffekt nach der Schrödingerschen Theorie. (The Compton effect according to Schrödinger's theory.) *Zeit. Phys*., 40, 117-33[; Heisenberg and Schrödinger provided alternative methods for determination of quantum *frequencies* and *intensities*, Compton effect already calculated by Dirac (June, 1926) using Heisenberg method, here the same problem treated by Schrödinger method, starts with the same *classic relativistic equation for kinetic energy* in terms of *momentum* and *energy*, which is *Hamiltonian equation* for the system, introduces same imaginary variables for *time* and *energy* to create same space-time symmetric form, applies in same way to *electron in electromagnetic field* described in terms of *vector potential* and *scalar potential*, and introduces same imaginary variable for scalar potential, adds the same *field energy* to the *kinetic energy* resulting in the same *classical relativistic Hamiltonian equations for a point electron moving in an electromagnetic field*, in accordance with Schrödinger's rules Gordon then substitutes the classical *quantum differential operators* for the momentum vector in the amended *Hamiltonian equation* and applies resulting differential operator to the *wave function* ψ to obtain the *Klein-Gordon equation*, $1/c^2\, \partial^2/\partial t^2\, \psi - \nabla^2\, \psi + m^2c^2/h^2\, \psi = 0$, (Dirac [February, 1928). The Quantum Theory of the Electron.] objected to this substitution on grounds of the interpretation of the wave function, and solutions with negative probabilities, negative energy, and positive charge for the electron); calculates radiation from *current density* and *charge density*, applies to Compton effect.

that *the operator of the wave equation of the quantum theory should be obtained from this F by the same procedure as in non-relativity theory, namely, by putting*

$$W = ih\, \partial/\partial t$$
$$p_r = -\, ih\, \partial/\partial x_r, \qquad r = 1, 2, 3,$$

in it.

This gives the *wave equation*

$$F\psi = \{(ih\, \partial/c\partial t + e/c\, A_0)^2 + \Sigma_r\, (-\, ih\, \partial/\partial x_r + e/c\, A_r)^2 + m^2c^2\}\ \psi = 0, \ (1)$$

the *wave function* ψ being a function of x_1, x_2, x_3, t. *This gives rise to two difficulties.*

The first is in connection with the physical interpretation of ψ. Gordon, and also independently Klein*,

* Klein, O. (October, 1927). Elektrodynamik und Wellenmechanik vom Standpunkt des Korrespondenzprinzips. (Electrodynamics and wave mechanics from the standpoint of the correspondence principle.) *Zeit. Phys.*, 41, 10, 407-42[; alternative calculation of Compton effect restricted to the *one-electron problem*, starts from Maxwell-Lorentz field equations, describes motion of an electron in an electromagnetic field by *four-potential* and *scalar potential*, regards *Hamilton-Jacobi differential equation* for the action function (Klein–Gordon equation) as expression for motion of the electron, following de Broglie and Schrodinger replaces this first order equation with a second-order linear equation representing *relativistic* generalization of Schrödinger's wave equation for one-electron problem, evaluates equations determining the electromagnetic field with the help of wave mechanics using the correspondence principle to determine wave-mechanical expressions for *electric density* and *current vector*, after neglecting relativity results in the same expressions as those obtained by Schrodinger, applies to a "bound" electron moving in an axially symmetric electrostatic field over which a weak homogeneous magnetic field is superimposed to derive normal Zeeman effect, applies to scattered radiation from a light wave on a "force-free" electron to obtain the Compton effect, five-dimensional wave mechanics].

from considerations of the conservation theorems, make the assumption that if ψ_m, ψ_n are two solutions

$$\rho_{mn} = -\, e/2mc^2\, \{ih\, (\psi_m \partial\psi_n/\partial t - \psi_n^- \, \partial\psi_m/\partial t) + 2eA_0\, \psi_m\psi_n^-\},$$

and

$$I_{mn} = -\, e/2m\, \{-\, ih\, (\psi_m \text{ grad } \psi_n^- - \psi_n^- \text{ grad } \psi_m) + 2\, e/c\, A_m\, \psi_m\psi_n^-\}$$

are to be interpreted as the *charge* and *current* associated with the transition *m* →*n.*

This appears to be satisfactory so far as *emission* and *absorption* of radiation are concerned, but is not so general as the interpretation of the *non-relativity quantum mechanics*, which has been developed[#] sufficiently to enable one to answer the question: *What is the probability of any dynamical variable at any specified time having a value lying between any specified limits*, when the system is represented by a given *wave function* ψ_n?

> [#] Jordan, P. (November, 1927). Über eine neue Begründung der Quantenmechanik. (On a new justification of quantum mechanics.) *Zeit. Phys.*, 40, 809-38; https://doi.org/10.1007/BF01390903; Dirac, P. A. M. (January, 1927). The Physical Interpretation of the Quantum Dynamics. *Roy. Soc. Proc., A*, 113, 765, 621-41[; *non-relativistic* matrix mechanics, Heisenberg's original matrix mechanics assumed that the elements of the diagonal matrix that represents the energy are the *energy levels* of the system, and the elements of the matrix that represents the total polarization, which are periodic functions of the time, determine the *frequencies* and *intensities* of the spectral lines in analogy to classical theory, in *Schrodinger's wave representation* physical results are based on assumption that the square of the *amplitude* of the wave function can be interpreted as a probability, enables probability of a *transition* being produced in a system by an arbitrary external perturbing force to be worked out, this paper provides a *general theory of obtaining physical results from quantum theory*, it shows all the physical information that one can hope to get from quantum dynamics and provides a general method for obtaining it, replaces special assumptions previously used, requires a theory of the more general schemes of matrix representation in which the rows and columns refer to any set of constants of integration that commute and of the laws of transformation from one such scheme to another, *does not take relativity mechanics into account*, counts time variable wherever it occurs as a parameter (a c-number), *transformation equations* that satisfy *quantum conditions* and *equations of motion*, *eigenfunctions* of Schrodinger's wave equation as *transformation functions* that enable transformation from scheme of matrix representation to scheme in which Hamiltonian is a diagonal matrix, dynamical variables represented by matrices whose rows and columns refer to the initial values of the *action variables* or to the *final values*, coefficients that enable transformation from one set of matrices to the other are those that determine the *transition probabilities*].

The Gordon-Klein interpretation can answer such questions if they refer to the position of the electron (by the use of ρ_{nm}*), but not if they refer to its momentum, or angular momentum or any other dynamical variable*. We should expect the interpretation of the *relativity* theory to be just as general as that of the *non-relativity* theory.

The general interpretation of *non-relativity quantum mechanics* is based on the *transformation theory*, and is made possible by the *wave equation* being of the form

$$(H - W)\,\psi = 0, \qquad (2)$$

i.e., being linear in W or $\partial/\partial t$, so that the *wave function* at any time determines the *wave function* at any later time. *The wave equation of the relativity theory must also be linear in W if the general interpretation is to be possible.*

The second difficulty in Gordon's interpretation arises from the fact that if one takes the *conjugate imaginary* of equation (1), one gets

$$\{(-W/c + e/c\, A_0)^2 + (-\mathbf{p} + e/c\, \mathbf{A})^2 + m^2c^2\}\ \psi = 0,$$

which is the same as one would get if one put – e for e. *The wave equation (1) thus refers equally well to an electron with charge e as to one with charge – e.* If one considers for definiteness the limiting case of large *quantum* numbers one would find that some of the solutions of the *wave equation* are *wave packets* moving in the way a particle of charge – e would move on the classical theory, while others are *wave packets* moving in the way a particle of charge e would move classically. *For this second class of solutions W has a negative value.*

Introduction of *special relativity* into the wave equation for the electron results in solutions with negative energy.

One gets over the difficulty on the classical theory by arbitrarily excluding those solutions that have a negative W. *One cannot do this on the quantum theory, since in general a perturbation will cause transitions from states with W positive to states with W negative.* Such a transition would appear experimentally as the electron suddenly changing its charge from – e to e, a phenomenon which has not been observed. *The true relativity wave equation should thus be such that its solutions split up into two non-combining sets, referring respectively to the charge – e and the charge e.*

In the present paper we shall be concerned only with the removal of the first of these two difficulties. The resulting theory is therefore still only an approximation, but it appears to be good enough to account for all the duplexity phenomena without arbitrary assumptions.

§ 2. *The Hamiltonian for no field.*

Our problem is to obtain a wave equation of the form (2)

$$[(H - W)\ \psi = 0, \qquad (2)]$$

which shall be invariant under a Lorentz transformation and shall be equivalent to (1)

$$[F\psi = \{(ih\ \partial/c\partial t + e/c\, A_0)^2 + \Sigma_r\, (-\,ih\ \partial/\partial x_r + e/c\, A_r)^2 + m^2c^2\}\ \psi = 0, \qquad (1)]$$

in the limit of large quantum numbers.

We shall consider first *the case of no field* [$A_0 = A_r = 0$], when [the *wave*] *equation* (1) reduces to

$$(-\,p_0^2 + p^2 + m^2c^2)\ \psi = 0 \qquad (3)$$

if one puts

$$p_0 = W/c = ih\ \partial/c\partial t\ [\text{and } p_r = -\,ih\ \partial/\partial x_r,\ (r = 1, 2, 3) \text{ so } p = \Sigma_r\, (-\,ih\ \partial/\partial x_r)].$$

The symmetry between p_0 and p_1, p_2, p_3 required by relativity shows that, since the Hamiltonian we want is linear in p_0, it must also be linear in p_1, p_2 and p_3. Our *wave equation* is therefore of the form

$$(p_0 + \alpha_1 p_1 + \alpha_2 p_2 + \alpha_3 p_3 + \beta)\,\psi = 0 \qquad (4)$$

where for the present all that is known about the dynamical variables or operators α_1, α_2, α_3, β is that they are independent of p_0, p_1, p_2, p_3, i.e. that they commute with t, x_1, x_2, x_3. Since we are considering the case of a particle moving in empty space, so that all points in space are equivalent, *we should expect the Hamiltonian not to involve t, x_1, x_2, x_3.* This means that α_1, α_2, α_3, β are independent of t, x_1, x_2, x_3, i.e., that they commute with p_0, p_1, p_2, p_3. *We are therefore obliged to have other dynamical variables besides the coordinates and momenta of the electron, in order that α_1, α_2, α_3, β may be functions of them.* The *wave function* ψ must then involve more variables than merely x_1, x_2, x_3, t.

Equation (4)

$$[(p_0 + \alpha_1 p_1 + \alpha_2 p_2 + \alpha_3 p_3 + \beta)\,\psi = 0 \qquad (4)]$$

leads to

$$0 = (-p_0 + \alpha_1 p_1 + \alpha_2 p_2 + \alpha_3 p_3 + \beta)(p_0 + \alpha_1 p_1 + \alpha_2 p_2 + \alpha_3 p_3 + \beta)\,\psi$$
$$= \{-p_0^2 + \sum \alpha_1^2 p_1^2 + \sum(\alpha_1\alpha_2 + \alpha_1\alpha_2)p_1 p_2 + \beta^2 + \sum(\alpha_1\beta + \beta\alpha_1)p_1\}\,\psi, \quad (5)$$

where the $\sum$ refers to cyclic permutation of the suffixes 1, 2, 3. This agrees with (3)

$$[(-p_0^2 + p^2 + m^2c^2)\,\psi = 0 \qquad (3)]$$

if

$$\alpha_r^2 = 1, \qquad \alpha_r\alpha_s + \alpha_s\alpha_r = 0, \quad (r \neq s) \qquad r, s = 1, 2, 3.$$
$$\beta^2 = m^2c^2, \qquad \alpha_r\beta + \beta\alpha_r = 0.$$

If we put $\beta = \alpha_4 mc$, these conditions become

$$\alpha_\mu^2 = 1 \qquad \alpha_\mu\alpha_\nu + \alpha_\nu\alpha_\mu = 0 \quad (\mu \neq \nu) \qquad \mu, \nu = 1, 2, 3, 4. \qquad (6)$$

We can suppose the α_μ's to be expressed as matrices in some matrix scheme, the matrix elements of α_μ being, say, $\alpha_\mu(\zeta'\,\zeta'')$. The *wave function* ψ must now be a function of ζ as well as x_1, x_2, x_3, t. The result of α_μ multiplied into ψ will be a function $(\alpha_\mu\,\psi)$ of x_1, x_2, x_3, t, ζ defined by

$$(\alpha_\mu\,\psi)\,(x, t, \zeta) = \Sigma_{\zeta'}\,\alpha_\mu(\zeta\,\zeta')\,\psi\,(x, t, \zeta').$$

We must now find four matrices α_μ to satisfy the conditions (6). We make use of the matrices

$$\sigma_1 = \begin{pmatrix} 0 & 1 \\ 1 & 0 \end{pmatrix} \qquad \sigma_2 = \begin{pmatrix} 0 & -i \\ i & 0 \end{pmatrix} \qquad \sigma_3 = \begin{pmatrix} 1 & 0 \\ 0 & 1 \end{pmatrix}$$

which Pauli introduced* to describe the three components of *spin angular momentum.*

* Pauli, *loc. cit.* [Pauli, W. (September, 1927). Zur Quantenmechanik des magnetischen Elektrons. (On the quantum mechanics of magnetic electrons.) *Zeit. Phys.*, 43, 601-23.]

These matrices have just the properties

$$\alpha_r^2 = 1 \qquad \alpha_r\alpha_s + \alpha_s\alpha_r = 0 \qquad (r \neq s), \tag{7}$$

that we require for our α's. We cannot, however, just take the α's to be three of our α's, because then it would not be possible to find the fourth. We must extend the α's in a diagonal manner to bring in two more rows and columns, so that we can introduce three more matrices ρ_1, ρ_2, ρ_3 of the same form as σ_1, σ_2, σ_3 but referring to different rows and columns, thus:

$$\sigma_1 = \begin{pmatrix} 0 & 1 & 0 & 0 \\ 1 & 0 & 0 & 0 \\ 0 & 0 & 0 & 1 \\ 0 & 0 & 1 & 0 \end{pmatrix} \qquad \sigma_2 = \begin{pmatrix} 0 & -i & 0 & 0 \\ i & 0 & 0 & 0 \\ 0 & 0 & 0 & -i \\ 0 & 0 & i & 0 \end{pmatrix} \qquad \sigma_3 = \begin{pmatrix} 1 & 0 & 0 & 0 \\ 0 & -1 & 0 & 0 \\ 0 & 0 & 1 & 0 \\ 0 & 0 & 0 & -1 \end{pmatrix}$$

$$\rho_1 = \begin{pmatrix} 0 & 0 & 1 & 0 \\ 0 & 0 & 0 & 1 \\ 1 & 0 & 0 & 0 \\ 0 & 1 & 0 & 0 \end{pmatrix} \qquad \rho_2 = \begin{pmatrix} 0 & 0 & -i & 0 \\ 0 & 0 & 0 & -i \\ i & 0 & 0 & 0 \\ 0 & i & 0 & 0 \end{pmatrix} \qquad \rho_3 = \begin{pmatrix} 1 & 0 & 0 & 0 \\ 0 & 1 & 0 & 0 \\ 0 & 0 & -1 & 0 \\ 0 & 0 & 0 & -1 \end{pmatrix}$$

The ρ's are obtained from the σ's by interchanging the second and third rows, and the second and third columns. We now have, in addition to equations (7)

$$\rho_r^2 = 1 \qquad \rho_r\rho_s + \rho_s\rho_r = 0 \qquad (r \neq s), \tag{7'}$$

and also

$$\rho_r\sigma_t = \sigma_s\rho_r.$$

If we now take

$$\alpha_1 = \rho_1\sigma_1, \qquad \alpha_2 = \rho_1\sigma_2, \qquad \alpha_3 = \rho_1\sigma_3, \qquad \alpha_4 = \rho_3,$$

all the conditions (6)

$$[\alpha_\mu^2 = 1 \qquad \alpha_\mu\alpha_\nu + \alpha_\nu\alpha_\mu = 0 \quad (\mu \neq \nu) \qquad \mu, \nu = 1, 2, 3, 4. \qquad (6)]$$

are satisfied, e.g.,

$$\alpha_1^2 = \rho_1\sigma_1\rho_1\sigma_1 = \rho_1^2\,\sigma_1^2 = 1$$
$$\alpha_1\alpha_2 = \rho_1\sigma_1\rho_1\sigma_2 = \rho_1^2\,\sigma_1\sigma_2 = -\,\rho_1^2\sigma_2\sigma_1 = -\,\alpha_2\alpha_1.$$

The following equations are to be noted for later reference

$$\rho_1\rho_2 = i\rho_3 = -\,\rho_2\rho_1$$
$$\sigma_1\sigma_2 = i\sigma_3 = -\,\sigma_2\sigma_1 \tag{8}$$

together with the equations obtained by cyclical permutation of the suffixes.

The wave equation (4)

$$[(p_0 + \alpha_1 p_1 + \alpha_2 p_2 + \alpha_3 p_3 + \beta)\ \psi = 0 \qquad (4)]$$

now takes the form [the *Dirac equation*]

$$\{p_0 + \rho_1 (\boldsymbol{\sigma}, \mathbf{p}) + \rho_3 mc\}\ \psi = 0 \qquad (9)$$

where [**p** is the *momentum* vector, $p_0 = ih/c\ \partial/\partial t$, $p_r = -\ ih\ \partial/\partial x_r$, r = 1, 2, 3; and] **σ** denotes the vector $(\sigma_1, \sigma_2, \sigma_3)$

[where $\sigma_1, \sigma_2, \sigma_3$ are the matrices

$$\sigma_1 = \begin{pmatrix} 0 & 1 & 0 & 0 \\ 1 & 0 & 0 & 0 \\ 0 & 0 & 0 & 1 \\ 0 & 0 & 1 & 0 \end{pmatrix} \qquad \sigma_2 = \begin{pmatrix} 0 & -i & 0 & 0 \\ i & 0 & 0 & 0 \\ 0 & 0 & 0 & -i \\ 0 & 0 & i & 0 \end{pmatrix} \qquad \sigma_3 = \begin{pmatrix} 1 & 0 & 0 & 0 \\ 0 & -1 & 0 & 0 \\ 0 & 0 & 1 & 0 \\ 0 & 0 & 0 & -1 \end{pmatrix}];$$

ρ_1 and ρ_3 are the matrices

$$\rho_1 = \begin{pmatrix} 0 & 0 & 1 & 0 \\ 0 & 0 & 0 & 1 \\ 1 & 0 & 0 & 0 \\ 0 & 1 & 0 & 0 \end{pmatrix} \qquad \rho_3 = \begin{pmatrix} 1 & 0 & 0 & 0 \\ 0 & 1 & 0 & 0 \\ 0 & 0 & -1 & 0 \\ 0 & 0 & 0 & -1 \end{pmatrix};$$

m is the mass of the electron; c is the speed of light].

§ 3. *Proof of Invariance under a Transformation.*

Multiply equation (9) [*Dirac equation*]

$$[\{p_0 + \rho_1 (\boldsymbol{\sigma}, \mathbf{p}) + \rho_3 mc\}\ \psi = 0 \qquad (9)]$$

by ρ_3 on the left-hand side. It becomes, with the help of (8)

$$[\rho_1\rho_2 = i\rho_3 = -\ \rho_2\rho_1$$
$$\sigma_1\sigma_2 = i\sigma_3 = -\ \sigma_2\sigma_1, \qquad (8)]$$

$$\{\rho_3 p_0 + i\rho_2 (\sigma_1 p_1 + \sigma_2 p_2 + \sigma_3 p_3) + mc\}\ \psi = 0$$

Putting

$$p_0 = ip_4,$$
$$\rho_3 = \gamma_4, \qquad \rho_2\sigma_r = \gamma_r, \qquad r = 1, 2, 3, \qquad (10)$$

we have [the *Dirac equation*]

$$\{i\Sigma\gamma_\mu p_\mu + mc\}\ \psi = 0, \quad \mu = 1, 2, 3, 4. \qquad (11)$$

The p_μ transform under a *Lorentz transformation* according to the law

$p_\mu' = \Sigma_\nu \alpha_{\mu\nu} p_\nu$,

where the coefficients $\alpha_{\mu\nu}$ are c-numbers satisfying

$\Sigma_\nu \alpha_{\mu\nu}\alpha_{\mu\tau} = \delta_{\nu\tau}$, $\quad\quad \Sigma_\tau \alpha_{\mu\tau}\alpha_{\nu\tau} = \delta_{\mu\nu}$.

The *wave equation* therefore transforms into

$\{i\Sigma\gamma_\mu' p_\mu' + mc\}\ \psi = 0,$ (12)

where

$\gamma_\mu' = \Sigma_\nu \alpha_{\mu\nu} \gamma_\nu$.

...

Thus, by a succession of *canonical transformations*, which can be combined to form a single *canonical transformation*, the ρ''s and σ''s can be brought into the form of the ρ's and σ's. The new *wave equation* (12) can in this way be brought back into the form of the original *wave equation* (11) or (9) [*Dirac equation*]

[$\{i\Sigma\gamma_\mu p_\mu + mc\}\ \psi = 0,\ \mu = 1, 2, 3, 4.$ (11)

$\{p_0 + \rho_1 (\boldsymbol{\sigma}, \mathbf{p}) + \rho_3 mc\}\ \psi = 0,$ (9)]

so that the results that follow from this original *wave equation* must be *independent of the frame of reference used.*

§ 4. *The Hamiltonian for an Arbitrary Field.*

To obtain the Hamiltonian for an electron in an *electromagnetic field* with *scalar potential* A_0 and *vector potential* **A**, we adopt the usual procedure of substituting $p_0 + e/c\ .\ A_0$ for p_0 and $\mathbf{p} + e/c\ .\ \mathbf{A}$ for **p** in the Hamiltonian for no field. From equation (9) we thus obtain [the *Dirac equation*]

$\{p_0 + e/c\ .\ A_0 + \rho_1 (\boldsymbol{\sigma}, \mathbf{p} + e/c\ \mathbf{A}) + \rho_3 mc\}\ \psi = 0.$ (14)

[*The Dirac equation*:

Alternative forms of the *Dirac equation*;

$\{p_0 + \rho_1 (\boldsymbol{\sigma}, \mathbf{p}) + \rho_3 mc\}\ \psi = 0;$ (9)

$\{i\Sigma\gamma_\mu p_\mu + mc\}\ \psi = 0,\ (\mu = 1, 2, 3, 4),$ (11)

where $p_0 = ip_4$, $\rho_3 = \gamma_4$, $\rho_2\sigma_r = \gamma_r$, $(r = 1, 2, 3)$;

$\{p_0 + e/c\ .\ A_0 + \rho_1 (\boldsymbol{\sigma}, \mathbf{p} + e/c\ \mathbf{A}) + \rho_3 mc\}\ \psi = 0.$ (14)

Other formulations include:

(a) $\{p_0 + e/c\ A_0 + \alpha_1 (p_1 + e/c\ A_1) + \alpha_2 (p_2 + e/c\ A_2) + \alpha_3 (p_3 + e/c\ A_3) + \alpha_4 mc\}\ \psi = 0.$

See Dirac, P. A. M. (March, 1928). The quantum theory of the Electron. Part II. *Roy. Soc. Proc., A*, 118, 779, 351-61, formula (1);

(b) $\{-i\hbar\gamma^\mu\delta_\mu + mc\}\ \psi = 0,$

where $\gamma^\mu\delta_\mu = (\beta^2/c\ \delta_t + \beta\alpha_1\delta_x + \beta\alpha_2\delta_y + \beta\alpha_3\delta_z) = \beta\ (\beta/c\ \delta_t + \sum^3_{n=1} \alpha_n\delta_n)$;

(c) $(\beta mc^2 + c \sum^3_{n=1} \alpha_n p_n)\, \psi(x, t) = i\hbar\, \partial\psi(x, t)/\partial t$, or

$\{- i\hbar/c\, \partial\psi(x, t)/\partial t + \sum^3_{n=1} \alpha_n p_n)\, \psi(x, t) + \beta mc\, \psi(x, t)\} = 0$,

where,

$\psi = \psi(x, t)$ is the wave function for the electron of rest mass

m is the rest mass of the electron

x, t are the space-time coordinates

$p_n = -\, ih\, \partial/\partial x_n$, (n = 1, 2, 3) are the momentum components (momentum operator in the Schrödinger equation)

c is the speed of light

$\hbar$ is the reduced Planck constant

α_1, α_2, α_3 and β are 4 x 4 matrices.

The new elements in this equation are the four 4 × 4 matrices α_1, α_2, α_3 and β, and the four-component wave function ψ. There are four components in ψ because the evaluation of it at any given point in configuration space is a *bispinor*. ***It is interpreted as a superposition of a spin-up electron, a spin-down electron, a spin-up positron, and a spin-down positron***. The 4 × 4 matrices α_k and β are all *Hermitian* and are *involutory*, and they all mutually *anticommute.*

[A *Hermitian matrix* (or self-adjoint matrix) is a complex square matrix that is equal to its own *conjugate transpose*—that is, the element in the i-th row and j-th column is equal to the complex *conjugate* of the element in the j-th row and i-th column, for all indices i and j.

A square matrix that is its own inverse, i.e. multiplication by the matrix A is an *involution* if and only if $A^2 = I$, where I is the n × n identity matrix:

$\alpha_i^2 = \beta^2 = I_4$

Two matrices M and N *anticommute* if M N = – N M:

$\alpha_i\alpha_j + \alpha_j\alpha_i = 0 \qquad (i \neq j)$

$\alpha_i\beta + \beta\alpha_i = 0$.]

The *Dirac equation* is superficially similar to the *Schrödinger equation* for a free particle with mass:

$-\, h^2/2m\, \nabla^2\phi = ih\, \partial/\partial t\, \phi$.

The left side represents the square of the momentum operator divided by twice the mass, which is the *non-relativistic kinetic energy*. Because *relativity* treats space and time as a whole, *a relativistic generalization of this equation requires that space and time derivatives must enter symmetrically as they do in the Maxwell equations that govern the behavior of light — the equations must be differentially of the same order in space and time*. In *relativity*, the momentum and the energies are the space and time parts of a *space-time* vector, the *four-momentum*, and they are related by the *relativistically* invariant relation

$$E^2 = m^2c^4 + p^2c^2$$

which says that the length of this *four-vector* is proportional to the *rest mass* m.]

This wave equation appears to be sufficient to account for all the duplexity phenomena. On account of the matrices ρ and σ containing four rows and columns, it will have four times as many solutions as the *non-relativity wave equation*, and twice as many as the previous *relativity wave equation* (1)

$$[F\psi = \{(ih\, \partial/c\partial t + e/c\, A_0)^2 + \Sigma_r (-\, ih\, \partial/\partial x_r + e/c\, A_r)^2 + m^2c^2\}\ \psi = 0. \quad (1)]$$

Since half the solutions must be rejected as referring to the charge + e on the electron, the correct number will be left to account for duplexity phenomena. The proof given in the preceding section of invariance under a Lorentz transformation applies equally well to the more general *wave equation* (14)

$$[\{p_0 + e/c\, .\, A_0 + \rho_1\, (\boldsymbol{\sigma}, \mathbf{p} + e/c\, \mathbf{A}) + \rho_3 mc\}\ \psi = 0. \qquad (14)]$$

We can obtain a rough idea of how (14) differs from the previous *relativity wave equation* (1) by multiplying it up analogously to (5). This gives, if we write e' for e/c …

…

Thus (15) becomes

$$0 = \{-\, (p_0 + e'A_0)2 + (\mathbf{p} + e'\mathbf{A})2 + m^2c^2 + e'h(\boldsymbol{\sigma}, \text{curl}\ \mathbf{A})$$
$$-\, ie'h\rho_1\, (\boldsymbol{\sigma}, \text{grad}\ A_0 + 1/c\ \partial\mathbf{A}/\partial t)\}\ \psi$$

$$= \{-\, (p_0 + e'A_0)2 + (\mathbf{p} + e'\mathbf{A})2 + m^2c^2 + e'h(\boldsymbol{\sigma}, \mathbf{H})$$
$$-\, ie'h\rho_1\, (\boldsymbol{\sigma}, \mathbf{E})\}\ \psi$$

where **E** and **H** are the *electric and magnetic vectors of the field.*

This differs from (1) by the two extra terms

$$eh/c\ (\boldsymbol{\sigma}, \mathbf{H}) + ieh/c\ \rho_1\ (\boldsymbol{\sigma}, \mathbf{E})$$

in F [where **E** and **H** are the *electric* and *magnetic vectors* of the field].

These two terms, when divided by the factor 2m can be regarded as the additional *potential energy* of the electron due to its new degree of freedom. The electron will therefore behave as though it has a *magnetic moment* eh/2mc. **σ** and an *electric moment* ieh/2mc. $\rho_1\boldsymbol{\sigma}$. *This magnetic moment is just that assumed in the spinning electron model.* The *electric moment*, being *a pure imaginary*, we should not expect to appear in the model. *It is doubtful whether the electric moment has any physical meaning*, since the Hamiltonian in (14) that we started from is real, and *the imaginary part only appeared when we multiplied it up in an artificial way in order to make it resemble the Hamiltonian of previous theories.*

§ 5. *The Angular Momentum Integrals for Motion a Central Field.*

We shall consider in greater detail the motion of an electron in a central field of force. We put **A** = 0 and $e'A_0 = V(r)$, an arbitrary function of the radius r, so that the Hamiltonian in (14)

$$[\{p_0 + e/c \,.\, A_0 + \rho_1 (\boldsymbol{\sigma}, \mathbf{p} + e/c\, \mathbf{A}) + \rho_3 mc\}\ \psi = 0. \qquad (14)]$$

becomes

$$F = p_0 + V + \rho_1 (\boldsymbol{\sigma}, \mathbf{p}) + \rho_3 mc$$

We shall determine the periodic solutions of the *wave equation* $F\ \psi = 0$, which means that p_0 is to be counted as a parameter instead of an operator; it is, in fact, just 1/c times the *energy* level.

We shall first find the *angular momentum* integrals of the motion. The *orbital angular momentum* **m** is defined by

$$\mathbf{m} = \mathbf{x} \times \mathbf{p},$$

and satisfies the following "Vertauschungs" relations

$$\begin{aligned} &m_1x_1 - x_1m_1 = 0, && m_1x_2 - x_2m_1 = ihx_3, \\ &m_1p_1 - p_1m_1 = 0, && m_1p_2 - p_2m_1 = ihp_3, \\ &\mathbf{m} \times \mathbf{m} = ih\mathbf{m}, && \mathbf{m}^2m_1 - m_1\mathbf{m}^2 = 0, \end{aligned} \qquad (17)$$

together with similar relations obtained by permuting the suffixes. Also, **m** commutes with r, and with p_r, the *momentum* canonically conjugate to r.

We have

$$\begin{aligned} m_1F - Fm_1 &= \rho_1 \{m_1 (\boldsymbol{\sigma}, \mathbf{p}) - (\boldsymbol{\sigma}, \mathbf{p})\, m_1\} \\ &= \rho_1 (\boldsymbol{\sigma}, m_1\mathbf{p} - \mathbf{p}m_1) \\ &= ih\rho_1 (\sigma_2p_3 - \sigma_3p_2) \end{aligned}$$

and so

$$\mathbf{m}F - F\mathbf{m} = ih\rho_1\, \boldsymbol{\sigma} \times \mathbf{p}. \qquad (18)$$

Thus, **m** is not a constant of the motion. We have further

$$\begin{aligned} \sigma_1F - F\sigma_1 &= \rho_1 \{\sigma_1 (\boldsymbol{\sigma}, \mathbf{p}) - (\boldsymbol{\sigma}, \mathbf{p})\, \sigma_1\} \\ &= \rho_1 (\sigma_1\boldsymbol{\sigma} - \boldsymbol{\sigma}\sigma_1, \mathbf{p}) \\ &= 2i\rho_1 (\sigma_3p_2 - \sigma_2p_3), \end{aligned}$$

with the help of (8), and so

$$\boldsymbol{\sigma}F - F\boldsymbol{\sigma} = -\, 2i\rho_1\, \boldsymbol{\sigma} \times \mathbf{p}.$$

Hence

$$(m + \tfrac{1}{2}\, h\boldsymbol{\sigma})\, F - F\, (m + \tfrac{1}{2}\, h\boldsymbol{\sigma}) = 0.$$

Thus m + ½ h$\boldsymbol{\sigma}$ (= **M** say) is a constant of the motion. *We can interpret this result by saying that the electron has a spin angular momentum of ½ h$\boldsymbol{\sigma}$*, which, added to the *orbital angular momentum* **m**, gives the *total angular momentum* M, which is a constant of the motion.

The *Vertauschungs* relations (17) all hold when M's are written for the m's. In particular

$$\mathbf{M} \times \mathbf{M} = ih\mathbf{M} \qquad \text{and} \qquad \mathbf{M}^2 M_3 = M_3 \mathbf{M}^2.$$

M_3 will be an *action variable* of the system. *Since the characteristic values of m_3 must be integral multiples of h in order that the wave function may be single-valued, the characteristic values of M_3 must be half odd integral multiples of h.* If we put

$$\mathbf{M}^2 = (j^2 - ¼)\, h^2, \qquad (19)$$

j will be another quantum number, and the characteristic values of M_3 will extend from $(j - ½)$ h to $(-j + ½)$ h*. Thus, *j* takes integral values.

One easily verifies from (18)

$$[\mathbf{m}F - F\mathbf{m} = ih\, \rho_1\, \boldsymbol{\sigma} \times \mathbf{p}. \qquad (18)]$$

that $\boldsymbol{m}^2$ *does not commute with F, and is thus not a constant of the motion. This makes a difference between the present theory and the previous spinning electron theory*, in which $\mathbf{m}^2$ is constant, and defines the *azimuthal quantum number k* by a relation similar to (19). We shall find that our *j* plays the same part as the *k* of the previous theory.

§ 6. *The Energy Levels for Motion in a Central Field.*

We shall now obtain the wave equation as a differential equation in r with the variables that specify the orientation of the whole system removed. We can do this by the use only of elementary non-commutative algebra in the following way.

…

If one neglects the last term, which is small on account of B being large, *this equation becomes the same as the ordinary Schrödinger equation for the system, with relativity correction included.* Since *j* has, from its definition, both positive and negative integral characteristic values, our equation will give twice as many *energy* levels when the last term is not neglected.

We shall now compare the last term of (26), which is of the same order of magnitude as the *relativity correction*, with the *spin correction* given by Darwin and Pauli.

…

The present theory will thus, in the first approximation, lead to the same energy levels as those obtained by Darwin, which are in agreement with experiment.

Dirac, P. A. M. (March, 1928). The quantum theory of the Electron. Part II.

Roy. Soc. Proc., A, 118, 779, 351-61; https://doi.org/10.1098/rspa.1928.0056.

Communicated by R. H. Fowler, F.R.S.—Received February 2, 1928.

St. John's College, Cambridge.

Application of the *Dirac equation* to the conservation theorem, the selection principle, the relative intensities of the lines of a multiplet, and the Zeeman effect.

In a previous paper by the author*

* Dirac, P. A. M. (February, 1928). The Quantum Theory of the Electron. *Roy. Soc. Proc., A*, 117, 778, 610–24. This is referred to later by *loc. cit.*

it is shown that the general theory of quantum mechanics together with *relativity* require the *wave equation for an electron moving in an arbitrary electromagnetic field* of potentials, A_0, A_1, A_2, A_3 to be of the form [the *Dirac equation*]

$$\{p_0 + e/c\ A_0 + \alpha_1 (p_1 + e/c\ A_1) + \alpha_2 (p_2 + e/c\ A_2) + \alpha_3 (p_3 + e/c\ A_3) + \alpha_4 mc\}\ \psi = 0. \quad (1)$$

[Dirac, P. A. M. (February, 1928). The Quantum Theory of the Electron. *Loc. cit.*: "To obtain the Hamiltonian for an electron in an electromagnetic field with scalar *potential* A_0 and vector *potential* **A**, we adopt the usual procedure of substituting $p_0 + e/c\ .\ A_0$ for p_0 and $\mathbf{p} + e/c\ .\ \mathbf{A}$ for $\mathbf{p}$ in the Hamiltonian for no field. From equation (9)

$$\{p_0 + \rho_1 (\boldsymbol{\sigma}, \mathbf{p}) + \rho_3 mc\}\ \psi = 0$$

we thus obtain

$$\{p_0 + e/c\ .\ A_0 + \rho_1 (\boldsymbol{\sigma}, \mathbf{p} + e/c\ \mathbf{A}) + \rho_3 mc\}\ \psi = 0. \quad (14)]$$

The α's are new dynamical variables which it is necessary to introduce in order to satisfy the conditions of the problem. They may be regarded as describing some internal motion of the electron, which for most purposes may be taken to be the spin of the electron postulated in previous theories. We shall call them the *spin variables*.

The α's must satisfy the conditions

$$\alpha_\mu^2 = 1, \qquad \alpha_\mu\alpha_\nu + \alpha_\nu\alpha_\mu = 0, \qquad (\mu \neq \nu.)$$

They may conveniently be expressed in terms of six variables ρ_1, ρ_2, ρ_3, σ_1, σ_2, σ_3 that satisfy

$$\rho_r^2 = 1, \qquad \sigma_r^2 = 1, \qquad \rho_r\sigma_s = \sigma_s\rho_r, \qquad (r, s = 1, 2, 3)$$

and

$$\rho_1\rho_2 = i\rho_3 = -\rho_2\rho_1, \qquad \sigma_1\sigma_2 = i\sigma_3 = -\sigma_2\sigma_1, \qquad (2)$$

together with the relations obtained from these by cyclic permutation of the suffixes, by means of the equations

$$\alpha_1 = \rho_1\sigma_1, \qquad \alpha_2 = \rho_1\sigma_2, \qquad \alpha_3 = \rho_1\sigma_3, \qquad \alpha_4 = \rho_3.$$

The variables σ_1, σ_2, σ_3 now form the three components of a vector, which corresponds (apart from a constant factor) to the *spin angular momentum* vector that appears in Pauli's theory of the spinning electron. The ρ's and σ's vary with the time, like other dynamical variables. Their *equations of motion*, written in the Poisson Bracket notation
[ ], are

$$\rho^{\cdot}_r = c\,[\rho_r, F], \qquad \sigma^{\cdot}_r = c\,[\sigma_r, F].$$

It should be observed that these *equations of motion* are consistent with the conditions (2)

$$[\rho_r^2 = 1, \qquad \sigma_r^2 = 1, \qquad \rho_r\sigma_s = \sigma_s\rho_r, \qquad (r, s = 1, 2, 3)$$
$$\rho_1\rho_2 = i\rho_3 = -\rho_2\rho_1, \qquad \sigma_1\sigma_2 = i\sigma_3 = -\sigma_2\sigma_1, \qquad (2)]$$

so that if the conditions are satisfied initially, they always remain satisfied. For example, we have

$$ih/c\,.\,\sigma^{\cdot}_1 = \sigma_1F - F\sigma_1 = 2i\rho_1\sigma_3\,(p_2 + e/c\,A_2) - 2i\rho_1\sigma_2\,(p_3 + e/c\,A_3).$$

Thus $\sigma^{\cdot}_1$ anticommutes with σ_1, so that

$$d\sigma_1^2/dt = \sigma^{\cdot}_1\sigma_1 + \sigma_1\sigma^{\cdot}_1 = 0.$$

The ρ's and σ's, and therefore also any function of them, can be represented by matrices with four rows and columns. A possible representation, in which ρ_3 and σ_3 are diagonal matrices, is given in (*loc. cit.*) § 2. *Such a representation can apply only to a single instant of time, since the ρ's and σ's vary with the time*. To get a scheme of representation which holds for all times, so that the equations of motion are valid in it, *we should have to have only constants of the motion as diagonal matrices*. It is, however, quite correct for the purpose of solving the *wave equation* (1)

$$[\{p_0 + e/c\,A_0 + \alpha_1\,(p_1 + e/c\,A_1) + \alpha_2\,(p_2 + e/c\,A_2) + \alpha_3\,(p_3 + e/c\,A_3) + \alpha_4 mc\}\,\psi = 0. \;(1)]$$

to take a matrix representation for the ρ's and σ's which holds only for a single instant of time (as was done in *loc. cit.*), since the *wave function* is then the *transformation function* connecting the ρ's, σ's and x's at this particular time with a set of variables that are constants of the motion, as is required for the general interpretation of quantum mechanics.

Before we proceed with the theory of atoms with single electrons that was begun in *loc. cit.*, the proof will be given of the *conservation theorem*, which states that the change in the *probability* of the electron being in a given volume during a given time is equal to the *probability* of its having crossed the boundary. This proof is supplementary to the work of *loc. cit.* § 3, and *is necessary before one can infer that the theory will give consistent results that are invariant under a Lorentz transformation.*

§ 1. The Conservation Theorem.

We shall first make a slight generalization of the usual interpretation of wave mechanics to apply to cases when the Hamiltonian is not *Hermitian.*

> [A *Hermitian matrix* (or self-adjoint matrix) is a complex square matrix that is equal to its own *conjugate transpose*—that is, the element in the i-th row and j-th column is equal to the *complex conjugate* of the element in the j-th row and i-th column, for all indices i and j.
>
> The *complex conjugate* of a complex number is the number with an equal real part and an imaginary part equal in magnitude but opposite in sign. That is, (if a and b are real, then) the complex conjugate of $a + b$i is equal to $a - b$i.
> The *complex conjugate* of z is often denoted as overline z (or, here, by z*).
>
> *Hermitian matrices* can be understood as the complex extension of real symmetric matrices.]

Let the *wave equation*, written in certain variables q, be

$$(H - W)\,\psi = 0. \qquad \text{(i)}$$

Consider also the equation

$$(H^\wedge - W^\wedge)\,\phi = 0$$

or

$$(H^\wedge + W)\,\phi = 0, \qquad \text{(ii)}$$

where the symbol $a^\wedge$ denotes the matrix obtained from the matrix a by transposing rows and columns. If ψ_m, ϕ_n are suitably normalized solutions of (i) and (ii) respectively, referring to the states m and n, we take to be the corresponding matrix element of the *probability* of the q's having specified values. If H is *Hermitian,* H^ is the *conjugate imaginary* of H (obtained by writing – i for i) and the solutions of (ii) are just the *conjugate imaginaries* to the solutions of (i), so that in this case our *probability* becomes the usual one $\psi_n{}^*\psi_m$. In the general case it is necessary to use the *transposed Hamiltonian* instead of the *conjugate imaginary Hamiltonian* in (ii) in order to secure that if ϕ_n, ψ_m are initially orthogonal or mutually normalized (i.e., $\int \phi_n\psi_m\, dq = 1$), they always remain orthogonal or mutually normalized respectively.

Our *wave equation* for an electron in an *electromagnetic field* is

$$\{p_0 + e'A_0 + \rho_1\,(\boldsymbol{\sigma}, \mathbf{p} + e'\mathbf{A}) + \rho_3 mc\}\,\psi = 0 \qquad (3)$$

where $e' = e/c$.

[cf *loc. cit.*
$\{p_0 + e/c\ .\ A_0 + \rho_1\,(\boldsymbol{\sigma}, \mathbf{p} + e/c\,\mathbf{A}) + \rho_3 mc\}\ \psi = 0.$ (14)]

The Hamiltonian here will be *Hermitian* if a matrix scheme for the spin variables is chosen in which they are *Hermitian*. However, if one now applies a Lorentz transformation to this *wave equation* and divides out by the coefficient of the new p_0, the resulting new Hamiltonian will not, in general, be Hermitian, although, as shown in *loc. cit.*, § 3, it may be brought back to its original Hermitian form by a *canonical transformation* of the matrix scheme for the spin variables. In the following work we require to have the same matrix representation of the *spin* variables *for all frames of reference*, so we cannot assume our Hamiltonian is *Hermitian*, and must use the above generalized interpretation.

The equation obtained by transposing rows and columns in the operator of (3)

$[\{p_0 + e'A_0 + \rho_1\,(\boldsymbol{\sigma}, \mathbf{p} + e'\mathbf{A}) + \rho_3 mc\}\ \psi = 0$ (3)]

is

$$[-\,p_0 + e'A_0 + \rho^\wedge_1\,(\boldsymbol{\sigma}^\wedge, -\,\mathbf{p} + e'\mathbf{A}) + \rho^\wedge_3 mc]\ \psi = 0. \qquad (4)$$

The *probability* per unit volume of the electron being in the neighborhood of any point is given, according to the above assumption, by $\phi\psi$, where this product must now be understood to mean the sum of the products of each of the four components of ϕ (referring respectively to the four rows or columns of the matrices ρ, σ) into the corresponding component of ψ. We have to prove that this *probability* is the time component of a 4-vector, and that the divergence of this 4-vector vanishes.

From (3)

$$\{\rho_3(p_0 + e'A_0) + \rho_1\rho_3\,(\boldsymbol{\sigma}, \mathbf{p} + e'\mathbf{A}) + mc\}\ \rho_3\psi = 0$$

or

$$\{\gamma_0(p_0 + e'A_0) + \Sigma_{r=1,2,3}\ \gamma_r(p_r + e'A_r) + mc\}\ \chi = 0, \qquad (5)$$

where

$$\gamma_0 = \rho_3, \qquad \gamma_r = \rho_1\rho_3\sigma_r, \qquad \chi = \rho_3\psi$$

Equation (5) is symmetrical between the four dimensions of space and time, and shows that $\gamma_0, -\gamma_1, -\gamma_2, -\gamma_3$ are the contravariant components of a 4-vector. If we multiply (4) by $\rho^\wedge_3$ on the left-hand side, we get

$$\gamma^\wedge_0\,(-\,p_0 + e'A_0) + \Sigma_{r=1,2,3}\ \gamma^\wedge_r(-\,p_r + e'A_r) + mc]\ \phi = 0, \qquad (6)$$

since

$$\gamma^\wedge_0 = \rho_3, \qquad \gamma^\wedge_r = \sigma^\wedge_r\rho^\wedge_3\rho^\wedge_1 = \rho^\wedge_3\rho^\wedge_1\sigma^\wedge_r.$$

The operator in this equation is just the transposed operator of (5). The *probability* per unit volume of the electron being in any place is now given by

$$\phi\psi = \phi\rho_3\chi = \phi\gamma_0\chi, \qquad (7)$$

where $\phi\alpha\chi$ denotes the sum of the products of each component of ϕ into the corresponding component of $\alpha\chi$, α being any function of the *spin variables*, represented by a matrix with four rows and columns. [Note that quite generally $\phi\alpha\chi = \chi\alpha\phi$.] Expression (7) is the time component of a 4-vector, whose special components, namely

$$-\phi\gamma_1\chi, \qquad -\phi\gamma_2\chi, \qquad -\phi\gamma_3\chi,$$

must give 1/c times the *probability* per unit time of the electron crossing unit area perpendicular to each of the three axes respectively.

We must now show that the *divergence* of this 4-vector vanishes, i.e. that

$$1/c\ (\phi\gamma_0\chi) - \Sigma_r\ \delta/\delta x_r\ (\phi\gamma_r\chi) = 0. \qquad (8)$$

Multiplying (5)

$$[\{\gamma_0(p_0 + e'A_0) + \Sigma_{r=1,2,3}\ \gamma_r(p_r + e'A_r) + mc\}\ \chi = 0, \qquad (5)]$$

by ϕ and (6)

$$[\gamma\hat{}_0\ (-p_0 + e'A_0) + \Sigma_{r=1,2,3}\ \gamma\hat{}_r(-p_r + e'A_r) + mc]\ \phi = 0, \qquad (6)]$$

by χ and subtracting, we get

$$\phi\ [\gamma_0 p_0 + \Sigma_{r=1,2,3}\ \gamma_r p_r]\ \chi + \chi\ [\gamma\hat{}_0 p_0 + \Sigma_r\ \gamma\hat{}_r p_r]\ \phi = 0,$$

which gives

$$\phi\ [\gamma_0\ \delta/c\delta t - \Sigma_r\ \delta/\delta x_r]\ \chi + \chi\ [\gamma\hat{}_0\ \delta/c\delta t - \Sigma_r\ \gamma\hat{}_r\ \delta/\delta x_r]\ \phi = 0,$$

or

$$\phi\ [\gamma_0\ \delta/c\delta t - \Sigma_r\ \delta/\delta x_r]\ \chi + 1/c\ \delta\phi/\delta t\ \gamma_0\chi - \Sigma_r\ \delta\phi/\delta x_r\ \gamma_r\chi = 0.$$

This gives immediately the *conservation equation* (8)

$$[1/c\ (\phi\gamma_0\chi) - \Sigma_r\ \delta/\delta x_r\ (\phi\gamma_r\chi) = 0. \qquad (8)]$$

as the γ's are here constant matrices.

§ 2. The Selection Principle.

In *loc. cit.* the *quantum number j* was introduced, which determines the magnitude of the resultant *angular momentum* for an electron moving in a central field of force. *j* can take both positive and negative integral values. Again, the *magnetic quantum number u* = M_3/h, say, that determines the component of the *total angular momentum* in some specified direction, was shown to take half odd integral values from $-|j| + ½$ to $|j| - ½$. The state $j = 0$ is thus excluded, and the weight of any state *j* is $2|j|$. The equation obtained to determine the energy levels, i.e., equation (25) or (26) [in *loc. cit.*], involves *j* only through the combination $j\ (j + 1)$ except in the last term, which represents the *spin correction*. Thus, two values of j which give the same value for $j\ (j + 1)$ form a *spin doublet*, so that $j + j'$ and $j = -(j' + 1)$ form a *spin doublet* when $j' > 0$. The connection between j-values and the usual notation for *alkali spectra* is therefore given by the following scheme:

j =	– 1	1 – 2	2 – 3	3 – 4 ...
	S	P	D	F

There is no *azimuthal quantum number k* in the present theory, an orbit for an *electron* in an atom being defined by three quantum numbers n, j, u only. One might on this account expect the *selection rules*, the relative *intensities* of the lines of a multiplet, etc., in the usual derivation of which k plays an important part, to be different in the present theory, but it will be found that they do just happen to be the same.

We shall first determine the *selection rule* for j. We use the following two theorems:

(i) If a dynamical variable X anticommutes with j, its matrix elements all refer to transitions of the type $j \rightarrow -j$.

(ii) If a dynamical variable Y satisfies

$$[[\mathrm{Y}, jh], jh] = -\mathrm{Y}, \qquad (9)$$

its matrix elements all refer to transitions of the type $j \rightarrow +/-j$.

To prove (i) we observe that the condition $j\,\mathrm{X} + \mathrm{X}\,j = 0$ gives

$$j' \,.\, \mathrm{X}\,(j'\,j'') + \mathrm{X}\,(j'\,j'') \,.\, j'' = 0$$

or

$$(j' + j'') \,.\, \mathrm{X}\,(j'\,j'') = 0.$$

Hence $\mathrm{X}\,(j'\,j'') = 0$ unless $j'' = -j'$.

A proof of (ii) involving *angle variables* has been given in a previous paper*.

* Dirac, P. A. M. (May, 1926). The elimination of the *nodes* in quantum mechanics. *Roy. Soc. Proc., A*, 111, 757, 281-305, § 3[; the laws of classical mechanics must be generalized when applied to atomic systems, *the commutative law of multiplication* as applied to dynamical variables is replaced by certain *quantum conditions* which are just sufficient to enable one to evaluate xy – yx when x and y are given, it follows that the dynamical variables cannot be ordinary numbers expressible in the decimal notation (which numbers will be called *c-numbers*), but may be considered to be numbers of a special kind (which will be called *q-numbers*), whose nature cannot be exactly specified, but which can be used in the algebraic solution of a dynamical problem in a manner closely analogous to the way the corresponding classical variables are used, the object of this paper is to simplify the *non-relativistic* quantum treatment by the introduction of *quantum variables*, in the

classical treatment of the dynamical problem of a number of particles or electrons moving in a central field of force and disturbing one another one always begins by making the initial simplification known as the *elimination of the nodes*, this consists in obtaining a *contact transformation* from the Cartesian co-ordinates and momenta of the electrons to a set of canonical variables of which all except three are independent of the orientation of the system as a whole while these three determine the orientation, introduces *action variables and their canonical conjugate angle variables*, *transformation equations*, substitutes set of *c-numbers* for *action variables* to fix *stationary state* and obtain physical results, applies to *anomalous Zeeman effect*, showed that *non-relativistic* theory gave the correct g-formula for *energy* of stationary states and Kronig's results for the relative intensities of the lines of a multiplet and their components in a weak magnetic field].

A simple proof analogous to the foregoing proof of (i) is as follows. Equation (9)

$$[[Y, jh], jh] = -Y, \qquad (9)]$$

gives

$$Yj^2 - 2jYj + j^2Y = Y$$

or

$$Y(j'\,j'')\,.\,j''^2 - 2j\,.\,Y(j'\,j'')\,.\,j'' + j'^2\,.\,Y(j'\,j'') = Y(j'\,j'').$$

Hence $Y(j'\,j'') = 0$ except when

$$j''^2 - 2j'\,j'' + j'^2 = 1,$$

i.e., when $j'' = j' \pm 1$.

We shall now evaluate $[[x_3, jh], jh]$. The definition of j is

$$jh = \rho_3\,\{(\boldsymbol{\sigma}, \mathbf{m} + h).$$

Hence

$$[x_3, jh] = \rho_3\,\{\sigma_1\,[x_3, m_1] + \sigma_2\,[x_3, m_2]\}$$
$$= \rho_3\,(\sigma_1 x_2 - \sigma_2 x_1), \qquad (10)$$

so that

$$[[x_3, jh], jh] = [\sigma_1 x_2 - \sigma_2 x_1, (\boldsymbol{\sigma}, \mathbf{m})].$$

Now

$$ih\,[\sigma_1, (\boldsymbol{\sigma}, \mathbf{m})] = \sigma_1\,(\boldsymbol{\sigma}, \mathbf{m}) - (\boldsymbol{\sigma}, \mathbf{m})\,\sigma_1 = 2i\,(\sigma_3, m_2 - \sigma_2, m_3]$$

or

$$\tfrac{1}{2}\,h\,[\sigma_1, (\boldsymbol{\sigma}, \mathbf{m})] = \sigma_3 m_2 - \sigma_2 m_3,$$

and similarly

$$\tfrac{1}{2}\,h\,[\sigma_2, (\boldsymbol{\sigma}, \mathbf{m})] = \sigma_1 m_3 - \sigma_3 m_1.$$

Hence …

…

so that

$$[[x_3, jh], jh] = -2u\,(\boldsymbol{\sigma}, \mathbf{x}) - x_3.$$

Thus, x_3 does not quite satisfy the condition that Y satisfies in (9)

$$[[[Y, jh], jh] = -Y, \qquad (9)]$$

owing to the extra term $-2u\,(\boldsymbol{\sigma}, \mathbf{x})$. This extra term, however, anticommutes with j. If we now form the expression $x_3 - c\underline{u}\,(\boldsymbol{\sigma}, \mathbf{x})$, where c is some quantity that commutes with j, we can choose c so as to make this expression satisfy completely the condition that Y satisfies in (9)

$$[[[Y, jh], jh] = -Y \qquad (9)]$$

We have, in fact,

. . .

. . .

Hence x_3 can be expressed as the sum of two terms, namely,

$$u/2(j^2 - ¼)\,(\boldsymbol{\sigma}, \mathbf{x}) \qquad \text{and} \qquad x_3 - u/2(j^2 - ¼)\,(\boldsymbol{\sigma}, \mathbf{x}),$$

of which the first anticommutes with j and therefore contains only matrix elements referring to *transitions* of the type $j \to -j$ while the second satisfies the condition that Y satisfies in (9), and therefore contains only matrix elements referring to *transitions* of the type $j \to j \pm 1$. A similar result holds for x_1 and x_2. Hence the selection rule for j is

$$j \to -j \qquad \text{or} \qquad j \to j \pm 1.$$

Thus, from states with $j = 2$ *transitions* can take place to states with = 1, – 2 or 3. Comparing this *selection rule* with the above scheme connecting j-values with the S, P, D notation, we see that it is exactly equivalent to the two *selection rules* for j and k of the usual theory, *and is therefore in agreement with experiment.*

§ 3. The Relative Intensities of the Lines of a Multiplet.

The relative *intensities* of the various components into which a line is split up in a weak magnetic field *must be the same on the present theory as on previous theories, as they depend only on the Vertauschungs relations connecting the co-ordinates x_r with the components of total angular momentum M_r*

$$[m_1x_1 - x_1m_1 = 0, \qquad m_1x_2 - x_2m_1 = ihx_3,$$
$$m_1p_1 - p_1m_1 = 0, \qquad m_1p_2 - p_2m_1 = ihp_3,$$
$$\mathbf{m} \times \mathbf{m} = ih\mathbf{m}, \qquad \mathbf{m}^2m_1 - m_1\mathbf{m}^2 = 0, \qquad \textit{loc. cit.}\ (17)]$$

which are taken over unchanged into the present theory. It will therefore be sufficient, for determining the relative *intensities* of the lines of a multiplet, to consider only one *Zeeman component* of each line, say, the component for which $\Delta u = 0$, i.e., the component that comes from x_3.

We shall determine the matrix elements of x_3, when expressed as a matrix in a scheme in which r, j, u and ρ_3 are diagonal. x_3 is diagonal in (i.e., commutes with) all of these variables except j. The part of x_3 referring to transitions $j \to -j$ we found to be

$$u/2(j^2 - ¼)\,(\boldsymbol{\sigma}, \mathbf{x}) = u/2(j^2 - ¼)\,\varepsilon\rho_1 r,$$

using the ε introduced in *loc. cit.* § 6.

[$r\varepsilon = \rho_1\,(\boldsymbol{\sigma}, \mathbf{x})$. *loc. cit.* (23)]

$\varepsilon\rho_1$ anticommutes with j, so that it can contain only matrix elements of the type $\varepsilon\rho_1(j, -j)$, and from the condition $(\varepsilon\rho_1)^2 = 1$ we must have

$$|\,\varepsilon\rho_1(j, -j)\,| = 1.$$

Hence

$$|\,x_3\,(j, -j)\,| = u/2(j^2 - ¼)\,r\,|\,\varepsilon\rho_1(j, -j)\,| = u/2(j^2 - ¼)\,r. \qquad (12)$$

Again, we have from (10)

[$[x_3, jh] = \rho_3\,\{\sigma_1\,[x_3, m_1] + \sigma_2\,[x_3, m_2]\}$

$= \rho_3\,(\sigma_1 x_2 - \sigma_2 x_1)$, (10)]

. . .

which gives

$$\{(j + 1)x_3 - x_3 j\}\{x_3(j + 1) - j x_3\} = r^2.$$

If we equate the (j, j) matrix elements of each side of this equation, we get on the left-hand side the sum of three terms, namely, the $(j, -j)$ matrix element of the first { } bracket times the $(-j, j)$ element of the second, the $(j, j + 1)$ element of the first times the $(j + 1, j)$ element of the second, and the $(j, j - 1)$ element of the first times the $(j - 1, j)$ element of the second. The second of these three terms vanishes, leaving

. . .

Hence

. . . (13)

Writing $-j$ for j, we get

. . . (14)

The three matrix elements of x_3 given in (12), (13) and (14) are associated with the three components of the multiplet formed by the combination of two doublets. The ratios of these matrix elements will, to a first approximation, remain unchanged when one makes a *transformation* from the matrix scheme in which r, j, u, ρ_3 are diagonal to a scheme in which the Hamiltonian is diagonal, and will therefore give the relative *intensities* of the Zeeman components $\Delta u = 0$ of the lines in a combination doublet. *These ratios are in agreement with those of previous theories based on the spinning electron model.*

§ 4. The Zeeman Effect.

If there is a uniform *magnetic field* of *intensity* H in the direction of the x_3 axis, we can take the *magnetic potentials* to be

$$A_1 = -\tfrac{1}{2} Hx_3, \qquad A_2 = \tfrac{1}{2} Hx_1, \qquad A_3 = 0.$$

The additional terms appearing in the Hamiltonian F will now be

$$\Delta F = \rho_1 e' (\boldsymbol{\sigma}, A) = -\tfrac{1}{2} He'\rho_1 (\sigma_1 x_2 - \sigma_2 x_1).$$

From (10)

$$[[x_3, jh] = \rho_3 \{\sigma_1 [x_3, m_1] + \sigma_2 [x_3, m_2]\}$$
$$= \rho_3 (\sigma_1 x_2 - \sigma_2 x_1), \qquad (10)]$$

it follows that $\rho_3 (\sigma_1 x_2 - \sigma_2 x_1)$ or $(\sigma_1 x_2 - \sigma_2 x_1)$, like x_3, contains only matrix elements of the type $(j, -j)$ or $(j, j \pm 1)$. Now ρ_1 anticommutes with j, and therefore contains only matrix elements of the type $(j, -j)$. Hence ΔF contains only matrix elements of the type (j, j) or $(j, -j \pm 1)$.

In *loc. cit.*, § 6, it was found [see equation (24)] that the Hamiltonian could be expressed as

$$F = p_0 + V + \varepsilon p_r + i\varepsilon\rho_3 jh/r + \rho_3 mc. \qquad (15)$$

It follows from (10) that $(\sigma_1 x_2 - \sigma_2 x_1)$ anti-commutes with $(\boldsymbol{\sigma}, x)$, and therefore also with ε. Thus, if we put

$$\Delta F = i\varepsilon\rho_3 r,$$

so that

$$\eta = \tfrac{1}{2} He/ch \,.\, \varepsilon\rho_2 (\sigma_1 x_2 - \sigma_2 x_1)/r,$$

η commutes with ε. Further, η commutes with ρ_3, r and p_r, so that it commutes with all the variables occurring in (15) except j. If we now express η as a matrix in j, we shall have obtained an expression for ΔF in terms of the variables occurring in (15). We have from (10) and (13)

...

and similarly

...

We have seen that the matrix elements of $\varepsilon\rho_1$, all of which are of the type $(j, -j)$, must be of modulus unity. Hence

... (16)

and similarly

...

Again, from (10) and (11)

...

so that

$$\eta\,(j, j) = He/2ch\; uj/(j^2 - ¼). \qquad (17)$$

If we now write down in full, as in *loc. cit.*, the *wave equation* corresponding to (15)

$$[F = p_0 + V + \varepsilon p_r + i\varepsilon\rho_3\, jh/r + \rho_3 mc, \qquad (15)]$$

and include the extra term ΔF, we shall have

$$[(F + \Delta F)\,\psi]_\alpha = (p_0 + V)\,\psi_\alpha - h\,\delta/\delta r\,\psi_\beta - (j/r + \eta r)\,h\psi_\beta + mc\psi_\alpha = 0,$$
$$[(F + \Delta F)\,\psi]_\beta = (p_0 + V)\,\psi_\beta + h\,\delta/\delta r\,\psi_\alpha - (j/r + \eta r)\,h\psi_\alpha + mc\psi_\beta = 0,$$

where η is now an operator, operating on ψ_α and ψ_β that commutes with everything except j. On eliminating ψ_α this gives, corresponding to (25) of *loc. cit.*,

...

We can neglect the $\eta^2 r^2$ term, which is proportional to the square of the *field strength*, and also the ηr term in the last bracket, which is of the order of magnitude of *field strength* times *spin correction*. The only first order effect of the field is the insertion of the terms $\eta - \eta j - j\eta$ in the first bracket. This bracket may now be written as

$$[2mE/h^2 + E^2/c^2h^2 + \{2(E + mc^2)/ch^2\}V + V^2/h^2 - j(j + 1)/r^2 + \eta - \eta j - j\eta \quad (18)$$

where E is the *energy* level, equal to $p_0c - mc^2$.

If the field is weak compared with the doublet separation, we can obtain a first approximation to the change in the *energy* levels by neglecting the non-diagonal matrix elements of ΔF or of η. The extra terms $\eta - \eta j - j\eta$ in (18) are now a constant instead of an operator, namely, the constant

$$-(2j - 1)\eta\,(j, j) = -\,He/ch\; uj/(j + ½)$$

from (17). *The energy levels will be reduced by $h^2/2m$ times this constant, if we neglect the fact that the characteristic E occurs in (18) in other places besides the term $2mE/h^2$,* which means *neglecting the interaction of the magnetic field with the relativity variation of mass with velocity.* The increase in the *energy levels* caused by the *magnetic field* is thus

$$He/2mc\; j/(j + ½)\; uh = \varpi g u h$$

where ϖ is the *Larmor frequency* He/2mc, and g, the *Lande splitting factor*, has the value

$$g = j/(j + ½)$$

For the succession of j-values, –1, 1, –2, 2, –3 ... g has the values, 2, 2/3, 4/3, 4/5, 6/5, ... *in agreement with Lande's formula for alkali spectra.*

We now take the case of a *magnetic field* that is strong compared with the doublet separation, but weak compared with the separations of terms of different series. This requires that the matrix elements of η of the type $\eta\ (j, -j-1)$ with $j > 0$ shall be taken into account, although those of the type $\eta\ (j, -j+1)$ can still be neglected. The reduction in the *energy* levels will now be approximately $h^2/2m$ times one or other of the characteristic values of the extra terms $\eta - \eta j - j\eta$ in (18). These characteristic values are the roots ξ of the equation

. . .

or

. . .

This gives, with the help of (16) and (17)

. . .

which reduces to

$$\xi^2 + He/ch\ 2u\xi + (He/ch)^2\ (u^2 - 1/4) = 0.$$

Hence

$$\xi^2 = -\ He/ch\ uj/(j \pm 1/2).$$

The increase in the *energy* levels due to the *magnetic field* is therefore

$$-\ h^2/2m\ \xi = h^2/2m\ He/ch\ (u \pm 1/2) = \varpi\ (u \pm 1/2)\ h,$$

in agreement with the previous spinning electron theory of the *Paschen-Back effect.*

One might expect that with still stronger magnetic fields the matrix elements $(j, -j+1)$ of η would come into play, and would cause interference between the Zeeman patterns of terms whose *quantum numbers k* in the usual notation differ by 2. The matrix elements $(j, -j+1)$ of $\eta - \eta j - j\eta$, however, vanish for arbitrary η, *so that no effect of this nature occurs.*

Pauli, W. (1940). The Connection Between Spin and Statistics[1].

Phys. Rev., 58, 8, 716–22; doi:10.1103/PhysRev.58.716.

Physikalisches Institut, Eidg. Technischen Hochschule, Zurich, Switzerland and Institute for Advanced Study, Princeton, New Jersey.

Received August 19, 1940.

[1] This paper is part of a report which was prepared by the author for the Solvay Congress 1939 and in which slight improvements have since been made. *In view of the unfavorable times*, the Congress did not take place, and the publication of the reports has been postponed for an indefinite length of time. The relation between the present discussion of the connection between spin and statistics, and the somewhat less general one of Belinfante, based on the concept of charge invariance, has been cleared up by Pauli, W. & Belinfante, F. J. (1940). *Physica* 7, 177.

Pauli derived that for a *relativistically invariant wave equation* for free particles: from postulate (I), *according to which the energy must be positive, the necessity of Fermi-Dirac statistics for particles with arbitrary half-integral spin*; from postulate (II), *according to which observables on different space-time points with a space-like distance are commutable, the necessity of Einstein-Bose statistics for particles with arbitrary integral spin.* Postulate I was introduced because "in case of *half-integral spin, …, a positive definite energy density, as well as a positive definite total energy, is impossible*". Similarly, postulate II was introduced so that "all physical quantities at finite distances exterior to the light cone … are commutable. … The justification for our postulate lies in the fact that measurements at two space points with a space-like distance can never disturb each other, *since no signals can be transmitted with velocities greater than that of light*". [???]

Abstract

In the following paper we conclude for the *relativistically invariant wave equation* for free particles: from postulate (I), according to which the energy must be positive, *the necessity of Fermi-Dirac statistics for particles with arbitrary half-integral spin*; from postulate (II), according to which observables on different space-time points with a space-like distance are commutable, *the necessity of Einstein-Bose statistics for particles with arbitrary integral spin.* It has been found useful to divide the quantities which are irreducible against Lorentz transformations into four symmetry classes which have a commutable multiplication like +1, –1, +e, –e with c =1.

---

§1. UNITS AND NOTATIONS

Since the requirements of the *relativity theory* [???] and the quantum theory are fundamental for every theory, it is natural to use as units the *vacuum velocity of light* c, and *Planck's constant* divided by 2π which we shall simply denote by $\hbar$. This convention means that all quantities are brought to the dimension of the power of a length by multiplication with powers of $\hbar$ and c. The reciprocal length corresponding to the *rest mass* m is denoted by $\kappa = mc/\hbar$.

As time coordinate, we use accordingly the length of the light path. In specific cases, however, we do not wish to give up the use of the *imaginary time coordinate*. Accordingly, a tensor index denoted by small Latin letters *i*, refers to the *imaginary time coordinate* and runs from 1 to 4. A special convention for denoting the *complex conjugate* seems desirable.

> [The *complex conjugate* of a complex number is the number with an equal real part and an imaginary part equal in magnitude but opposite in sign. That is, if a and b are real numbers then the *complex conjugate* of $a + bi$ is $a - bi$. The product of a complex number and its conjugate is a real number: $a^2 + b^2$.]

Whereas for quantities with the index 0 an asterisk signifies the *complex-conjugate* in the ordinary sense (e.g., for the current vector S_i the quantity $S_0{}^*$ is the *complex conjugate* of the charge density S_0), in general $U^*{}_a$ signifies the *complex-conjugate* of U_a ... multiplied with $(-1)^n$, where n is the number of occurrences of the digit 4 among the *i*, *k*, (e.g. $S_4 = iS_0$, $S_4{}^* = iS_0{}^*$).

Dirac's spinors u_ρ with $\rho = 1, .., 4$ have always a Greek index running from 1 to 4, and $u_\rho{}^*$ means the *complex-conjugate* of u_ρ, in the ordinary sense.

Wave functions, insofar as they are ordinary vectors or tensors, are denoted in general with capital letters, U_i, U_{ik} … . The symmetry character of these tensors must in general be added explicitly. As classical fields the *electromagnetic* and the *gravitational* fields, as well as fields with rest mass zero, take a special place, and are therefore denoted with the usual letters φ_i, $f_{ik} = - f_{ki}$, and $g_{ik} = - g_{ki}$, respectively.

The *energy-momentum tensor* T_{ik} is so defined, that the *energy-density* W and the *momentum density* G_k are given in natural units by $W = - T_{44}$ and $G_k = - iT_{k4}$ with k =1, 2, 3.

§2. IRREDUCIBLE TENSORS. DEFINITION OF SPINS

We shall use only a few general properties of those quantities which transform according to irreducible representations of the *Lorentz group*[2].

[2] See Waerden, B. L. v. d. (1932). *Die gruppentheoretische Methode in der Quantentheorie*. Berlin.

The proper *Lorentz group* is that continuous linear group the transformations of which leave the form

$$\Sigma_{k=1}^{4} x_k^2 = x^2 - x_0^2$$

invariant and in addition to that satisfy the condition that they have the determinant +1 and do not reverse the time. A tensor or spinor which transforms irreducibly under this group can be characterized by two integral positive numbers (p, g). (The corresponding "*angular momentum quantum numbers*" (j, k) are then given by p =2j + 1, q = 2k + i, with integral or half-integral j and k.*)

* In the spinor calculus this is a spinor with 2j undotted and 2k dotted indices.

The quantity U(j,k) characterized by (j,k) has p . q = (2j + 1)(2k + 1) independent components. Hence to (0, 0) corresponds the scalar, to (½, ½) the vector, to (1, 0) the self-dual skew-symmetrical tensor, to (1, 1) the symmetrical tensor with vanishing spur, etc. *Dirac's spinor* u, reduces to two irreducible quantities (½, 0) and (0, ½) each of which consists of two components. If U(j,k) transforms according to the representation

$$U_r' = \Sigma_{r=1}^{(2j+1)(2k+1)} \Lambda_{rs} U_s$$

then U*(k, j) transforms according to the complex-conjugate representation Λ*. Thus, for k = j, Λ* = Λ. This is true only if the components of U(j,k) and U(k,j) are suitably ordered. For an arbitrary choice of the components, a similarity transformation of Λ and Λ* would have to be added. In view of §1 we represent generally with U* the quantity the transformation of which is equivalent to Λ* if the transformation of U is equivalent to Λ.

The most important operation is the reduction of the product of two quantities

$$U_1(j_1, k_1)\, U_2(j_2, k_2)$$

which, according to the well-known *rule of the composition of angular momenta*, decompose into several U(j,k) where, independently of each other j, k run through the values

$$j = j_1 + j_2, j_1 + j_2 - 1, \ldots, | j_1 - j_2 |$$
$$k = k_1 + k_2, k_1 + k_2 - 1, \ldots, | k_1 - k_2 |$$

By limiting the transformations to the subgroup of *space rotations* alone, the distinction between the two numbers j and k disappears and U(j,k) behaves under this group just like the product of two irreducible quantities U(j)U(k) which in turn reduces into several irreducible U(l) each having $2l + 1$ components, with

$$l = j + k, j + k - 1, \ldots, | j - k |$$

Under the *space rotations* the U(l) with integral l transform according to single-valued representation, whereas those with half-integral l transform according to double-valued representations. Thus the unreduced quantities T(j,k) with integral (half-integral) j + k are single-valued (double-valued).

If we now want to determine the spin value of the particles which belong to a given field it seems at first that these are given by $l = j + k$. Such a definition would, however, not correspond to the physical facts, for there then exists no relation of the spin value with the number of independent plane waves, which are possible *in the absence of interaction* for given values of the components k; in the *phase factor* exp i(kx). *In order to define the spin in an appropriate fashion*[3]*, we want to consider first the case in which the rest mass m of all the particles is different from zero.*

[3] See Fierz, M. (1939). *Helv. Phys. Acta*, 12, 3; also Broglie, L. de (1939). *Comptes rendus*, 208, 1697; (1939). *Loc. cit.*, 209, 265.

In this case *we make a transformation to the rest system of the particle*, where all the space components of k_i are zero, and *the wave function depends only on the time*. In this system we reduce the *field components*, which according to the *field equations* do not necessarily vanish, into parts irreducible against *space rotations*. To each such part, with r = 2s + i components, belong r different eigenfunctions which under *space rotations* transform among themselves and *which belong to a particle with spin s*. If the field equations describe particles with only one *spin* value there then exists in the rest system only one such irreducible group of components. From the *Lorentz invariance*,

[In *relativistic* physics, Lorentz symmetry or *Lorentz invariance*, named after the Dutch physicist Hendrik Lorentz, *is an equivalence of observation or observational symmetry due to special relativity implying that the laws of physics stay the same for all observers that are moving with respect to one another within an inertial frame*. It has also been described as "the feature of nature that says experimental results are independent of the orientation or the boost velocity of the laboratory through space".]

it follows, for an arbitrary system of reference, that r or Σr eigenfunctions always belong to a given arbitrary k_i. The number of quantities U(j,k) which enter the theory is, however, in a *general coordinate system* more complicated, since these quantities together with the vector k_i have to satisfy several conditions.

In the case of zero rest mass there is a special degeneracy because, as has been shown by Fierz, this case permits a *gauge transformation* of the second kind*.

> * By "gauge-transformation of the first kind" we understand a transformation $U \to Ue^{i\alpha}$ $U^* \to U^* e^{-i\alpha}$ with an arbitrary space and time function α. By "gauge-transformation of the second kind" we understand a transformation of the type
>
> $\varphi_k \to \varphi_k - 1/e\, i\, \partial\alpha/\partial x_k$
>
> as for those of the electromagnetic potentials.

If the field now describes only one kind of particle with the rest mass zero and a certain spin value, *then there are for a given value of k; only two states, which cannot be transformed into each other by a gauge transformation. The definition of spin may, in this case, not be determined so far as the physical point of view is concerned because the total angular momentum of the field cannot be divided up into orbital and spin angular momentum by measurements.* But it is possible to use the following property for a definition of the *spin*. If we consider, in the *q number theory*, states where only one particle is present, then not all the eigenvalues j(j + 1) of the square of the angular momentum are possible. But j begins with a certain minimum value s and takes then the values s, s + 1, ... [4].

> [4] The general proof for this has been given by Fierz, M. (1940). *Helv. Phys. Acta*, 13, 45.

This is only the case for m = 0. For photons, s = 1; j = 0 is not possible for one single photon[5].

> [5] See for instance W. Pauli in the article "Wellenmechanik" in the *Handbuck der Physik*, Vol. 24/2, p. 260.

For *gravitational* quanta s = 2 and the values j = 0 and j = 1 do not occur.

In an arbitrary system of reference and for arbitrary rest masses, the quantities U all of which transform according to double-valued (single-valued) representations with half-integral (integral) j + k describes only particles with half-integral (integral) spin. *A special investigation is required only when it is necessary to decide whether the theory describes particles with one single spin value or with several spin values.*

§3. PROOF OF THE INDEFINITE CHARACTER OF THE CHARGE IN CASE OF INTEGRAL AND OF THE ENERGY IN CASE OF HALF-INTEGRAL SPIN

We consider first a theory which contains only U with integral j + k, i.e., *which describes particles with integral spins only*. It is not assumed that only particles with one single spin value will be described, but all particles shall have integral spin.

We divide the quantities U into two classes: (1) the "+1 class" with j integral, k integral; (2) the "–1 class" with j half-integral, k half-integral.

The notation is justified because, according to the indicated rules about the reduction of a product into the irreducible constituents under the Lorentz group, the product of two quantities of the +1 class or two quantities of the –1 class contains only quantities of the +1 class, whereas the product of a quantity of the +1 class with a quantity of the –1 class contains only quantities of the –1 class. *It is important that the complex conjugate U* for which j and k are interchanged belong to the same class as U.* As can be seen easily from the multiplication rule, tensors with even (odd) number of indices reduce only to quantities of the +1 class (–1 class). The propagation vector k_i we consider as belonging to the –1 class, since it behaves after multiplication with other quantities like a quantity of the –1 class.

We consider now a homogeneous and linear equation in the quantities U which, however, does not necessarily have to be of the first order. Assuming a plane wave, we may put k_l for $-i\,\partial/\partial x_l$. *Solely on account of the invariance against the proper Lorentz group* it must be of the typical form

$$\Sigma\, kU^+ = \Sigma\, U^-, \qquad \Sigma\, kU^- = \Sigma\, U^+. \tag{1}$$

This typical form shall mean that there may be as many different terms of the same type present, as there are quantities U^+ and U^-. Furthermore, among the U^+ may occur the U^+ as well as the $(U^+)^*$, whereas other U may satisfy reality conditions $U = U^*$. Finally, we have omitted an *even* number of k factors. These may be present in arbitrary number in the term of the sum on the left- or right-hand side of these equations. It is now evident that these equations remain invariant under the substitution

$$k_i \to -k_i\,; \qquad U^+ \to U^+, \qquad [(U^+)^* \to (U^+)^*]; \tag{2}$$
$$U^- \to -U^-, \qquad [(U^-)^* \to -(U^-)^*]$$

Let us consider now tensors T of even rank (scalars, skew-symmetrical or symmetrical tensors of the 2nd rank, etc.), which are composed quadratically or bilinearly of the U's.

They are then composed solely of quantities with even j and even k and thus are of the typical form

$$T \sim \Sigma\, U^{+}U^{+} + \Sigma\, U^{-}U^{-} + \Sigma\, U^{+}kU^{-}, \qquad (3)$$

where again a possible even number of k factors is omitted and no distinction between U and U* is made. Under the substitution (2) they remain unchanged, T→T.

The situation is different for tensors of odd rank S (vectors, etc.) which consist of quantities with half-integral j and half-integral k. These are of the typical form

$$S \sim \Sigma\, U^{+}kU^{+} + \Sigma\, U^{-}kU^{-} + \Sigma\, U^{-}, \qquad (4)$$

and hence *change the sign under the substitution (2), S→– S.* Particularly is this the case for the *current* vector s_i. To the transformation $k_i \to -k_i$ belongs for arbitrary wave packets the transformation $x_i \to -x_i$ and *it is remarkable that from the invariance of Eq. (1) against the proper Lorentz group alone there follows an invariance property for the change of sign of all the coordinates.* In particular, the indefinite character of the *current density* and the total charge for even spin follows, since to every solution of the field equations belongs another solution for which the components of s_k change their sign. *The definition of a definite particle density for even spin which transforms like the 4-component of a vector is therefore impossible.*

We now proceed to a discussion of the somewhat less simple case of *half-integral spins.* Here we divide the quantities U, which have half-integral j + k, in the following fashion: (3) the "+e class" with *j integral k half-integral*, (4) the "–e class" with *j half-integral k integral.*

The multiplication of the classes (1), …, (4), follows from the rule $e^2 = 1$ and the *commutability of the multiplication.* This law remains unchanged if e is replaced by –e. We can summarize the multiplication law between the different classes in the following multiplication table:

…

We notice that these classes have the multiplication law of Klein's "four-group".

It is important that here the *complex-conjugate* quantities for which j and k are interchanged do not belong to the same class, so that

U^{+e}, $(U^{-e})^*$ belong to the +e class
U^{-e}, $(U^{+e})^*$ belong to the –e class.

We shall therefore cite the *complex-conjugate* quantities explicitly. (One could even choose the U^{+e} suitably so that all quantities of the –e class are of the form $(U^{+e})^*$.)

Instead of (1)

$$[\Sigma\, kU^+ = \Sigma\, U^-, \qquad \Sigma\, kU^- = \Sigma\, U^+. \qquad (1)]$$

we obtain now as typical form

$$\Sigma\, kU^{+e} + \Sigma\, k(U^{-e})^* = \Sigma\, U^{-e} + \Sigma\, (U^{+e})^* \qquad (5)$$
$$\Sigma\, kU^{-e} + \Sigma\, k(U^{+e})^* = \Sigma\, U^{+e} + \Sigma\, (U^{-e})^*$$

since a factor k or $-i\, \partial/\partial x_l$ always changes the expression from one of the classes +e or –e into the other. As above, an even number of k factors have been omitted.

Now we consider instead of (2)

$$[k_i \to -k_i\,; \qquad U^+ \to U^+, \qquad [(U^+)^* \to (U^+)^*]; \qquad (2)$$
$$U^- \to -U^-, \qquad [(U^-)^* \to -(U^-)^*]\,]$$

the substitution

$$k_i \to -k_i\,; \qquad U^{+e} \to i\, U^{+e}, \quad (U^{-e})^* \to i\, (U^{-e})^*; \qquad (6)$$
$$(U^{+e})^* \to -i\, (U^{+e})^*; \quad U^{-e} \to -i\, U^{-e}.$$

This is in accord with the algebraic requirement of the passing over to the *complex conjugate*, as well as with the requirement that quantities of the same class as U^{+e}, $(U^{-e})^*$ transform in the same way. Furthermore, it does not interfere with possible r*eality conditions* of the type $U^{+e} = (U^{-e})^*$ or $U^{-e} = (U^{+e})^*$. Equations (5) remain unchanged under the substitution (6).

We consider again tensors of even rank (scalars, tensors of 2nd rank, etc.), which are composed bilinearly or quadratically of the U and their *complex-conjugate*. For reasons similar to the above they must be of the form

$$T \sim \Sigma\, U^{+e} U^{+e} + \Sigma\, U^{-e} U^{-e} + \Sigma\, U^{+e}\, kU^{-e} + \ldots \qquad (7)$$

Furthermore, the tensors of odd rank (vectors, etc,) must be of the form

$$S \sim \Sigma\, U^{+e} kU^{+e} + \Sigma\, U^{-e} kU^{-e} + \Sigma\, U^{+e} U^{-e} + \ldots. \qquad (8)$$

The result of the substitution (6) is now the opposite of the result of the substitution (2): the tensors of even rank change their sign; the tensors of odd rank remain unchanged:

$$T \to -T; \qquad S \to +S. \qquad (9)$$

In case of half-integral spin, therefore, a positive definite energy density, as well as a positive definite total energy, is impossible. The latter follows from the fact, that, under the above substitution, the *energy density* in every spacetime point changes its sign as a result of which the *total energy* changes also its sign.

*It may be emphasized that it was not only unnecessary to assume that the wave equation is of the first order**,

* But we exclude operations like $(k^2 + \kappa^2)^{1/2}$, which operate at finite distances in the coordinate space.

but also that *the question is left open whether the theory is also invariant with respect to space reflections* ($x' = -x$, $x_0' = -x_0$). *This scheme covers therefore also Dirac's two component wave equations. (with rest mass zero)*.

These considerations do not prove that for integral spins there always exists a definite *energy density* and for half-integral spins a definite *charge density*. In fact, it has been shown by Fierz [6] that this is not the case for spin >1 for the *densities*.

[6] Fierz, M. (1939). *Helv. Phys. Acta*, 12, 3.

There exists, however (in the *c number theory*), a definite *total charge* for half-integral *spins* and a definite *total energy* for the integral *spins*.

[The term *c-number* (short for classical number) was introduced by Dirac. It refers to real and complex numbers. These numbers are considered classical because they are not quantum operators that act on elements of the Hilbert space of states in a quantum system. In other words, c-numbers are the familiar numbers we encounter in everyday mathematics, such as real numbers (like 3.14) and complex numbers (like 2 + 3i).]

The *spin value* ½, is discriminated through the possibility of a definite *charge density*, and the spin values 0 and 1 are discriminated through the possibility of defining a definite *energy density*. Nevertheless, the present theory permits arbitrary values of the *spin quantum numbers* of elementary particles as well as arbitrary values of the *rest mass*, the *electric charge*, and the *magnetic moments* of the particles.

§4. QUANTIZATION OF THE FIELDS IN THE ABSENCE OF INTERACTIONS. CONNECTION BETWEEN SPIN AND STATISTICS

The impossibility of defining in a physically satisfactory way the *particle density* in the case of *integral spin* and the *energy density* in the case of *half-integral spins* in the c-number theory is an indication that a satisfactory interpretation of the theory within the limits of the one-body problem is not possible*.

> * The author therefore considers as not conclusive the original argument of Dirac, according to which the *field equation* must be of the first order.

In fact, all *relativistically invariant* theories lead to particles, which in external fields can be emitted and absorbed in pairs of opposite charge for electrical particles and singly for neutral particles. The fields must, therefore, undergo a second quantization. For this we do not wish to apply here the canonical formalism, in which time is unnecessarily sharply distinguished from space, and which is only suitable if there are no supplementary conditions between the canonical variables[7].

> [7] On account of the existence of such conditions the canonical formalism is not applicable for spin >1 and therefore the discussion about the connection between spin and statistics by J. S. de Wet, (1940). *Phys. Rev.*, 57, 646, which is based on that formalism is not general enough.

Instead, we shall apply here a generalization of this method which was applied for the first time by Jordan and Pauli to the electromagnetic field[8].

> [8] The consistent development of this method leads to the "*many-time formalism*" of Dirac, which has been given by Dirac, P. A. M. (1935). *Quantum Mechanics*. (Oxford, second edition).

This method is especially convenient in the absence of interaction, where all fields $U^{(r)}$ satisfy the wave equation of the second order …

We shall, however, *expressively postulate in the following that all physical quantities at finite distances exterior to the light cone (for* $|x_0' - x_0''| < |x' - x''|$*) are commutable**.

> * For the canonical quantization formalism this postulate is satisfied implicitly. But this postulate is much more general than the canonical formalism.

…

The justification for our postulate lies in the fact that measurements at two space points with a space-like distance can never disturb each other, since no signals can be transmitted with velocities greater than that of light. …

…

So far, we have not distinguished between the two cases of Bose statistics and the exclusion principle. …

...

Hence, we come to the result: *For integral spin the quantization according to the exclusion principle is not possible. ...*

On the other hand, it is formally possible to quantize the theory for half-integral spins according to Einstein-Bose-statistics, but according to the general result of the preceding section *the energy of the system would not be positive*. Since for physical reasons it is necessary to postulate this, *we must apply the exclusion principle in connection with Dirac's hole theory.*

For the positive proof that a theory with a *positive total energy* is possible by quantization according to Bose-statistics (*exclusion principle*) for integral (half-integral) *spins*, we must refer to the already mentioned paper by Fierz. In another paper by Fierz and Pauli[11] the case of an external *electromagnetic field* and also the connection between the special case of *spin* 2 and the *gravitational theory* of Einstein has been discussed.

[11] Fierz, M. & Pauli, W. (1939). *Proc. Roy. Soc.*, A173, 211.

In conclusion we wish to state, that according to our opinion the connection between *spin* and statistics is one of the most important *applications* of the *special relativity theory*. [???]

PART II Exchange Interaction.

Exchange interaction is a quantum mechanical constraint on the states of indistinguishable particles. While sometimes called an exchange force, or, in the case of fermions, Pauli repulsion, its consequences cannot always be predicted based on classical ideas of force. *Both bosons and fermions can experience exchange interaction.*

The wave function of indistinguishable particles is subject to exchange symmetry: the wave function either changes sign (for fermions) or remains unchanged (for bosons) when two particles are exchanged. The *exchange symmetry* alters the expectation value of the distance between two indistinguishable particles when their *wave functions* overlap. For fermions the expectation value of the distance increases, and for bosons it decreases (compared to distinguishable particles).

The *exchange interaction* arises from the combination of *exchange symmetry* and the *Coulomb interaction.*

> [*Exchange symmetry* is a fundamental concept in quantum physics that applies to identical particles. It states that:
>
> - The *wave function* of indistinguishable particles either changes sign (for *fermions*) or remains unchanged (for *bosons*) when two particles are exchanged.
> - *For fermions, the combined wave function is antisymmetric under swapping, preventing the particles from occupying the same state.*
> - The outcomes of measurements should be unaffected if identical particles are swapped.
>
> The *Coulomb interaction*, or Coulomb force, is the force of attraction or repulsion of particles or objects because of their electric charge.]

For an *electron* in an electron gas, the *exchange symmetry* creates an "exchange hole" in its vicinity, which other electrons with the *same spin* tend to avoid due to the Pauli exclusion principle. This decreases the energy associated with the Coulomb interactions between the electrons with same spin. *Since two electrons with different spins are distinguishable from each other and not subject to the exchange symmetry, the effect tends to align the spins. Exchange interaction* is the main physical effect responsible for *ferromagnetism*, and has no classical analogue.

For *bosons*, the *exchange symmetry* makes them bunch together, and the *exchange interaction* takes the form of an effective attraction that causes identical particles to be found closer together, as in Bose–Einstein condensation.

Exchange interaction effects were discovered independently by physicists Werner Heisenberg and Paul Dirac in 1926. [Heisenberg, W. (June, 1926). Mehrkörperproblem und Resonanz in der Quantenmechanik. (Multibody Problem and Resonance in Quantum Mechanics.) *Zeit. Phys.*, 38, 6–7, 411–26; doi//10.1007/ BF01397160; Dirac, P. A. M. (October, 1926). On the Theory of Quantum Mechanics. *Roy. Soc. Proc., A*, 112, 762, 661-77 (copy below).]

Exchange symmetry

Quantum particles are fundamentally indistinguishable. Wolfgang Pauli demonstrated that this is a type of symmetry: states of two particles must be either *symmetric* or *antisymmetric* when coordinate labels are exchanged. In a simple one-dimensional system with two identical particles in two states ψ_a and ψ_b the system *wavefunction* can therefore be written two ways:

$$\psi_a(x_1)\ \psi_b(x_2) \pm \psi_a(x_2)\ \psi_b(x_1).$$

Exchanging x_1 and x_2 gives either a *symmetric* combination of the states ('plus') or an *antisymmetric* combination ('minus'). *Particles that give symmetric combinations are called bosons*; *those with antisymmetric combinations are called fermions*.

The two possible combinations imply different physics. For example, the expectation value of the square of the distance between the two particles is:

$$\langle(x_1 - x_2)^2\rangle_\pm = \langle x^2\rangle a + \langle x^2\rangle b - 2\langle x\rangle_a \langle x\rangle_b \mp 2\,|\,\langle x\rangle_{ab}\,|^2$$

The last term *reduces the expected value for bosons* and *increases the value for fermions* but *only when the states ψ_a and ψ_b physically overlap* ($\langle x\rangle_{ab} \neq 0$).

The physical effect of the exchange symmetry requirement is not a force. Rather *it is a significant geometrical constraint*, increasing the curvature of *wavefunctions* to prevent the overlap of the states occupied by indistinguishable fermions. The terms "*exchange force*" and "*Pauli repulsion*" for fermions are sometimes used as an intuitive description of the effect but this intuition can give incorrect physical results.

Exchange interactions between localized electron magnetic moments

Quantum mechanical particles are classified as bosons or fermions. The spin–statistics theorem of quantum field theory demands that all particles with half-integer spin behave as fermions and all particles with integer spin behave as bosons. Multiple bosons may occupy the same *quantum state*; however, *by the Pauli exclusion principle, no two fermions can occupy the same state. Since electrons have spin 1/2, they are fermions.* This means that *the overall wave function of a system must be antisymmetric when two electrons are exchanged,* i.e. interchanged with respect to both *spatial* and *spin* coordinates. First, however, exchange will be explained with the neglect of spin.

Exchange of spatial coordinates

Taking a hydrogen molecule-like system (i.e. one with two electrons), one may attempt to model the *state of each electron* by first assuming the electrons behave independently, and taking *wave functions in position space* of $\Phi_a(r_1)$ for the first electron and $\Phi_b(r_2)$ for the second electron. The functions Φ_a and Φ_b are orthogonal, and *each corresponds to an energy eigenstate*. Two *wave functions* for the overall system in position space can be constructed. One uses an *antisymmetric* combination of the *product wave functions in position space*:

$$\Psi_A(\vec{r}_1,\vec{r}_2) = 1/\sqrt{2}\ [\Phi_a(\vec{r}_1)\ \Phi_b(\vec{r}_2) - \Phi_b(\vec{r}_1)\ \Phi_a(\vec{r}_2)] \qquad (1)$$

The other uses a *symmetric* combination of the *product wave functions in position space*:

$$\Psi_S(\vec{r}_1,\vec{r}_2) = 1/\sqrt{2}\ [\Phi_a(\vec{r}_1)\ \Phi_b(\vec{r}_2) + \Phi_b(\vec{r}_1)\ \Phi_a(\vec{r}_2)] \qquad (2)$$

To treat the problem of the hydrogen molecule perturbatively, the overall Hamiltonian is decomposed into a unperturbed Hamiltonian of the non-interacting hydrogen atoms $\mathcal{H}^{(0)}$ and a perturbing Hamiltonian, which accounts for interactions between the two atoms $\mathcal{H}^{(1)}$. The full Hamiltonian is then:

$$\mathcal{H} = \mathcal{H}^{(0)} + \mathcal{H}^{(1)}$$

where

$$\mathcal{H}^{(0)} = -\ \hbar^2/2m\ \nabla_1^2 - \hbar^2/2m\ \nabla_2^2 - e^2/r_{a1} - e^2/r_{b2}$$

and

$$\mathcal{H}^{(1)} = e^2/R_{ab} + e^2/r_{12} - e^2/r_{a2} - e^2/r_{b1}$$

The first two terms of $\mathcal{H}^{(0)}$ denote the *kinetic energy of the electrons*. The remaining terms account for the *attraction between the electrons and their host protons* ($r_{a1}/_{b2}$). The terms in $\mathcal{H}^{(1)}$ account for the *potential energy* corresponding to: *proton–proton repulsion* (R_{ab}), *electron–electron repulsion* (r_{12}), and *electron–proton attraction between the electron of one host atom and the proton of the other* ($r_{a2/b1}$). All quantities are assumed to be *real*.

Two *eigenvalues* for the system *energy* are found:

$$E_{\pm} = E_{(0)} + (C \pm J_{ex})/(1 \pm S^2) \qquad (3)$$

where the E_+ is the *spatially symmetric solution* and E_- is the *spatially antisymmetric solution*, corresponding to Ψ_S and Ψ_A respectively. A variational calculation yields similar results. $\mathcal{H}$ can be diagonalized by using the position–space functions given by Eqs. (1) and (2). In Eq. (3), C is the two-site two-electron *Coulomb integral* (It may be interpreted as the *repulsive potential* for electron-one at a particular point $\Phi_a(\vec{r}_1)^2$ in an *electric field* created by electron-two distributed over the space with the *probability density* $\Phi_b(\vec{r}_2)^2$), S[a] is the *overlap integral*, and J_{ex} is the *exchange integral*, which is similar to the two-site *Coulomb integral* but *includes the exchange of the two electrons*. It has no simple physical interpretation, but *it can be shown to arise entirely due to the anti-symmetry requirement*. These integrals are given by:

$$C = \int \Phi_a(\vec{r}_1)^2 \, (1/R_{ab} + 1/r_{12} - 1/r_{a1} - 1/r_{b2}) \, \Phi_b(\vec{r}_2)^2 \, d^3r_1 \, d^3r_2 \qquad (4)$$
$$S = \int \Phi_b(\vec{r}_2) \, \Phi_a(\vec{r}_2) \, d^3r_2 \qquad (5)$$
$$J_{ex} = \int \Phi_a^*(\vec{r}_1) \, \Phi_b^*(\vec{r}_2) \, (1/R_{ab} + 1/r_{12} - 1/r_{a1} - 1/r_{b2}) \, \Phi_b(\vec{r}_1) \, \Phi_a(\vec{r}_2) \, d^3r_1 \, d^3r_2. \qquad (6)$$

Although *in the hydrogen molecule the exchange integral, Eq. (6), is negative, Heisenberg first suggested that it changes sign at some critical ratio of internuclear distance to mean radial extension of the atomic orbital.*

Inclusion of spin

The *symmetric* and *antisymmetric* combinations in Equations (1) and (2) did not include the *spin* variables (α = spin-up; β = spin-down); there are also *antisymmetric* and *symmetric* combinations of the *spin* variables:

$$\alpha(1)\beta(2) \pm \alpha(2)\beta(1) \qquad (7)$$

To obtain the *overall wave function*, these spin combinations have to be coupled with Eqs. (1) and (2). The resulting overall *wave functions*, called *spin-orbitals*, are written as *Slater determinants*. When the *orbital wave function* is symmetrical the *spin* one must be anti-symmetrical and vice versa. Accordingly, E_+ above corresponds to the *spatially symmetric/spin-singlet* solution and E_- to the *spatially antisymmetric/spin-triplet* solution.

J. H. Van Vleck presented the following analysis: "The *potential energy* of the interaction between the two electrons in *orthogonal orbitals* can be represented by a matrix, say E_{ex}. From Eq. (3), the *characteristic values* [eigenvalues] of this matrix are $C \pm J_{ex}$. The *characteristic values* of a matrix are its diagonal elements after it is converted to a diagonal matrix. Now, the *characteristic values* of the square of the magnitude of the resultant *spin*, …

and, hence,

$$E_{ex} = C - ½ J_{ex} - 2J_{ex}\langle \vec{s}_a \cdot \vec{s}_b \rangle \qquad (9)$$

$$[V_{12} = K_{12} - ½ J_{12} - 2J_{12}\, \mathbf{s}_1 \,.\, \mathbf{s}_2, \qquad (9) \text{ in Part III below}]$$

where the *spin momenta* are given as $\langle \vec{s}_a \rangle$ and $\langle \vec{s}_b \rangle$.

Dirac pointed out that the critical features of the *exchange interaction* could be obtained in an elementary way by neglecting the first two terms on the right-hand side of Eq. (9), thereby considering the two *electrons* as simply having their *spins* coupled by a potential of the form:

$$-2J_{ab}\,\langle \vec{s}_a \cdot \vec{s}_b \rangle \qquad (10)$$

It follows that the *exchange interaction* Hamiltonian between two *electrons* in *orbitals* Φ_a and Φ_b can be written in terms of their *spin momenta* $\vec{s}_a$ and $\vec{s}_b$. This interaction is named the *Heisenberg exchange Hamiltonian* or the *Heisenberg–Dirac Hamiltonian* in the older literature:

$$\mathscr{H}_{Heis} = -2J_{ab}\,\langle \vec{s}_a \cdot \vec{s}_b \rangle \qquad (11)$$

J_{ab} is not the same as the quantity labeled J_{ex} in Eq. (6). Rather, J_{ab}, which is termed the *exchange constant*, is a function of Eqs. (4), (5), and (6), namely,

$$J_{ab} = ½\,(E_+ - E_-) = (J_{ex} - CS^2)/(1 - S^4). \qquad (12)$$

However, with *orthogonal orbitals* (in which $S = 0$), for example with *different orbitals in the same atom*, $J_{ab} = J_{ex}$.

Effects of exchange

If J_{ab} is positive the exchange energy favors electrons with parallel spins; *this is a primary cause of ferromagnetism in materials in which the electrons are considered localized in the Heitler–London model of chemical bonding*, but this model of *ferromagnetism* has severe limitations in solids (see below). If J_{ab} is *negative*, the interaction favors electrons with *antiparallel spins*, potentially causing *antiferromagnetism*. The sign of J_{ab} is essentially determined by the relative sizes of J_{ex} and the product of CS. This sign can be deduced from the expression for the difference between the energies of the triplet and singlet states, $E_- - E_+$:

$$E_{-} - E_{+} = 2(CS^2 - J_{ex})/(1 - S^4) \qquad (13)$$

Although these consequences of the *exchange interaction* are *magnetic* in nature, the cause is not; *it is due primarily to electric repulsion and the Pauli exclusion principle. In general, the direct magnetic interaction between a pair of electrons (due to their electron magnetic moments) is negligibly small compared to this electric interaction.*

Exchange energy splittings are very elusive to calculate for molecular systems at large internuclear distances. However, analytical formulae have been worked out for the hydrogen molecular ion.

Normally, exchange interactions are very short-ranged, confined to electrons in orbitals on the same atom (*intra-atomic exchange*) or *nearest neighbor atoms* (*direct exchange*) but longer-ranged interactions can occur via intermediary atoms and this is termed *superexchange*.

Direct exchange interactions in solids

In a crystal, generalization of the *Heisenberg Hamiltonian* in which the sum is taken over the *exchange Hamiltonians* for all the (i,j) pairs of atoms of the many-electron system gives:

$$\mathcal{H}_{Heis} = ½\,(-\,2J\,\Sigma_{i,j}\,\langle \vec{S}_i \cdot \vec{S}_j\rangle) = -\,\Sigma_{i,j}\,J\,\langle \vec{S}_i \cdot \vec{S}_j\rangle \qquad (14)$$

The 1/2 factor is introduced because the interaction between the same two atoms is counted twice in performing the sums. Note that the J in Eq. (14) is the *exchange constant* J_{ab} above not the *exchange integral* J_{ex}. The *exchange integral* J_{ex} is related to yet another quantity, called the *exchange stiffness constant* (A) which serves as a characteristic of a *ferromagnetic* material. The relationship is dependent on the crystal structure. For a simple cubic lattice with lattice parameter a

$$A_{sc} = J_{ex}\langle S^2\rangle/a. \qquad (15)$$

For a body-centered cubic lattice,

$$A_{bcc} = 2J_{ex}\langle S^2\rangle/a \qquad (16)$$

and for a face-centered cubic lattice,

$$A_{fcc} = 4J_{ex}\langle S^2\rangle/a. \qquad (17)$$

The form of Eq. (14) corresponds identically to the *Ising model* of *ferromagnetism* except that in the *Ising model*, the dot product of the two *spin angular momenta* is replaced by the scalar product $S_{ij}S_{ji}$. The *Ising model* was invented by Wilhelm Lenz in 1920 and solved for the one-dimensional case by his doctoral student Ernst Ising in 1925. The energy of the *Ising model* is defined to be:

$$E = -\sum_{i \neq j} J_{ij} \langle S_{iz} S_{jz} \rangle. \quad (18)$$

Limitations of the Heisenberg Hamiltonian and the localized electron model in solids

Because the Heisenberg Hamiltonian presumes the electrons involved in the exchange coupling are localized in the context of the Heitler–London, or valence bond (VB), theory of chemical bonding, *it is an adequate model for explaining the magnetic properties of electrically insulating narrow-band ionic and covalent non-molecular solids* where this picture of the bonding is reasonable. Nevertheless, theoretical evaluations of the *exchange integral* for non-molecular solids that display metallic conductivity in which the electrons responsible for the *ferromagnetism* are itinerant (e.g. iron, nickel, and cobalt) have historically been either of the wrong sign or much too small in magnitude to account for the experimentally determined *exchange constant* (e.g. as estimated from the Curie temperatures via $T_C \approx 2\langle J \rangle / 3k_B$ where $\langle J \rangle$ is the *exchange interaction* averaged over all sites). The Heisenberg model thus cannot explain the observed *ferromagnetism* in these materials. In these cases, a delocalized, or Hund–Mulliken–Bloch (molecular orbital/band) description, for the electron wave functions is more realistic. Accordingly, the *Stoner model of ferromagnetism* is more applicable. [???]

> [The *Stoner model* is a simplified model of a solid which is formulated in terms the dimensionless *density of spin up (down) electrons* and the *dispersion relation of spinless electrons* where the electron-electron interaction is disregarded. It can be used to calculate the total energy of the system as a function of its polarization.
>
> The *Heisenberg model* describes magnetic interactions in terms of *spin exchange* between neighboring magnetic moments. It focuses on the alignment of spins in a crystal lattice, considering the *exchange interaction energy*.]
>
> [Stuart, R.; Marshall, W. (1960). Direct Exchange in Ferromagnets. *Phys. Rev.*, 120, 2, 353–7; doi:10.1103/physrev.120.353: "The direct exchange integral which occurs in the *Heisenberg theory of ferromagnetism* is evaluated for all internuclear spacings. We find that it is always positive, whereas Bethe originally suggested it would be positive only at large spacing and more recently it has been suggested that the integral should always be negative. However, at the observed internuclear separation the magnitude calculated is of the order of 70 times too small to explain

the experimentally determined *exchange constant* in *ferromagnetic metals*, and we therefore conclude that direct exchange is not responsible for ferromagnetism in these metals." ???]

[This article in Expedia surprisingly fails to reference Heisenberg, W. (September, 1928). *Zur Theory of Ferromagnetismus*. (On the theory of ferromagnetism.). (See above.) in which Heisenberg correctly demonstrated that two conditions were necessary for the appearance of *ferromagnetism* in a sold: the crystal lattice must be a type such that *any atom has at least 8 neighbors*; and the *principal quantum number* of the electrons that are responsible for magnetism must be $n \geq 3$, which is the case for Fe, Co, Ni, whose lattices were all cubic, some of which were space-centered ($z = 8$) and some of which were face-centered ($z = 12$). In this article he applied Heitler-London's calculations in place of his earlier resonance theory in a 1926 paper, which was referenced.

In his 1928 paper, Heisenberg noted that empirical results exhibit *ferromagnetism* as an entirely similar state of affairs to what was previously observed in the spectrum of the helium atom; and it seemed to follow from the levels in the helium atoms that a powerful interaction prevailed between the spin directions of two electrons that led to the splitting of the level structure into systems of singlets and triplets. He also noted that this was closely related to explaining ferromagnetic phenomena as being implied by the *exchange phenomenon* (resulting from *quantum entanglement*). He proceeded to show that the *Coulomb interaction*, together with the *Pauli principle*, succeeded in evoking the same effects as the molecular field that Weiss postulated, noting that it was only in recent times that mathematical methods were developed for the treatment of such a complicated problem in the important investigations of Wigner, Hund, and Heitler and London. As a first approximation, Heisenberg assumed that the lattice separations were very large, and that every electron belonged to its own atom, and applied Heitler-London's calculations to the case of 2n electrons in a state of interaction, finding 2n electrons in 2n positionally different quantum cells. Due to their smallness, he was able to leave the magnetic interactions outside of consideration, and showed that *the spin moments of all electrons become partly parallel and partly anti-parallel as a result of the exchange processes*. By adding the fundamental Pauli principle to this, viz., that the eigenfunctions of the total system should be *anti-symmetric* in all electrons, he showed that an entirely well-defined *total magnetic moment* belonged to each level value of the perturbed system, and there were (2n)! levels in the unperturbed system (ignoring the Pauli principle and spin). He then showed that a *statistical treatment of ferromagnetism was possible when all energy values had been calculated.* Heisenberg concluded that *an atom in a lattice can only one be exchanged with its "neighbors"*; exchanges with atoms that lie further away that the "neighboring atoms" could then be neglected. The *number of "neighbors" of an atom* is, e.g., 1 in a molecular lattice of diatomic molecules, 2 in a linear chain, 4 in a quadratic surface lattice, 6 in a simple cubic lattice, 8 *in a cubic, space-centered lattice*, and 12 in a cubic, face-centered lattice. By assuming a distribution of energy values about the mean had the approximate form of a Gaussian error curve, Heisenberg showed that small or negative values of the constant β [= zJ_0/kT)] resulted in *paramagnetism*; and that *ferromagnetism was only possible for lattice types for which an atom had at least eight neighbors*, which was the case for Fe, Co, Ni, whose lattices were all cubic, some of which were space-centered ($z = 8$) and some of which were face-centered ($z = 12$). He concluded that two conditions were necessary for the appearance of *ferromagnetism*: the crystal lattice must be a type such that *any atom has at least 8 neighbors*; and the *principal quantum number* of the electrons that are responsible for magnetism must be $n \geq 3$.]

Dirac, P. A. M. (October, 1926). On the Theory of Quantum Mechanics.

Roy. Soc. Proc., A, 112, 762, 661-77; https://doi.org/10.1098/rspa.1926.0133.JSTOR 94692.

Communicated by R. H. Fowler, F.R.S.

Received August 26, 1926.

St. John's College, Cambridge.

In this paper, Dirac developed a *relativistic* treatment of Schrodinger's wave theory from a more general point of view in which the time t and its conjugate momentum – W were treated from the beginning on the same footing as the other variables. He applied his *relativistic formulation* to a system containing an atom with two electrons and found that *if the positions of the two electrons were interchanged the new state of the atom was physically indistinguishable from the original one*. In order that that the theory only enabled calculation of *observable quantities* it was necessary to treat (*mn*) and (*nm*) as only one *state*. *Unsymmetrical* functions of the co-ordinates (and momenta) of the two electrons could not be represented by matrices. *Symmetrical* functions such as the total *polarizations* of the atom could be considered to be represented by matrices without inconsistency. These matrices were by themselves sufficient to determine all the physical properties of the system. The *theory of uniformizing variables* introduced by the author *could no longer apply*. The new theory allowed two solutions satisfying the necessary conditions; one led to Pauli's principle that not more than one electron can be in any given orbit, and the other, when applied to the analogous problem of the ideal gas, led to the Einstein-Bose statistical mechanics. *With neglect of relativity mechanics this accounted for the absorption and stimulated emission of radiation* and showed that the elements of the matrices representing the total polarization determined the transition probabilities. *This could not be applied to spontaneous emission.*

---

§ 1. *Introduction and Summary.*

The new mechanics of the atom introduced by Heisenberg* may be based on the assumption that the variables that describe a dynamical system do not obey the commutative law of multiplication, but satisfy instead certain quantum conditions.

* Heisenberg, W. (July, 1925). Über quantentheoretische Umdeutung kinematischer und mechanischer Beziehungen. (On the quantum-theoretical re-interpretation of kinematic and mechanical relations.) *Zeit. Phys.*, 33, 879-93; Heisenberg proposes a quantum mechanics in which only relationships among observable quantities occur, not possible to assign to the electron a point in space as a function of time, builds on Kramer's dispersion theory and instead assigns to the electron an *emitted radiation*, substitutes *frequencies* and *amplitudes* of Fourier components of emitted radiation of electron, instead of reinterpreting x(t) as a *sum* over transition components represents position by *set* of transition

components, assigns *transition frequencies* and *transition amplitudes* as observables, replaces classical component by *transition* component corresponding to the quantum jump from state n to state $n - \alpha$, translates the old *quantum condition* that fixes the properties of the *states* to a new condition to calculate the amplitude of a *transition* between two states by replacing the differential by a difference, in quantum case *frequencies* do not combine in same way as classical harmonics but in accordance with the *Ritz combination principle* under which spectral lines of any element include frequencies that are either the sum or the difference of the frequencies of two other lines, in quantum case frequencies combine by multiplying *transition amplitudes* (equivalent to matrix multiplication), results in non-commutativity of kinematical quantities, shows simple quantum theoretical connection to Kramers' dispersion theory, the *equation of motion* $\ddot{x} + f(x) = 0$ and the *quantum condition* $h = 4\pi m \sum_{\alpha=-\infty}^{+\infty} \{|a(n, n+\alpha)|^2\omega(n, n+\alpha) - |a(n, n-\alpha)|^2\omega(n, n-\alpha)\}$ together contain if solvable *a complete determination not only of the frequencies and energies but also of the quantum theoretical transition probabilities*].

One can build up a theory without knowing anything about the dynamical variables except the algebraic laws that they are subject to, and (can show) that they may be represented by matrices whenever a set of *uniformizing variables* for the dynamical system exists. It may be shown, however (see § 3), that *there is no set of uniformizing variables for a system containing more than one electron, so that the theory cannot progress very far on these lines*.

A new development of the theory has recently been given by Schrodinger[#].

[#] See various papers in the *Ann. Physik* , beginning with Schrodinger, E. (March, 1926). Quantisierung als Eigenwertproblem. (Erste Mitteilung) (Quantization as an eigenvalue problem. (First communication).) *Ann. Physik*, 384, 4, 79, 361-376; https://doi.org/10.1002/andp.19263840404; the first of 4 papers published by Schrodinger in German during 1926]

[Also see Schrodinger, E. (December, 1926). A Wave Theory of the Mechanics of Atoms and Molecules. *Phys. Rev.*, 28, 1049-70; *non-relativistic* development of de Broglie's *relativistic* wave mechanics in which *phase-waves* associated with motion of material points, in particular with motion of an electron or proton, assumes material points are wave-systems, *wave-equation* $\Delta\psi + 8\pi^2 m(E - V)\psi/h^2 = 0$, *laws of motion* and *quantum conditions* deduced simultaneously from Hamiltonian principle, *wave function* converts atom into system of fluctuating charges spread out continuously in space, generates electric moment that changes in time, discrepancy between frequency of motion and frequency of emission disappears, frequency of emission coincides with differences of frequency of motion, superposition of frequencies, definite localization of electric charge in space and time associated with the wave-system, solutions of *wave equation* for simplified hydrogen atom or one body problem correspond to Bohr's stationary energy levels of the elliptic orbits, the selected values called "*eigenvalues*" and the solutions that belong to them "*eigenfunctions*", the charge of the electron is spread out through space but the *wave-phenomenon* is restricted to a small sphere of a few Angstroms diameter constituting the atom, also possible to calculate *amplitudes* of harmonic components of the *electric moment* for any direction in space, in the case of the *Stark effect* (perturbation of the hydrogen-atom caused by an external homogeneous electric field) parallel to the electric field or perpendicular to the field, shows that squares of these *amplitudes* are proportional to *intensities* of the several line components polarized in either direction, *wave mechanics has been developed without reference to relativity modifications of classical mechanics or to action of a magnetic field on the atom, not been possible to extend the relativistic theory to a system of more than one electron, relativistic theory of hydrogen atom in grave contradiction with experiment*, how to take into account *electron spin* is yet unknown.]

Starting from the idea that an atomic system cannot be represented by a trajectory, i.e., by a point moving through the co-ordinate space, but must be represented by a wave in this space, Schrodinger obtains from a variation principle a differential equation which the *wave function* ψ must satisfy. This differential equation turns out to be very closely connected with the Hamiltonian equation which specifies the system, namely, if

$$H\ (q_r, p_r) - W = 0$$

is the Hamiltonian equation of the system, where the q_r, p_r are canonical variables, then the *wave equation* for ψ is

$$\{H\ (q_r, ih\ \delta/\delta q_r) - W\}\ \psi = 0, \qquad (1)$$

where h is $(2\pi)^{-1}$ times the usual Planck's constant.

[Schrodinger (1926):

$$\Delta\psi + 8\pi^2 m(E - V)\psi/h^2 = 0, \qquad (16)]$$

Each *momentum* p_r in H is replaced by the operator ih $\delta/\delta q_r$, and is supposed to operate on all that exists on its right-hand side in the term in which it occurs. Schrodinger takes the values of the parameter W for which there exists a ψ satisfying (1) that is continuous, single-valued and bounded throughout the whole of q-space to be the *energy* levels of the system, and shows that when the general solution of (1) is known, matrices to represent the p_r and q_r may easily be obtained, satisfying all the conditions that they have to satisfy according to Heisenberg's matrix mechanics, and consistent with the energy levels previously found. *The mathematical equivalence of the theories is thus established.*

In the present paper, Schrodinger's theory is considered in § 2 from a slightly more general point of view, in which *the time t and its conjugate momentum –W are treated from the beginning on the same footing as the other variables*. A more general method, requiring only elementary symbolic algebra, of obtaining matrix representations of the dynamical variables is given.

In § 3 the problem is considered of a system containing several similar particles, such as an atom with several electrons. *If the positions of two of the electrons are interchanged, the new state of the atom is physically indistinguishable from the original one*. In such a case one would expect only symmetrical functions of the co-ordinates of all the electrons to be capable of being represented by matrices. *It is found that this allows one to obtain two solutions of the problem satisfying all the necessary conditions, and the theory is incapable of deciding which is the correct one.* One of the solutions leads to Pauli's principle that *not more than one electron can be in any given orbit*, and the other, when applied to the analogous problem of the ideal gas, leads to the *Einstein-Bose statistical mechanics*.

The effect of an arbitrarily varying perturbation on an atomic system is worked out in § 5 with the help of a new assumption. The theory is applied to the *absorption* and stimulated *emission* of radiation by an atom. A generalization of the description of the phenomena by Einstein's B coefficients is obtained, in which the *phases* play their proper parts. *This method cannot be applied to spontaneous emission.*

§ 2. *General Theory.*

According to the new point of view introduced by Schrodinger, we no longer leave unspecified the nature of the dynamical variables that describe an atomic system, but count the q's and t as ordinary mathematical variables (this being permissible since they commute with one another) and take the p's and W to be the differential operators

$$p_r = - ih\ \delta/\delta q_r, \qquad - W = - \delta/\delta q_r, \qquad (2)$$

Whenever a p_r or W occurs in a term of an equation, it must he considered as meaning the corresponding differential operator operating on all that occurs on its right-hand side in the term in question. Thus, by carrying out the operations, one can reduce any function of the p's, q s, W and t to a function of the q's and t only.

The relations (2) require two obvious modifications to be made in the algebra governing the dynamical variables. Firstly, *only rational integral functions of the p's and W have a meaning*, and, secondly, *one can multiply up an equation by a factor (integral in the p's and W) on the left-hand side, but one cannot, in general, multiply up by factor on the right-hand side*. Thus, if one is given the equation a = b, one can infer from it that Xa = Xb, where X is arbitrary, but one cannot in general infer that aX = bX.

There are, however, certain equations a = b for which it is true that aX = bX for any X, and these equations we call *identities*. The *quantum conditions*

$$q_r p_s - p_s q_r = ih\delta_{rs}, \qquad p_r p_s - p_s p_r = 0,$$

with the similar relations involving –W and t, are *identities*, as it can easily be verified (and has been verified by Schrodinger) that the relations

$$(q_r p_s - p_s q_r)\, X = ih\delta_{rs} X,$$

etc., hold for any X. These relations form the main justification for the assumptions (2)

$$[p_r = -\, ih\ \delta/\delta q_r, \qquad -\, W = -\, \delta/\delta q_r. \qquad (2)]$$

If a = b is an identity, we can deduce, since aX = bX and Xa = Xb, that

$$aX - Xa = bX - Xb,$$

or

$$[a, X] = [b, X].$$

Thus, we can equate the Poisson bracket of either side of an identity with an arbitrary quantity, and so our quantum identity is the analogue of an *identity* on the classical theory. We assume the general equation xy – yx = ih [x, y] and the *equations of motion* of a dynamical system to be *identities*.

A dynamical system is specified by a Hamiltonian equation between the variables

$$H\,(q_r, p_r, t) - W = 0, \qquad (3)$$

or more generally

$$F\,(q_r, p_r, t, W) = 0, \qquad (4)$$

and the *equations of motion* are

$$dx/ds = [x, F],$$

where x is any function of the dynamical variables, and s is a variable which depends on the form in which (4) is written, and, in particular, is just t if (4)

$$[F\,(q_r, p_r, t, W) = 0, \qquad (4)]$$

is written in the form (3)

$$[H\,(q_r, p_r, t) - W = 0. \qquad (3)]$$

On the new theory we consider the equation

$$F\psi = 0, \qquad (5)$$

which, if we take ψ to be a function of the q's and t only, is an ordinary differential equation for ψ. From the general solution of this differential equation the matrices that form the solution of the mechanical problem may be very easily obtained.

Since (5) is linear in its general solution is of the form

$$\psi = \Sigma\, c_n \psi_n, \qquad (6)$$

where the c_n's are arbitrary constants and the ψ_n's are a set of independent solutions, which may be called *eigenfunctions.* Only solutions that are continuous, single-valued and bounded throughout the whole domain of the q's and t are recognized by the theory. Instead of a discreet set of *eigenfunctions* ψ_n there may be a continuous set $\psi(\alpha)$, depending on a parameter α, and satisfying the differential equation for all values of α in a certain range, in which case the sum in (6) must be replaced by an integral $\int c\alpha\ \psi(\alpha)\ d\alpha$, or both a discreet set and a continuous set may occur together. For definiteness, however, we shall write down explicitly only the discreet sum in the following work.

We shall now show that any constant of integration of the dynamical system (either a first integral or a second integral) can be represented by a matrix whose elements are constants, there being one row and column of the matrix corresponding to each *eigenfunction* ψ_n.

...

The matrix representation we have obtained is not unique, since any set of independent eigenfunctions ψ_n will do. *To obtain the matrices of Heisenberg's original quantum mechanics, we must choose the* ψ_n*'s in a particular way.* We can always, by a linear transformation, obtain a set of ψ_n's which makes the matrix representing any given constant of integration of the dynamical system a diagonal matrix. Suppose now that the Hamiltonian F does not contain the time explicitly, so that W is a constant of the system, and is the *energy*, and we choose the ψ_n's so as to make the matrix representing W a diagonal matrix, i.e., so as to make

$$W\psi_n = W_n \psi_n, \qquad (7)$$

where W_n is a numerical constant. Let x be any function of the dynamical variables *that does not involve the time explicitly*, and put

$$x\psi_n = \Sigma_m\, x_{mn} \psi_m,$$

where the x_{mn}'s are functions of the time only. We shall now show that the x_{mn}'s are of the form

$$x_{mn} = a_{mn} e^{\,i(W_m - W_n)t/h} \qquad (8)$$

where the a_{mn}'s are constants, as on Heisenberg's theory. ...

...

We have thus shown that with the ψ_n's chosen in this way the matrices satisfy all the conditions of Heisenberg's matrix mechanics, except the condition that the matrices that represent real quantities are Hermitic (i.e., have their *mn* and *nm* elements conjugate imaginaries). There does not seem to be any simple general proof that this is the case, as the proof would have to make use of the fact that the ψ_n's are bounded. It is easy to prove the particular case that the matrix representing W is Hermitic, i.e., that the W_n's are real, since from (7)

$$[W\psi_n = W_n\psi_n, \qquad (7)]$$

must be of the form

$$\psi_n = u_n e^{-iW_n t/h},$$

where u_n is independent of t, and if W_n contains an imaginary part, ψ_n would not remain bounded as t becomes infinite. In general, the matrices representing real quantities could be Hermitic only if the arbitrary numerical constants by which the ψ_n's may be multiplied are chosen in a particular way.

We may regard an *eigenfunction* as being associated with definite numerical values for some of the constants of integration of the system. Thus, if we find constants of integration *a*, *b*, . . . such that

$$a\psi_n = a_n\psi_n, \qquad b\psi_n = b_n\psi_n, \qquad \ldots \qquad (11)$$

where a_n, b_n, . . . are numerical constants, we can say that ψ_n represents a *state* of the system in which *a*, *b*, . . . have the numerical values a_n, b_n, . . . (Note that *a*, *b*, ... must commute for (11) to be possible.) In this way we can have *eigenfunctions* representing *stationary states* of an atomic system with definite values for the *energy*, *angular momentum*, and other constants of integration.

It should be noticed that the choice of the time t as the variable that occurs in the elements of the matrices representing variable quantities is quite arbitrary, and any function of t and the q's that increases steadily would do. To determine accurately the radiation emitted by the system in the direction of the x-axis, one would have to use
(t – x/c) instead of t*

* Dirac, P. A. M. (June, 1926). Relativity Quantum Mechanics with an Application to Compton Scattering. *Roy. Soc. Proc., A*, 111, 758, 405-423 [; the object of this paper is to extend quantum mechanics to systems for which the Hamiltonian involves the time explicitly and to comply with the *theory of special relativity* by treating time on the same footing as the other variables, sets $x_4 = ict$ (so that $x_1^2 + x_2^2 + x_3^2 + x_4^2 = 0$ and $x_1^2 + x_2^2 + x_3^2 = c^2t^2$) and $p_4 = iW/c$ where W is the energy, shows that – W is the *momentum* conjugate to t, substitutes $(t - x_1/c)$ for t as *uniformizing variable* in order that its contribution to the exchange of energy with the radiation field may vanish, applies *relativistic* quantum

mechanics to Compton scattering and calculation of *frequency* and *intensity* of scattered radiation; *no improvement in agreement with experiments from relativistic formulation*].

It is probable that the representation of a constant of integration of the system by a matrix of constant elements is more fundamental than the representation of a variable quantity by a matrix whose elements are functions of some variable such as t or (t – x/c). It would appear to be possible to build up an electromagnetic theory in which the *potentials* of the field at a specified point x_0, y_0, z_0, t_0 in space-time are represented by matrices of constant elements that are functions of x_0, y_0, z_0, t_0.

§ 3. *Systems containing Several Similar Particles.*

In Heisenberg's matrix mechanics it is assumed that the elements of the matrices that represent the dynamical variables determine the *frequencies* and *intensities* of the components of radiation emitted. The theory thus enables one to calculate just those quantities that are of physical importance, and gives no information about quantities such as *orbital frequencies* that one can never hope to measure experimentally. We should expect this very satisfactory characteristic to persist in all future developments of the theory.

Consider now a system that contains two or more similar particles, say, for definiteness, *an atom with two electrons*. Denote by (*mn*) that *state* of the atom in which one electron is in an orbit labelled *m*, and the other in the orbit *n*. The question arises whether the two *states* (*mn*) and (*nm*), which are physically indistinguishable as they differ only by the interchange of the two electrons, are to be counted as two different *states* or as only one *state*, i.e., do they give rise to two rows and columns in the matrices or to only one? If the first alternative is right, then the theory would enable one to calculate the *intensities* due to the two *transitions* (*mn*) -> (*m'n'*) and (*mn*) -> (*n'm'*) separately, as the *amplitude* corresponding to either would be given by a definite element in the matrix representing the total *polarisation*. The two *transitions* are, however, physically indistinguishable, and only the sum of the intensities for the two together could be determined experimentally. Hence, *in order to keep the essential characteristic of the theory that it shall enable one to calculate only observable quantities*, one must adopt the second alternative that (*mn*) and (*nm*) count as only one *state*.

This alternative, though, also leads to difficulties. The symmetry between the two electrons requires that the *amplitude* associated with the transition (*mn*) -> (*m'n'*) of x_1, a *co-ordinate* of one of the electrons, shall equal the *amplitude* associated with the transition (*nm*) -> (*n'm'*) of x_2, the corresponding *coordinate* of the other electron, i.e.,

$$x_1\,(mn;\,m'n') = x_2\,(nm;\,n'm'). \qquad (12)$$

If we now count (*mn*) and (nm) as both defining the same row and column of the matrices, and similarly for (*m'n'*) and (*n'm'*), equation (12) shows that each element of the matrix x_1 equals the corresponding element of the matrix x_2, so that we should have the matrix equation $x_1 = x_2$. This relation is obviously impossible, as, amongst other things, it is inconsistent with the quantum conditions. *We must infer that unsymmetrical functions of the co-ordinates (and momenta) of the two electrons cannot be represented by matrices. Symmetrical* functions, such as the total *polarisation* of the atom, can be considered to be represented by matrices without inconsistency, and these matrices are by themselves sufficient to determine all the physical properties of the system.

One consequence of these considerations is that the theory of uniformizing variables introduced by the author can no longer apply. This is because, corresponding to any transition (*mn*) -> (*m'n'*), there would be a term $e^{i(aw)}$ in the Fourier expansions, and we should require there to be a unique state, (*m''n''*), say, such that the same term $e^{i(aw)}$ corresponds to the transition (*m'n'*) -> (*m''n''*), and $e^{2i(aw)}$ corresponds to (*mn*) -> (*m''n''*). If the *m*'s and *n*'s are *quantum numbers*, and we take the case of one *quantum number* per *electron* for definiteness, we should have to have

$$m'' - m' = m' - m, \qquad n'' - n' = n' - n.$$

Since, however, the *state* (*m'n'*) may equally well be called the *state* (*n'm'*), we may equally well take

$$m'' - n' = n' - m, \qquad n'' - m' = m' - n.$$

which would give a different *state* (*m''n''*). There is thus no unique *state* (*m''n''*) that the theory of uniformizing variables demands.

If we neglect the interaction between the two electrons, then we can obtain the *eigenfunctions* for the whole atom simply by multiplying the *eigenfunctions* for one electron when it exists alone in the atom by the *eigenfunctions* for the other electron alone, and taking the same time variable for each*.

* The same time variable t must be taken in each owing to the fact that we write the Hamiltonian equation for the whole system: $H(1) + H(2) - W = 0$, where $H(1)$ and $H(2)$ are the Hamiltonians for the two electrons separately, so that there is a common time t conjugate to minus the total energy W.

Thus, if ψ_n (x, y, z, t) is the *eigenfunction* for a single *electron* in the orbit *n*, then the *eigenfunction* for the whole atom in the state (*mn*) is

$$\psi_m (x_1, y_1, z_1, t)\, \psi_n (x_2, y_2, z_2, t) = \psi_m(1)\, \psi_n(2),$$

say, where x_1, y_1, z_1 and x_2, y_2, z_2 are the co-ordinates of the two electrons, and $\psi(r)$ means $\psi(x_r, y_r, z_r, t)$. The *eigenfunction* $\psi_m(2)\ \psi_n(1)$, however, also corresponds to the same *state* of the atom if we count the (*mn*) and (*nm*) *states* as identical. But two independent *eigenfunctions* must give rise to two rows and columns in the matrices. If we are to have only one row and column in the matrices corresponding to both (*mn*) and (*nm*), we must find a set of *eigenfunctions* of the form

$$\psi_{mn} = a_{mn}\ \psi_m(1)\ \psi_n(2) + b_{mn}\ \psi_m(2)\ \psi_n(1),$$

where the a_{mn}'s and b_{mn}'s are constants, which set must contain only one ψ_{mn} corresponding to both (*mn*) and (*nm*), and must be sufficient to enable one to obtain the matrix representing any *symmetrical* function A of the two electrons. This means the ψ_{mn}'s must be chosen such that A times any chosen ψ_{mn} can be expanded in terms of the chosen ψ_{mn}'s in the form

$$A\psi_{mn} = \Sigma_{m'n'}\ \psi_{m'n'}\ A_{m'n',mn}, \qquad (13)$$

where the $A_{m'n',mn}$'s are constants or functions of the time only.

There are two ways of choosing the set of ψ_{mn}'s to satisfy the conditions. We may either take $a_{mn} = b_{mn}$, which makes each ψ_{mn} a *symmetrical* function of the two electrons, so that the left-hand side of (13) is *symmetrical* and only *symmetrical eigenfunctions* will be required for its expansion, or we may take $a_{mn} = -\ b_{mn}$, which makes ψ_{mn} *antisymmetrical*, so that the left-hand side of (13) is *antisymmetrical* and only *antisymmetrical eigenfunctions* will be required for its expansion. Thus, *the symmetrical eigenfunctions alone or the antisymmetrical eigenfunctions alone give a complete solution of the problem. The theory at present is incapable of deciding which solution is the correct one*. We are able to get complete solutions of the problem which make use of less than the total number of possible *eigenfunctions* at the expense of being able to represent only *symmetrical* functions of the two electrons by matrices.

These results may evidently be extended to any number of *electrons*. For r non-interacting *electrons* with co-ordinates x_1, y_1, z_1..., x_r, y_r, z_r, ..., the *symmetrical eigenfunctions* are

$$\Sigma_{\alpha 1, \dots \alpha r}\ \psi_{n1}(\alpha_1),\ \psi_{n2}(\alpha_2)\ \dots\ \psi_{nr}(\alpha_r), \qquad (14)$$

where α_1, α_2 ... α_r are any permutation of the integers 1, 2 ... r, while the *antisymmetrical* ones may be written in the determinantal form

$$\begin{vmatrix} \psi_{n1}(1), & \psi_{n1}(2), \dots & \psi_{n1}(r) \\ \psi_{n1}(1), & \psi_{n1}(2), \dots & \psi_{n1}(r) \\ \dots & \dots \quad \dots & \dots \\ \psi_{n1}(1), & \psi_{n1}(2), \dots & \psi_{n1}(r) \end{vmatrix} \qquad (15)$$

If there is interaction between the *electrons*, there will still be *symmetrical* and *antisymmetrical eigenfunctions*, although they can no longer be put in these simple forms. In any case the *symmetrical* ones alone or the *antisymmetrical* ones alone give a complete solution of the problem.

An *antisymmetrical eigenfunction* vanishes identically when two of the *electrons* are in the same orbit. This means that in the solution of the problem *with antisymmetrical eigenfunctions there can be no stationary states with two or more electrons in the same orbit, which is just Pauli's exclusion principle*. The solution *with symmetrical eigenfunctions*, on the other hand, *allows any number of electrons to be in the same orbit, so that this solution cannot be the correct one for the problem of electrons in an atom.*

§ 4. *Theory of the Ideal Gas.*

The results of the preceding section apply to any system containing several similar particles, in particular to an assembly of gas molecules. There will be two solutions of the problem, in one of which the *eigenfunctions* are *symmetrical* functions of the co-ordinates of all the molecules, and in the other *antisymmetrical*.

The *wave equation* for a single *molecule* of *rest-mass* m moving in free space is

$$\{p_x^2 + p_y^2 + p_z^2 - W^2/c^2 + m^2c^2\}\ \psi = 0$$
$$\{\delta^2/\delta x^2 + \delta^2/\delta y^2 + \delta^2/\delta z^2 - 1/c^2\ \delta^2/\delta t^2 + m^2c^2/h^2\}\ \psi = 0$$

and its solution is of the form

$$\psi_{\alpha 1,\, \alpha 2,\, \alpha 3} = \exp.\ i\ (\alpha_1 x + \alpha_2 y + \alpha_3 z - Et)/h, \qquad (16)$$

where α_1, α_2, α_3 and E are constants satisfying

$$\alpha_1^2 + \alpha_2^2 + \alpha_3^2 - E^2/c^2 + m^2c^2 = 0.$$

The *eigenfunction* (16) represents an atom having the *momentum* components α_1, α_2, α_3 and the *energy* E.

We must now obtain some restriction on the possible *eigenfunctions* due to the presence of boundary walls. It is usually assumed that the *eigenfunction*, or *wave function* associated with a *molecule*, vanishes at the boundary, but we should expect to be able to deduce this, if it is true, from the general theory. We assume, as a natural generalization of the methods of the preceding section, that there must be only just sufficient *eigenfunctions* for one to be able to represent by a matrix any function of the co-ordinates that has a physical meaning. Suppose for definiteness that each *molecule* is confined between two boundaries at $x = 0$ and $x = 2\pi$. Then only those functions of x that are defined only for $0 < x < 2\pi$ have a physical meaning and must be capable of being represented by matrices. (This will require fewer *eigenfunctions* than if every function of x had to be capable of being represented by a matrix.) These functions f(x) can always be expanded as Fourier series of the form

$$f(x) = \Sigma_n \alpha_n e^{inx}, \qquad (17)$$

where the α_n's are constants and the n's integers. If we choose from the *eigenfunctions* (16)

$$[\psi_{\alpha 1,\, \alpha 2,\, \alpha 3} = \exp.\ i\,(\alpha_1 x + \alpha_2 y + \alpha_3 z - Et)/h, \qquad (16)]$$

those for which α_1/h is an integer, then f(x) times any chosen *eigenfunction* can be expanded as a series in the chosen *eigenfunctions* whose coefficients are functions of t only, and hence can be represented by a matrix. Thus, these chosen *eigenfunctions* are sufficient, and are easily seen to be only just sufficient, for the matrix representation of any function of x of the form (17)

$$[f(x) = \Sigma_n \alpha_n e^{inx}. \qquad (17)]$$

Instead of choosing those *eigenfunctions* with integral values for α_1/h, we could equally well take those with α_1/h equal to half an odd integer, or more generally with $\alpha_1/h = n + \varepsilon$, where n is an integer and ε is any real number. *The theory is incapable of deciding which are the correct ones.* For statistical problems, though, they all lead to the same results.

When y and z are also bounded by $0 < y < 2\pi$, $0 < z < 2\pi$, we find for the number of waves associated with *molecules* whose energies lie between E and E + dE the value

$$4\pi/c^3h^3\,(E^2 - m^2c^4)^{1/2}E dE.$$

This value is in agreement with the ordinary assumption that the *wave function* vanishes at the boundary. It reduces, *when one neglects relativity mechanics*, to the familiar expression

$$2\pi/h^3\,(2m)^{3/4}E_1^{\,1/2}dE_1, \qquad (18)$$

where $E_1 = E - mc^2$ is the *kinetic energy*. For an arbitrary volume of gas V the expression must be multiplied by $V/(2\pi)^3$.

To pass to the *eigenfunctions* for the *assembly of molecules*, between which there is assumed to be no interaction, we multiply the *eigenfunctions* for the separate *molecules*, and then take either the *symmetrical eigenfunctions*, of the form (14)

$$[\Sigma_{\alpha 1, \ldots \alpha r}\ \psi_{n1}(\alpha_1),\ \psi_{n2}(\alpha_2) \ldots\ \psi_{nr}(\alpha_r), \qquad (14)]$$

or the *antisymmetrical* ones, of the form (15)

$$[\begin{vmatrix} \psi_{n1}(1), & \psi_{n1}(2), \ldots & \psi_{n1}(r) \\ \psi_{n1}(1), & \psi_{n1}(2), \ldots & \psi_{n1}(r) \\ \ldots & \ldots \quad \ldots & \ldots \\ \psi_{n1}(1), & \psi_{n1}(2), \ldots & \psi_{n1}(r) \end{vmatrix}. \qquad (15)]$$

We must now make *the new assumption that all stationary states of the assembly (each represented by one eigenfunction) have the same a priori probability*. If now we adopt the solution of the problem that involves *symmetrical eigenfunctions*, we should find that all values for the number of *molecules* associated with any wave have the same *a priori* probability, which gives just the *Einstein-Bose statistical mechanics**.

* Bose, S. (December, 1924). Plancks Gesetz und Lichtquantenhypothese. (Planck's law and light quantum hypothesis.) *Zeit. Phys*., 26, 178-181; https://doi.org/10.1007/BF01327326; Einstein, A. (1924). Quantentheorie des einatomigen idealen Gases. (Quantum theory of the monatomic ideal gas.) *Sitzungsb. d. Preuss. Ac*., 261-7; (1925). Quantentheorie des einatomigen idealen Gases. 2. Abhandlung. (Quantum Theory of the Monatomic Ideal Gas, Part II.) *Sitzungsb. d. Preuss. Ac*., 3-14.

On the other hand, we should obtain a different statistical mechanics if we adopted the solution with *antisymmetrical eigenfunctions*, as we should then have either 0 or 1 *molecule* associated with each wave. *The solution with symmetrical eigenfunctions must be the correct one when applied to light quanta*, since it is known that the *Einstein-Bose statistical mechanics* leads to Planck's law of black-body radiation. *The solution with antisymmetrical eigenfunctions, though, is probably the correct one for gas molecules*, since it is known to be the correct one for electrons in an atom, and one would expect molecules to resemble electrons more closely than light quanta.

We shall now work out, according to well-known principles, the *equation of state* of the gas on the assumption that the solution with *antisymmetrical eigenfunctions* is the correct one, so that not more than one *molecule* can be associated with each wave. Divide the waves into a number of sets such that the waves in each set are associated with *molecules* of about the same energy. Let A_s be the number of waves in the *s*th set, and let E_s be the *kinetic energy* of a *molecule* associated with one of them. Then the probability of a distribution (or the number of *antisymmetrical eigenfunctions* corresponding to distributions) in which N_s molecules are associated with waves in the *s*th set is

$$W = \prod_s \{A_s!/[N_s!\,(A_s - N_s)!]\},$$

giving for the *entropy*

$$S = k \log W = k\Sigma_s\{A_s (\log A_s - 1) - N_s (\log N_s - 1) - (A_s - N_s)\,[\log (A_s - N_s) - 1]\}.$$

This is to be a maximum, so that

$$0 = \delta S = k\Sigma_s\{-\log N_s + \log (A_s - N_s)\}\delta N_s = k\Sigma_s \log (A_s/N_s - 1).\ \delta N_s,$$

for all variations δN_s that leave the *total number of molecules* $N = \Sigma_s N_s$ and the *total energy* $E = \Sigma_s E_s N_s$ unaltered, so that

$$\Sigma_s \delta N_s = 0, \qquad \Sigma_s E_s \delta N_s = 0,$$

We thus obtain

$$\log (A_s/N_s - 1) = \alpha + \beta E_s,$$

where α and β are constants, which gives

$$N_s = A_s/(e^{\alpha + \beta Es} + 1). \tag{19}$$

By making a variation in the *total energy* E and putting $\delta E/\delta S = T$, the *temperature*, we readily find that $\beta = 1/kT$, so that (19) becomes

$$N_s = A_s/(e^{\alpha + Es/kT} + 1),$$

[where N_s is the number of molecules associated with waves in the *s*th set].

This formula gives the *distribution in energy* of the *molecules*. On the Einstein-Bose theory the corresponding formula is

$$N_s = A_s/(e^{\alpha + Es/kT} - 1).$$

...

The saturation phenomenon of the Einstein-Bose theory does not occur in the present theory. The *specific heat* can easily be shown to tend steadily to zero as T -> 0, instead of first increasing until the saturation point is reached and then decreasing, as in the Einstein-Bose theory.

§ 5. *Theory of Arbitrary Perturbations.*

In this section we shall consider the problem of an atomic system subjected to a perturbation from outside (e.g., an incident *electromagnetic field*) which can vary with the time in an arbitrary manner. Let the *wave equation* for the undisturbed system be

$$(H - W)\,\psi = 0, \tag{20}$$

where H is a function of the p's and q's only. Its general solution is of the form

$$\psi = \Sigma_n\, c_n \psi_n, \tag{21}$$

where the c_n's are constants. We shall suppose the ψ_n's to be chosen so that one is associated with each *stationary state* of the *atom*, and to be multiplied by the proper constants to make the matrices that represent real quantities Hermitic.

Now suppose a perturbation to be applied, beginning at the time t = 0. The *wave equation* for the disturbed system will be of the form

$$(H - W + A)\,\psi = 0, \tag{22}$$

where A is a function of the p's, q's and t and is real. It will be shown that we can obtain a solution of this equation of the form

$$\psi = \Sigma_n a_n \psi_n, \qquad (23)$$

where the a_n's are functions of t only, which may have the arbitrary values c_n at the time t = 0.

We shall consider the general solution (21) of equation (20) to represent an *assembly* of the undisturbed *atoms* in which $|c_n|^2$ is the number of atoms in the *n*th state, and shall assume that (23) represents in the same way an assembly of the disturbed atoms, $|a_n(t)|^2$ being the number in the *n*th state at any time t. We take $|a_n|^2$ instead of any other function of a_n because, as will be shown later, this makes the total number of atoms remain constant.

The condition that ψ defined by equation (23) shall satisfy equation (22) is

$$\begin{aligned} 0 &= \Sigma_n (H - W + A)\, a_n \psi_n, \\ &= \Sigma_n a_n (H - W + A)\, \psi_n - ih\, \Sigma_n \dot{a}_n \psi_n, \end{aligned} \qquad (24)$$

since H and A commute with a_n,# while $W a_n - a_n W = ih\dot{a}_n$ identically.

The statement *a* commutes with *b* means *ab* = *ba* identically.

Suppose $A\psi_n$ to be expanded in the form

$$A\psi_n = \Sigma_m A_{mn} \psi_m,$$

where the coefficients A_{mn} are functions of t only, and satisfy $A_{mn}{}^* = A_{nm}$, where the * denotes the conjugate imaginary. Equation (24) now becomes, since $(H - W)\, \psi_n = 0$,

$$\Sigma_{mn}\, a_n A_{mn} \psi_m - ih \Sigma_m \dot{a}_m \psi_m = 0$$

Taking out the coefficient of ψ_m, we find

$$ih\dot{a}_m = \Sigma_n a_n A_{mn}, \qquad (25)$$

which is a simple differential equation showing how the a_m's vary with the time.

Taking conjugate imaginaries, we find

$$-ih\dot{a}_m{}^* = \Sigma_n a_n{}^* A_{mn}{}^* = \Sigma_n a_n A_{mn}.$$

Hence, if $N_m = a_m a_m{}^*$ is the number of *atoms* in the *m*th state, we have

$$\begin{aligned} ih\dot{N}_m &= ih\, (\dot{a}_m a_m{}^* + \dot{a}_m{}^* a_m) \\ &= \Sigma_n (a_n A_{mn}\, a_m{}^* - a_n{}^* A_{nm}\, a_m). \end{aligned}$$

This gives

$$ih \Sigma_m \dot{N}_m = \Sigma_{nm} (a_m{}^* A_{mn} a_n - a_n{}^* A_{nm}\, a_m) = 0.$$

as required.

If the perturbation consists of incident *electromagnetic radiation* moving in the direction of the x-axis and *plane polarized* with its *electric vector* in the direction of the y-axis, the perturbing term A in the Hamiltonian is, *with neglect of relativity mechanics*, $\kappa/c.\eta^{\cdot}$ #, where η is the *total polarization* in the direction of the y-axis and 0, κ, 0, 0 are the components of the *potential* of the incident radiation.

We have neglected a term involving κ^2. This approximation is legitimate, even though we later evaluate the number of transitions that occur in a time T to the order κ^2, provided T ie large compared with the periods of the atom.

. . .

$$[\Delta N_m =]\ 1/h^2c^2 \,.\, \Sigma_n \{|\,c_n\,|^2 - |\,c_m\,|^2\}|\,\eta^{\cdot}_{nm}\,|^2\ |\textstyle\int_0^T \kappa(t)\ e^{\,i(W_m - W_n)t/h}\,dt|^2 \qquad (29)$$

[where $|\,c_n\,|^2$ is the number of atoms in the *n*th state.]

This gives ΔN_m, the increase in the number of atoms in the state *m* from the time t = 0 to the time t = T. The term in the summation that has the suffix *n* may be regarded as due to transitions between the state *m* and the state *n*.

If we resolve the radiation from the time t = 0 to the time t = T into its harmonic components, we find for the *intensity* of *frequency* ν per unit frequency range [I_ν] the value

$$I_\nu = 2\pi\nu^2c^{-1}\ |\textstyle\int_0^T \kappa(t)\ e^{\,2\pi i\nu t/h}\,dt|^2.$$

Hence the term in expression (29) for ΔN_m due to transitions between state *m* and state *n* may be written

$$1/2\pi h^2\,\nu^2 c \,.\, \Sigma_n \{|\,c_n\,|^2 - |\,c_m\,|^2\}|\,\eta^{\cdot}_{nm}\,|^2\ I_\nu,$$

where

$$2\pi\nu = (W_m - W_n)/h,$$

or

$$2\pi/h^2c \,.\, \{|\,c_n\,|^2 - |\,c_m\,|^2\}|\,\eta_{nm}\,|^2\ I_\nu.$$

If one averages over all directions and states of *polarization* of the incident radiation, this becomes

$$2\pi/3h^2c \,.\, \{|\,c_n\,|^2 - |\,c_m\,|^2\}|\,P_{nm}\,|^2\ I_\nu,$$

where

$$|\,P_{nm}\,|^2 = |\,\xi_{nm}\,|^2 + |\,\eta_{nm}\,|^2 + |\,\zeta_{nm}\,|^2,$$

ξ, η and ζ being the three components of *total polarization*. Thus, one can say that the radiation has caused $2\pi/3h^2c \,.\, |\,c_n\,|^2\,|\,P_{nm}\,|^2\ I_\nu$ transitions from state *n* to state *m*, and $2\pi/3h^2c \,.\, |\,c_m\,|^2\,|\,P_{nm}\,|^2\ I_\nu$, transitions from state *m* to state *n*, the probability coefficient for either process being

$$B_{n\to m} = B_{m\to n} = 2\pi/3h^2c \,.\, | P_{nm} |^2,$$

in agreement with the ordinary Einstein theory.

The present theory thus accounts for the *absorption* and stimulated *emission* of radiation, and shows that the elements of the matrices representing the total *polarization* determine the transition probabilities. *One cannot take spontaneous emission into account without a more elaborate theory involving the positions of the various atoms and the interference of their individual emissions, as the effects will depend upon whether the atoms are distributed at random, or arranged in a crystal lattice, or all confined in a volume small compared with a wave-length.* The last alternative mentioned, which is of no practical interest, appears to be the simplest theoretically.

It should be observed that we get the simple Einstein results *only because we have averaged over all initial phases of the atoms*. The following argument shows, however, that *the initial phases are of real physical importance, and that in consequence the Einstein coefficients are inadequate to describe the phenomena except in special cases*. If initially all the atoms are in the normal state, then it is easily seen that the expression (29)

$$[\Delta N_m = 1/h^2c^2 \,.\, \Sigma_n \{| c_n |^2 - | c_m |^2\} | \eta^{\cdot}_{nm} |^2 \, |\textstyle\int_0^T \kappa(t)\, e^{\,i(W_m - W_n)t/h}\, dt|^2 \qquad (29)]$$

for ΔN_m holds without the averaging process, so that in this case the *Einstein coefficients* are adequate. If we now consider the case when some of the atoms are initially in an excited state, we may suppose that they were brought into this state by radiation incident on the atoms before the time $t = 0$. The effect of the subsequent incident radiation must then depend on its phase relationships with the earlier incident radiation, since *a correct way of treating the problem would be to resolve both incident radiations into a single Fourier integral*. If we do not wish the earlier radiation to appear explicitly in the calculation, we must suppose that it impresses certain phases on the atoms it excites, and that these phases are important for determining the effect of the subsequent radiation. It would thus not be permissible to average over these phases, but one would have to work directly from equation (28).

Walter Heinrich Heitler (January 2, 1904 – November 15, 1981).

Heitler was a German physicist who made contributions to quantum electrodynamics and quantum field theory. He brought chemistry under quantum mechanics through his theory of valence bonding.

In 1922, Heitler began his study of Physics at the Karlsruhe Technische Hochschule, in 1923 at the Humboldt University of Berlin, and in 1924 at the Ludwig Maximilian University of Munich(LMU), where he studied under both Arnold Sommerfeld and Karl Herzfeld. The latter was his thesis advisor when he obtained his doctorate in 1926; Herzfeld taught courses in theoretical physics and one in physical chemistry, and in Sommerfeld's absence often took over his classes. From 1926 to 1927, he was a Rockefeller Foundation Fellow for postgraduate research with Niels Bohr at the Institute for Theoretical Physics at the University of Copenhagen and with Erwin Schrödinger at the University of Zurich. He then became an assistant to Max Born at the Institute for Theoretical Physics at the University of Göttingen. Heitler completed his Habilitation, under Born, in 1929, and then remained as a Privatdozent until 1933. In that year, he was let go by the university because he was Jewish.

At the time Heitler received his doctorate, three Institutes for Theoretical Physics formed a consortium which worked on the key problems of the day, such as atomic and molecular structure, and exchanged both scientific information and personnel in their scientific quests. These institutes were located at the LMU, under Arnold Sommerfeld, the University of Göttingen, under Max Born, and the University of Copenhagen, under Niels Bohr. Furthermore, Werner Heisenberg and Born had just recently published their trilogy of papers which launched the *matrix mechanics* formulation of *quantum mechanics*. Also, in early 1926, Erwin Schrödinger, at the University of Zurich, began to publish his quintet of papers which launched the *wave mechanics* formulation of *quantum mechanics* and showed that the *wave mechanics* and *matrix mechanics* formulations were equivalent. These papers immediately put the personnel at the leading theoretical physics institutes onto applying these new tools to understanding atomic and molecular structure. It was in this environment that Heitler went on his Rockefeller Foundations Fellowship, leaving LMU and within a period of two years going to do research and study with the leading figures of the day in theoretical physics, Bohr's personnel in Copenhagen, Schrödinger in Zurich, and Born in Göttingen.

In Zurich, with Fritz London, Heitler applied the new quantum mechanics to deal with the saturable, nondynamic forces of attraction and repulsion, i.e., exchange forces, of the hydrogen molecule. Their valence bond treatment of this problem, was a landmark in that it brought chemistry under quantum mechanics. Furthermore, their work greatly influenced chemistry through Linus Pauling, who had just received his doctorate and on a

Guggenheim Fellowship visited Heitler and London in Zurich. Pauling spent much of his career studying the nature of the chemical bond. The application of quantum mechanics to chemistry would be a prominent theme in Heitler's career.

While Heitler was at Göttingen, Adolf Hitler came to power in 1933. With the rising prominence of anti-Semitism under Hitler, Born took it upon himself to take the younger Jewish generation under his wing.[18] In doing so, Born arranged for Heitler to get a position that year as a Research Fellow at the University of Bristol, with Nevill Francis Mott.

At Bristol, Heitler was a Research Fellow of the Academic Assistance Council, in the H. H. Wills Physics Laboratory. At Bristol, among other things, he worked on *quantum field theory* and *quantum electrodynamics* on his own, as well as in collaboration with other scientific refugees from Hitler, such as Hans Bethe and Herbert Fröhlich, who also left Germany in 1933.

With Bethe, he published a paper on pair production of gamma rays in the Coulomb field of an atomic nucleus, in which they developed the Bethe-Heitler formula for Bremsstrahlung.

In 1936, Heitler published his major work on quantum electrodynamics, *The Quantum Theory of Radiation*, which marked the direction for future developments in quantum theory. The book appeared in many editions and printings and has been translated into Russian.

Heitler also contributed to the understanding of cosmic rays, as well as predicted the existence of the electrically neutral pi meson. While developing the theory of cosmic ray showers in 1937, he became aware of the latest experimental work in the field: the observation of cosmic ray interactions in Nuclear emulsion by Austrian physicists Marietta Blau and Hertha Wambacher. He mentioned this to Bristol colleague Cecil Powell, saying that the method appeared so straightforward that 'even a theoretician might be able also to do it'. This intrigued Powell, and he convinced theoretician Heitler to travel to Switzerland with a batch of llford emulsions and expose them on the Jungfraujochat at 3500m. In a letter to 'Nature' in August 1939, Heitler and Powell were able to confirm the observations of Blau and Wambacher. Thus Heitler had some influence in setting Cecil Powell on the first step of his path to the 1950 Nobel Prize in Physics, "for his development of the photographic method of studying nuclear processes and his discoveries regarding mesons made with this method".

After the fall of France in 1940, Heitler was briefly interned on the Isle of Man for several months. Heitler remained at Bristol eight years, until 1941, when he became a professor at the Dublin Institute for Advanced Studies, which was arranged there by Erwin Schrödinger, Director of the School for Theoretical Physics. He has been described as the "unsung hero of DIAS in the 1940s".

At Dublin, Heitler's work with H. W. Peng on radiation damping theory and the meson scattering process resulted in the Heitler-Peng integral equation. During his stay in Dublin, he lived at 21 Seapark Road, Clontarf, down the road from Erwin Schrödinger.

During the 1942–1943 academic year, Heitler gave a course on elementary wave mechanics, during which W. S. E. Hickson took notes and prepared a finished copy. These notes were the basis for Heitler's book *Elementary Wave Mechanics: Introductory Course of Lectures*, first published in 1943. A new edition was published as *Elementary Wave Mechanics* in 1945. This version was revised and republished many times, as well as being translated into French and Italian and published in 1949 and in German in 1961. A further revised version appeared as *Elementary Wave Mechanics With Applications to Quantum Chemistry* in 1956, as well as in German in 1961.

Schrödinger resigned as Director of the School for Theoretical Physics in 1946, but stayed at Dublin, whereupon Heitler became Director. Heitler stayed at Dublin until 1949, when he accepted a position as Ordinarius Professor for Theoretical Physics and Director of the Institute for Theoretical Physics at the University of Zurich, where he remained until 1974, when he retired. In 1958, Heitler held the Lorentz Chair for Theoretical Physics at the University of Leiden.[38] While in Zurich, after some years, he began writing on the philosophical relationship of science to religion. His books were published in German, English, and French.

Heitler died in Zurich on November 15, 1981 at the age of 78.

Fritz Wolfgang London (March 7, 1900–March 30, 1954).

London was a German born physicist. His fundamental contributions to the theories of chemical bonding and of intermolecular forces (London dispersion forces) are now considered classic and are discussed in standard textbooks of physical chemistry. With his brother Heinz London, he made a significant contribution to understanding electromagnetic properties of superconductors with the London equations and was nominated for the Nobel Prize in Chemistry on five separate occasions.

London was born in Breslau, Germany (now Wrocław, Poland) as the son of Franz London (1863-1917).

London's early work with Walter Heitler on chemical bonding is now treated in any textbook on physical chemistry. This paper was the first to properly explain the bonding in a homonuclear molecule such as H_2. It is no coincidence that the Heitler–London work appeared shortly after the introduction of quantum mechanics by Heisenberg and Schrödinger, because *quantum mechanics was crucial in their explanation of the covalent bond.* Another necessary ingredient was *the realization that electrons are indistinguishable, as expressed in the Pauli principle.*

Other early work of London was in the area of intermolecular forces. He coined the expression "dispersion effect" for the attraction between two rare gas atoms at large (say about 1 nanometer) distance from each other. Nowadays this attraction is often referred to as "London force". In 1930 he gave (together with R. Eisenschitz) a unified treatment of the interaction between two noble gas atoms that attract each other at large distance, but repel each other at short distances. Eisenschitz and London showed that this repulsion is a consequence of enforcing the electronic wavefunction to be antisymmetric under electron permutations. This antisymmetry is required by the Pauli principle and the fact that electrons are fermions.

For atoms and nonpolar molecules, the London dispersion force is the only intermolecular force, and is responsible for their existence in liquid and solid states. For polar molecules, this force is one part of the van der Waals force, along with forces between the permanent molecular dipole moments.

London was the first theoretical physicist to make the fundamental, and at the time controversial, suggestion that superfluidity is intrinsically related to the Einstein condensation of bosons, a phenomenon now known as Bose–Einstein condensation. Bose recognized that the statistics of massless photons could also be applied to massive particles; he did not contribute to the theory of the condensation of bosons.

London was also one of the early authors (including Schrödinger) to have properly understood the principle of local gauge invariance (Weyl) in the context of the then new quantum mechanics.

London predicted the effect of flux quantization in superconductors and with his brother Heinz postulated that the electrodynamics of superconductors is described by a massive field. I.e. that whilst magnetic flux is expelled from a superconductor, this happens exponentially over a finite length with an exponent which is now called the London penetration depth.

London also developed a theory of a rotational response of a superconductor, pointing out that rotation of a superconductor generates magnetic field London moment. This effect is used in models of rotational dynamics of neutron stars.

Being a Jew, London lost his position at the University of Berlin after Hitler's Nazi Party passed the 1933 racial laws. He took visiting positions in England and France, and emigrated to the United States in 1939, of which he became a naturalized citizen in 1945. Later in his life, London was a professor at Duke University. He was awarded the Lorentz Medal in 1953. He died from a heart ailment in Durham, North Carolina, in 1954.

Heitler, W. & London, F. (June, 1927). Wechselwirkung neutraler Atome und homöopolare Bindung nach der Quantenmechanik. (Interaction of neutral atoms and homeopolar bonding according to quantum mechanics.)

Zeit. Phys., 44, 455–72. https://doi.org/10.1007/BF01397394; also at http://quantum-chemistry-history.com/Heitler_London_Dat/WechselWirk1927/WechselWirk1927.htm (in German); translation by T. G. Underwood.

Received 30 June 1927.

Zürich, Physikal. Institut der Universität.

[1] Presented at the conference of the German Physical Society Freiburg i. Br., June 12, 1927.

With 2 illustrations.

In this paper, Heitler and London examined the interaction between *neutral atoms* which resulted in *non-polar valance bonds*. They applied quantum mechanics to the calculation of the *interaction energy* of the atoms when they move closer together. Due to *quantum entanglement*, it was found that two neutral atoms could interact with each other in two ways. *The problem was twofold degenerate, corresponding to the two ways of assigning the electrons to the neutral atoms.* Examination of the different cases of two H atoms and two He atoms showed that by applying the *Pauli principle*, the selected *eigenfunctions* of the system changed or maintained their sign, respectively, when two *electrons* were swapped if the two electrons compared had the same, or different, *spin*. It was found that in the case of He there was only one solution, which yields about the right size of the He gas kinetic-radius, *due to the fact that 2 He atoms (and the same applied to all noble gases) could not differ in their spin* – in contrast to hydrogen (and all atoms with unfinished shells) – so that 2 He atoms had only one possible mode of behaving.

Abstract

The interplay of forces between *neutral atoms* shows a characteristic quantum mechanical ambiguity. This ambiguity is therefore capable of encompassing the various modes of relation which experience furnishes: in the case of hydrogen, for example, the possibility of a *homopolar bond* or elastic reflection, whereas in the case of noble gases only the latter — and this already as an effect of a first approximation of approximately the right magnitude. In the selection and discussion of the different modes of behavior, the Pauli principle also proves itself here, when applied to systems of several atoms.

The interaction between *neutral atoms* has given rise to a theoretic treatment which has so far caused considerable difficulties. Whilst the attractions of ions make a simple picture which has been known for a long time, the conditions in the case of neutral atoms, in particular the possibility of a *non-polar bond*, are quite difficult to understand, if one did not want to resort to very artificial explanations [2].

[2] For literature on this subject, see e.g. *Handb. d. Phys*., XXII, 1926, article Herzfeld.

The development of *quantum mechanics has provided new points of view for the treatment of these problems: first of all, in the new "models" the charge distribution is completely different* [3] *than in the Bohr models* (i.e. decay like e^{-r}), which would already result in a completely different interplay of forces between "neutral" atoms.

[3] As is well known, *the correspondence between classical and quantum mechanical quantities relates only to the electric moments,* i.e., to the focal points of the charge, not to the charge distribution itself, which difference does not come into play in the usual spectroscopic questions.

However, a characteristic quantum mechanical *vibrational phenomenon*, which is closely related to the *resonance* oscillations introduced by Heisenberg, is crucial for the understanding of the possible behaviors between *neutral atoms*. We will study these ratios using the example of two H-atoms (§ 1) and two He-atoms (§ 3). To anticipate the result (§ 2): for the *interaction energy*, one obtains a solution: one which is attractive at medium distances of the atoms, and one at small distances and which are necessary for a *homopolar* molecule formation (already in a first approximation, in which perturbations caused by polarization are to be avoided). *However, this solution is not permitted in the ground state due to the Pauli prohibition* (§ 4) for He. A second solution, which is the only one that can be considered for He, provides repulsion everywhere (van der Waals' b-forces).

§ 1. *Interaction of two hydrogen atoms.*

We set ourselves the task of determining the change in energy, which *two neutral hydrogen atoms* in the ground state experience when we approach them to the distance R (measured by the distance of the nuclei) from each other). Depending on whether this additional energy decreases or increases as the atoms gradually approach, we infer attraction or repulsion.[1]

[1] It is not superfluous to point out that the *connection between wave-mechanical frequency and mechanical energy assumed here is quite hypothetical*, for we know that it is precisely

the Lorentz force approach, which governs the action of the field on matter, that is *replaced in quantum mechanics by something quite different, namely, the wave equation*.

1. We designate the two nuclei, the distance of which is once and for all fixed and equal to R, by *a* and b, the two electrons by numbers 1 and 2, and finally the distances of the electrons from the nuclei and from each other by $r_{a1}, \ldots, r_{12}$.

The *wave equation of our problems - of the 2-hydrogen atom problems* - then reads (with $\Delta_{12} = \Delta_1 + \Delta 2 = \partial^2/\partial x_1^2 + \partial^2/\partial y_1^2 + \partial^2/\partial z_1^2 + \partial^2/\partial x_2^2 + \partial^2/\partial y_2^2 + \partial^2/\partial z_2^2$):

$$(\chi) \equiv \Delta_{12}\,\chi + 8\pi^2 m/h^2\,\{E - (\varepsilon^2/R + \varepsilon^2/r_{12} - \varepsilon^2/r_{a1} - \varepsilon^2/r_{a2} - \varepsilon^2/r_{b1} - \varepsilon^2/r_{b2})\}\,\chi = 0 \qquad (1)$$

We are interested in those solutions χ *which correspond to the perturbations of two neutral H-atoms in the ground state*, and accordingly we shall approximate them from the well-known *eigenfunctions* of the *hydrogen ground state*.

If the electron 1 is located at the nucleus *a*, then the well-known hydrogen *eigenfunction* is to be assigned to it

$$\psi(1) = 1/\sqrt{\pi}\,(1/a_0)^{3/2}\,e^{-ra1/a0} \qquad (2)$$

where a_0 means Bohr's *orbital radius* of the 1_1-orbit. The argument 1 denotes the coordinates of the electron 1 in q-space, we will write it in the future as an index, also ψ_1, (a measurement with the eigenvalue index is not to be feared, since we are always looking at atoms in their ground state). Accordingly,

$$\varphi_1 = 1/\sqrt{\pi}\,(1/a_0)^{3/2}\,e^{-rb1/a0} \qquad (2')$$

means that the first electron is located at nucleus b. The two *eigenfunctions* (2) and (2') are quite different, although they can be brought to coincide by a simple translation; because they take place in different areas of space[1].

[1] Accordingly, it should always be borne in mind in the following that the other conclusions remain unchanged, even if ψ and φ are different in their form, whereby our considerations are at once reduced to an essentially universal concept.

Corresponding *eigenfunctions* exist for electron 2:

$$\psi_2 = 1/\sqrt{\pi}\,(1/a_0)^{3/2}\,e^{-ra2/a0} \qquad \varphi_2 = 1/\sqrt{\pi}\,(1/a_0)^{3/2}\,e^{-rb2/a0} \qquad (2a)$$

which indicates that electron 2 is located at nucleus *a* or b. The *eigenfunctions* (2a) differ from the *eigenfunctions* (2) and (2') only in q-space.

2. As unperturbed *eigenfunctions* we shall have to choose those which say that an electron is in one nucleus and the other electron is in another.[2]

[2] The possibility of *ionization* is excluded here for the time being; to what extent this is justified, we will only show later (§ 5).

If one thinks of these two as yet uncoupled systems as one system, it is well known that the product of these two *eigenfunctions* must be regarded as the common *eigenfunction*. However, depending on the distribution of the two electron ions on the two nuclei, this is possible in two ways. First of all, you have:

$$\psi_1\varphi_2 \quad \text{(1 is at } a\text{, 2 is at b).} \tag{3a}$$

With the same right, however, one also obtains:

$$\psi_2\varphi_1 \quad \text{(2 is at } a\text{, 1 is at b).} \tag{3b}$$

Both possibilities belong to the same *energy* of the entire system (double hydrogen energy). It is a case of twofold degeneration: all pairs of orthogonal linear combinations of $\psi_1\varphi_2$ and $\psi_2\varphi_1$:

$$\begin{aligned} \alpha &= a\,\psi_1\varphi_2 + b\,\psi_2\varphi_1 \\ \beta &= c\,\psi_1\varphi_2 + d\,\psi_2\varphi_1 \end{aligned} \tag{4}$$

with the conditions of *normalization* and *orthogonality*

$$\left.\begin{aligned} a^2 + b^2 + 2abS &= 1 \\ c^2 + d^2 + 2cdS &= 1 \\ ac + bd + (ad + bc)S &= 0 \end{aligned}\right\} \tag{5}$$

(where $S = \int \psi_1\varphi_1\psi_2\varphi_2\, d\tau_1\, d\tau_2$)

are to be regarded as intrinsically perturbed *eigenfunctions* of these problems.

3. The exact *eigenfunctions* of the differential equation (1) are derived from linear combinations (4), the *eigenfunctions* of "zero approximation" and differ from the latter only by a small perturbation. The coefficients *a*, b, c, d in (4) for these *eigenfunctions* of zero approximation can be determined according to known rules. Although equation (1) lacks the character of a perturbation problem, because no term in the *potential* function can be separated as a "*perturbation potential*", the method for determining the said coefficients and the perturbation energies is exactly the same as if there were a *perturbation potential*. In order to make the calculations more transparent, we want to choose from the outset the

correct linear combinations that are standardized according to (Ö) (and verify their correctness afterwards);

$$\alpha = 1/\surd(2 + 2S)\ (\psi_1\varphi_2 + \psi_2\varphi_1) \quad \}$$
$$\beta = 1/\surd(2 - 2S)\ (\psi_1\varphi_2 - \psi_2\varphi_1). \quad \} \qquad (4a)$$

For the *perturbed eigenfunctions*, we will now have to start:

$$\chi_\alpha = \alpha + \upsilon_\alpha \quad \}$$
$$\chi_\beta = \beta + \upsilon_\beta. \quad \} \qquad (6)$$

With this approach, we enter into equation (1) and take into account that the functions ψ_1, ψ_2, φ_1 and φ_2 satisfy the four equations,

$$\Delta_1\psi_1 + 8\pi^2 m/h^2\ (E_0 + \varepsilon^2/ra_1)\ \psi_1 = 0, \quad \}$$
$$\Delta_2\psi_2 + 8\pi^2 m/h^2\ (E_0 + \varepsilon^2/ra_2)\ \psi_2 = 0, \quad \}$$
$$\Delta_1\varphi_1 + 8\pi^2 m/h^2\ (E_0 + \varepsilon^2/ra_1)\ \varphi_1 = 0, \quad \}$$
$$\Delta_2\varphi_2 + 8\pi^2 m/h^2\ (E_0 + \varepsilon^2/ra_2)\ \varphi_2 = 0, \quad \} \qquad (7)$$

where E_0 means – 13.5 volts, the *eigenvalue of the hydrogen ground state*. With the designation

$$E_1 = E - 2\ E_0 \qquad (8)$$

for the *eigenvalue perturbation* we get from (1) the two differential equations for υ_α and υ_β:

$$\surd(2 + 2S)\ L\upsilon_\alpha + (E_1 - \varepsilon^2/r_{12} - \varepsilon^2/R)\ (\psi_1\varphi_2 + \psi_2\varphi_1)$$
$$+ (\varepsilon^2/r_{a1} + \varepsilon^2/r_{b2})\ \psi_2\varphi_1 + (\varepsilon^2/r_{b1} + \varepsilon^2/r_{a2})\ \psi_1\varphi_2 \qquad (9\alpha)$$
$$\surd(2 - 2S)\ L\upsilon_\beta + (E_1 - \varepsilon^2/r_{12} - \varepsilon^2/R)\ (\psi_1\varphi_2 - \psi_2\varphi_1)$$
$$+ (\varepsilon^2/r_{a1} + \varepsilon^2/r_{b2})\ \psi_2\varphi_1 - (\varepsilon_2/r_{b1} + \varepsilon^2/r_{a2})\ \psi_1\varphi_2. \qquad (9\beta)$$

In order for these equations to be solved at all in υ inhomogeneous equations for an *eigenvalue* E of the homogeneous equation, it must be required that the inhomogeneities are orthogonal to all *eigenfunctions* of the homogeneous equations, i.e. to χ_α and χ_β, which belong to the same *eigenvalue* E.

First of all, we find that the inhomogeneity of (9α) is already orthogonal to β, and the inhomogeneity of (9β) is orthogonal to α (except for negligible quantities). This verifies the correctness of the approach (4a). The other two demands:

Inhomogeneity (9α) orthogonal to α,

Inhomogeneity (9β) orthogonal to β,

deliver the two *eigenvalue perturbations*

$$E_\alpha = E_{11} - (E_{11}S - E_{12})/(1 + S) \quad \}$$
$$E_\beta = E_{11} + (E_{11}S - E_{12})/(1 - S) \quad \} \qquad (10)$$

with the following designations:

$$E_{11} = \int [(\varepsilon^2/r_{12} + \varepsilon^2/R)\,(\psi_1{}^2\varphi_{22} + \psi_2{}^2\varphi_1{}^2)/2 - (\varepsilon^2/r_{a1} + \varepsilon^2/r_{b2})\,\psi_2{}^2\varphi_1{}^2/2 \qquad (11)$$
$$- (\varepsilon^2/r_{a2} + \varepsilon^2/r_{b1})\,\psi_1{}^2\varphi_2{}^2/2]\; d\tau_1\, d\tau_2,$$
$$E_{12} = \int (2\varepsilon^2/r_{12} + 2\varepsilon^2/R - \varepsilon^2/r_{a1} - \varepsilon^2/r_{a2} - \varepsilon^2/r_{b1} - \varepsilon^2/r_{b2})\,\psi_1\varphi_1\psi_2\varphi_2/2\; d\tau_1\, d\tau_2,$$
$$S = \int \psi_1\varphi_1\psi_2\varphi_2\; d\tau_1\, d\tau_2.$$

§ 2. *Discussion of the result.*

1. So we get two *perturbation energies* corresponding to the inputs (5a):

$$E_\alpha = E_{11} - (E_{11}S - E_{12})/(1 + S)$$

belongs to $1/\surd(2 + 2S)\,(\psi_1\varphi_2 + \psi_2\varphi_1)$ (*symmetrical* in 1 and 2),

$$E_\beta = E_{11} + (E_{11}S - E_{12})/(1 - S)$$

belongs to $1/\surd(2 - 2S)\,(\psi_1\varphi_2 - \psi_2\varphi_1)$ (*antisymmetrical* in 1 and 2).

It is a result that can only be described very artificially in classical terms, *that two neutral atoms can interact with each other in two ways.* We are still a long way from a real understanding of this fact. But it is desirable at least to be clear about how this curious ambiguity comes about mathematically. *The essential point is evidently that the problem is originally twofold degenerate* (la and lb), *corresponding to the two ways of assigning the electrons to the neutral atoms*[1].

[1] In classical mechanics, this problem has not degenerated: the same energy value belongs to the two electron arrangements. But the criterion of a degeneration of classical mechanics is not $E_k - E_l = 0$ but $\nu \equiv \partial E/\partial J = 0$ (a "proximity relationship" in terms of the effects of variable J). In fact, the electrons in this case also have their full degree of periodicity. Only when the energy is sufficient to overcome the potential visual wave between the two atoms (or when the two atoms are sufficiently approached), does now $E_k - E_l = 0$ result in $\partial E/\partial J = 0$ and the problem is degenerate. *The electrons now describe orbits around both nuclei and are constantly exchanging information.* In *quantum mechanics* this distinction does not occur [cf. Hund, F. (1926). *Zeit. Phys*., 40, 442].

> *No matter how high the (finite) potential threshold, there is always a certain probability of an exchange of electrons.* While in classical mechanics there is a way to attach a label to electrons (place each electron in a sufficiently deep potential well and keep away from a large amount of energy), something similar is impossible in quantum mechanics. Even if you know an electron in a potential well at an instant, you are never sure whether it has not already exchanged with another in the next moment. *For this reason, a statistic which in principle discriminates against an individualization of electrons and takes into account only their numerical distribution among the states, such as the Bose or Fermi statistics, is so immensely adapted to the possibilities of quantum mechanical description.*

The abolition of this degeneracy is linked as can be seen from (10), the non-disappearance of the quantum function of the atom *a* at the place of the atom b and vice versa (otherwise $\psi_1\varphi_1 = 0$ and consequently $E_{12} = 0$ and $S = 0$); but *this means that there must be a finite probability for the electron of a to belong to the atom b.* In fact, the exponential behavior of ψ and φ always satisfies this condition. The magnitude $1/\hbar\ (E_\beta - E_\alpha)$ will have to be interpreted as the frequency at which an *exchange of the two electrons* is effected on average. For long distances, this difference decreases as $\varepsilon^2/a_0\ e^{-2R/a0}$ so that distant atoms very rarely enter into an exchange [1].

> [1] "Large" distances are already the mean gas-kinetic distances ($3.3\ .\ 10^{-7}$ cm); with them the period of exchange is of the order of magnitude a_0/ε^2 h. $e^{-2/a0\ 3.3.10-7} \sim 10^{30}$ years; on the other hand, with the distances in the crystal lattice ($\sim 3\ .10^{-8}$ cm), an exchange takes 10^{-10} sec.

The whole phenomenon is almost intertwined with the quantum mechanical *resonance* phenomenon discussed by Heisenberg. But, while in *resonance*, the electrons of different motions of one and the same *eigenfunction* series exchange their energy, *here electrons of the same excitation levels (of the same energy), but at different eigenfunction systems (ψ and φ), exchange their energies.* There, the general occurrence of the same jump frequency twice is characteristic (*resonance* phenomenon); here, on the other hand, there is no question of resonance.

2. *Let us now turn to the quantitative discussion of interaction energies* (10). Even without evaluating the integrals that occur, it can be said that the indication of $E_{11}S - E_{12}$ is always positive. In a Sturm-Liouville *eigenvalue* problem of any number of dimensions with homogeneous boundary conditions, the natural oscillation that has no *nodes* has the lowest *eigenvalue*. Every other orthogonal natural oscillation has *nodes* and a higher *eigenvalue*. Now, however, the solution α is apparently *nodeless*, while β always has a *node* as an *antisymmetric eigenfunction.*

[In quantum mechanics, a *node* is a point or region where the probability of finding a particle is zero. *Nodes* are points of zero probability densities, and the regions in the neighborhood of *nodes* will have small probability densities. *Nodes* can be found in simple quantum mechanical systems, such as the behavior of a particle in a harmonic oscillator potential. In atomic orbitals, *nodes* are the regions or spaces around the nucleus where the probability of finding an electron is zero.]

So

$$E_\beta > E_\alpha. \qquad (12)$$

The meaning of E_{11} is immediately evident from (11): *it is the purely Coulomb interaction of the existing charge distribution*. The calculation results in

$$E_{11} = \varepsilon^2/a_0 \, e^{-2R/a0} \, (a_0/R + 5/8 - 3/4 \, R/a_0 - 1/6 \, R^2/a_0^2) \qquad (13)$$

...

The rest of the energies (11) are not so easy to interpret. Not all of the integrals found there could be evaluated. The calculation is:

$$S = (1 + R/a_0 + R^2/3a_0^2)^2 \, e^{-2R/a0} \qquad (14)$$

$$\int \psi_1\psi_2\varphi_1\varphi_2/2 \, (\varepsilon^2/r_{a1} + \varepsilon^2/r_{a2} + \varepsilon^2/r_{b1} + \varepsilon^2/r_{b2} - 2\varepsilon^2/R) \, d\tau_1 d\tau_2$$
$$= 2\varepsilon^2/a_0 \, (1 + 2R/a_0 + 4R^2/3a_0^2 + R^3/3a_0^3) \, e^{-2R/a0} - \varepsilon^2/R \, S,$$

while for the missing integral over $1/r_{12}$ we only have an upper limit[1].

[1] It is the determination of the self-potential of a charge distribution on confocal ellipsoids, which is much more complicated than Dirac's well-known determination for similarly located ellipsoids. In equation (15), the ellipsoid was replaced by a sphere with the same total charge.

It is

$$\int \psi_1\psi_2\varphi_1\varphi_2/r_{12} \, d\tau_1 d\tau_2 < \varepsilon^2/a_0 \, 5/8 \, (1 + R/a_0 + R^2/3a_0^2)^2 \, e^{-2R/a0}. \qquad (15)$$

With this upper limit, the curves E_α and E_β in Fig. 1 are obtained.

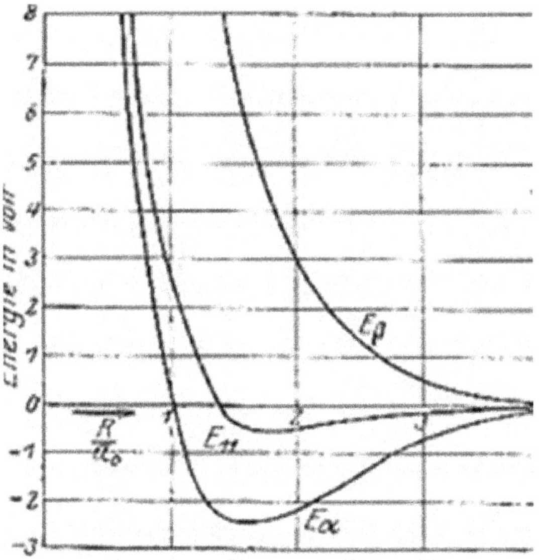

Potential zweier neutraler H-Atome.
(E_α = homöopolare Anziehung,
E_β = elastische Reflexion.)

E_β constantly shows *repulsion*; E_α shows *attraction at a great distance*, then a *minimum* approximately at R = 3/2 a_0 and *at a short distance repulsion*. The *non-polar attraction* that occurs here shows itself to be a characteristic *quantum mechanical effect*. It occurs without taking into account the disturbances caused by polarization. It is also noteworthy that the *repulsion* E_β, as can be seen, *is for the most part also due to this quantum mechanical effect*, is not based on the Colomb interaction E_{11}.

The physical significance of the two solutions, we are concerned with is the following: *The antisymmetric solution with the interaction energy E_β corresponds to the van der Waals repulsion (elastic reflection) of the two hydrogen atoms. But if α is excited, the two hydrogen atoms can join together to form a homopolar molecule, where the minimum of E_α indicates equilibrium.*

We will also show that in the interaction of two non-excited noble gas atoms (§ 3) the solution corresponding to the formation of molecules is prohibited by quantum theory. The substantiation of these allegations is set out in § 4.

The estimated curves give approximately the values of the respective atomic quantities: *atomic diameter*, *dissociation energy* and *moment of inertia* of the molecule. As far as the error caused by the estimation is concerned, it should be noted that in any case E_β is actually higher, E_α lower, whereby the typical character of the two *potentials* would be even more pronounced. Since it is not our goal to calculate the most accurate numerical values possible, but to gain insight into the physical conditions of *homopolar bonding*, let us content ourselves with this estimation here.

§ 3. *Interaction of two He-atoms.*

We use the same method to investigate the exchange of two neutral He atoms. The two nuclei are called again *a* and b and are kept at a fixed distance R. The electrons are numbered 1 to 4. For each nucleus there exists an *eigenfunction*, ψ for the nucleus *a*, φ for the nucleus b, which are now functions of two electrons each, and Heisenberg's He-theory has shown that the *eigenfunctions* of the *He-ground state* are *symmetrical* in the spatial coordinates of the two electrons[1].

[1] Heisenberg, W. (June, 1926). Mehrkörperproblem und Resonanz in der Quantenmechanik. (Multibody Problem and Resonance in Quantum Mechanics.) *Zeit. Phys*., 38, 411-426.

By creating an ψ and a φ-function, an *eigenfunction* is created (we write the arguments again as indices):

$$\psi_{ik} \cdot \varphi_{lj} \qquad (i \neq k \neq l \neq j = 1, 2, 3, 4) \tag{16}$$

of the unperturbed problem (mostly separated neutral atoms). By permutation of the four electrons (taking into account that $\psi_{ik} = \psi_{ki}$, $\varphi_{ik} = \varphi_{ki}$) $\binom{4}{2} = 6$ new *eigenfunctions*, which all belong to the same *eigenvalue*. They are:

$$\begin{array}{lll} \psi_{12} \cdot \varphi_{34}, & \varphi_{34} \cdot \psi_{12}, & \} \\ \psi_{13} \cdot \varphi_{42}, & \varphi_{42} \cdot \psi_{13}, & \} \\ \psi_{14} \cdot \varphi_{23}, & \varphi_{23} \cdot \psi_{14}, & \} \end{array} \tag{17}$$

The electrons of the helium ground state must be divergent from each other with respect to their spin. We choose the designation in such a way that the electrons 1 and 3 are of the same *spin*, also 2 and 4, but 1 and 2 have a different *spin*. Then, according to the *Pauli principle*, it is excluded that 1 and 3 or 2 and 4 are located at the same He nucleus. We can, therefore, exclude two of the six *eigenfunctions* (17) from the outset and limit ourselves to

$$\begin{array}{lll} \psi_{12} \cdot \varphi_{34}, & \varphi_{34} \cdot \psi_{12}, & \} \\ \psi_{14} \cdot \varphi_{23}, & \varphi_{23} \cdot \psi_{14}, & \} \end{array} \tag{17a}$$

The wave equation for the 2-He problem is:

$$\Sigma_{i=1}^{4} [\Delta_i\chi + 8\pi^2 m/h^2 \{E_0 - \varepsilon^2(4/R + \Sigma_{k=1}^{4} 1/r_{ik} - 1/r_{ai} - 1/r_{bi})\ \chi] = 0. \tag{18}$$

It is *symmetrical* in all four electrons. Here, too, we do not have a perturbation problem of the usual form, since it is just as impossible to separate a *perturbation potential* as in § 1;

but we infer from the *symmetry* of the differential equation that the *eigenfunctions* of the zero-approximation are the same as if there were a small *perturbation potential* in the many *symmetrical* equations. We call this H.

Of course, it is not the first-order *eigenvalues* that can be used to deduce your secular problem, but we can deduce the correct *eigenfunctions* of the first order, which are characterized solely by the *symmetry* character of the differential equation. ...
...

For the same reason as in the case of the *2-hydrogen problem*, E_α the lowest *eigenvalue*, must consequently be $H_{12} < 0$; then E_α belongs the *nodeless symmetric eigenfunction* α. In the following section, however, we shall show that this α, which would energetically necessitate the formation of molecules, is not permissible in quantum theory for neutral He, but that of the four solutions (20) only β can occur, which again means *elastic reflection.*

§ 4. ***The Pauli Principle and Molecular Formation.***

1. How to convince yourself easily; our solutions include (4a) or (20) systems which do not combine with each other[1].

[1] It is distinguished from a similar one, recently by Hund, F. (October, 1927). Zur Deutung der Molekelspektren. I. (On the interpretation of the molecular spectra. I.) *Zeit. Phys.*, 40, 742-64; https://doi.org/10.1007/BF01400234, namely the two-center problem.

1. This circumstance opens up the possibility of applying the Pauli principle, which has proven itself so well in the discussion of the electron configurations of the individual atoms, in a broader sense to the system of two atoms interacting, in order to achieve a narrower selection of the modes of behavior of two atoms permitted by quantum theory. The following formulation of this regulation is sufficient for our purposes[2]:

[2] *This formulation of the Pauli principle is not entirely correct.* But a really general version of it is not yet known. We refer to the work of Heisenberg, W. (1927). *Zeit. Phys.*, 41, 239, as well as especially on soon to be published reflections of F. Hund.

the selected eigenfunctions of the system should change/maintain their sign, when two electrons are swapped, if the two electrons compare/have different spin. (A so-called "*eigenfunction of the spin*" is not to be taken into account here.)

2. We apply this provision to the solutions of § 1 and 3. For hydrogen, the *symmetrical* α retains its sign when 1 and 2 are swapped, β changes. At α the two electrons have different

spin, at β the same *spin*. However, since there is no constraint on the *spin* of separated atoms, both solutions can occur - depending on chance.

For the solutions (20) of the *2-He problem*, however, we know that 1 and 3 and 2 and 4 each had the same *spin*, 1 and 2, and 3 and 4 had different ones. As a result, *there are limits to the interchangeability of the electrons from the outset*. If the single He-atom is to remain viable (Pauli principle, applied to the single He-atom), then only the electrons 1 for 3 and 2 for 4 may be exchanged. With respect to each of these swaps, the selected *eigenfunctions* must be *antisymmetric*. If both pairs are swapped, it must reproduce itself again (without a change of sign). The *eigenfunctions* α, β, γ, δ now merge into an electron in which solutions of the form $\varphi + \psi$ and $\varphi - \psi$ occur, which combine with each other.

We see, then, that in He's case the only solution to the demands suffices. To it, however, belongs the higher *eigenvalue*. ... Closer discussion shows that *it yields about the right size of the He gas kinetic-radius. The extraordinary difference that we find here in the case of He is evidently due to the fact that 2 He atoms (and the same applies to all noble gases) cannot differ in their spin* – in contrast to hydrogen (and all atoms with unfinished shells) -, so that 2 He atoms have only one possible mode of behaving.

3. We now want to show which of the solutions thus eliminated allow *molecule formation*. ...

§ 5. ***Hydrogen and ion formation.*** Our considerations in the first paragraph cannot be complete insofar as *the possibility of ion formation has not been taken into account*. In our approaches, we had taken the starting point from the functions (3):

$$\psi_1\varphi_2 \quad \text{or,} \quad \psi_2\varphi_1$$

which provide for an electron for each nucleus, — according to our question about the interaction of neutral atoms.

Now, if we are interested in the *interaction of two H-ions* (H^+ and H^-), we will look at the configurations of two electrons at the same nucleus:

$$\psi_1\psi_2 \quad \text{or,} \quad \varphi_2\varphi_1, \qquad (21)$$

that we have not yet taken into account. One might think that this neglect was unjustified, since the *eigenfunctions* (21) belong to the same *eigenvalue* as those which we have hitherto used alone (3),

[$\psi_1\varphi_2$ (1 is at *a*, 2 is at b). (3a)

$\psi_2\varphi_1$ (2 is at *a*, 1 is at b). (3b)]

so that "we should therefore assume a fourfold degeneration".

However, *this is not correct*: even for widely separated ions, the very significant perturbations of the two electrons to each other must be taken into account from the outset if both electrons are at the same nuclei (analogous to the perturbations in the He atom!). For separated ions, instead of the $2E_0$ (– 27 volts) of the neutral atoms, the energy increased by the ionization is (13.5 volts) (corrected for the amount of electron affinity, which is very small). Accordingly, instead of (21) He-like perturbed functions, the following functions are to be used as *eigenfunctions*:

ψ_{12} or, φ_{12},

which are *symmetrical* in 1 and 2 and differ only in their position in q-space, not in form.

If we now take into account the perturbation of the He-like ion H^- by the H^+ and, we obtain, *because of the symmetry character of the wave equation* (1),

$$[(\chi) \equiv \Delta_{12}\,\chi + 8\pi^2 m/h^2\ \{E - (\varepsilon^2/R + \varepsilon^2/r_{12} - \varepsilon^2/r_{a1} - \varepsilon^2/r_{a2} - \varepsilon^2/r_{b1} - \varepsilon^2/r_{b2})\}\ \chi = 0 \quad (1)$$

$$\text{with } \Delta_{12} = \Delta_1 + \Delta 2 = \partial^2/\partial x_1^2 + \partial^2/\partial y_1^2 + \partial^2/\partial z_1^2 + \partial^2/\partial x_2^2 + \partial^2/\partial y_2^2 + \partial^2/\partial z_2^2)]$$

in the usual way as *eigenfunctions* of zeroth order (without standardization):

$$\alpha^* = \psi_{12} + \varphi_{12}, \qquad (22)$$
$$\beta^* = \psi_{12} - \varphi_{12}.$$

When the two ions approach each other, the corresponding *eigenvalues* E_α* and E_β* first lose weight, about the same as the ionic potential $-\varepsilon^2/R$ (Fig. 2).

...

Fig. 2. Ion potential (E_α*, E_β*), compared to the potential of neutral atoms.

In this case, E_α*, as belonging to the *nodeless* α^*, will always be deeper than E_β*. The difference $E_\beta^* - E_\alpha^*$ is proportional to the frequency of the exchange of ionic charges.

As long as the *eigenvalues* E_α* and E_β* do not come close to our energy curves of § 2 (they are shown in Fig. 2[1]),

[1] On the basis of the following considerations, we use the assignment of the terms as proposed by F. Hund., i.e., Fig. 12, for it is evident from the considerations of § 4 that not all states of the 2-H problem can pass into states of H_2 molecules, let alone He.

the calculations of § 1 are to be regarded as *perturbations* of the *eigenvalue* $2E_0 = -27$ volts and are not objectionable. But something new occurs, as the new curves come very close to the old ones. For these values of R, the *eigenvalue* in question is multiplied, one has degeneracy, the "correct" *eigenfunctions* of zero order then become certain linear combinations of the *eigenfunctions* before the intersection, which results from a secular problem. The calculations of § 1 would therefore be objectionable if, on this occasion, they were to produce other linear combinations than those used therein. Let us look into that now.

1. First, E_α* and possibly E_β* will intersect the upper of the two energy curves E_β. Since α* and β* are *symmetrical*, while β is *anti-symmetric* in 1 and 2, which symmetry relations are not lost even in the case of any large perturbation because of the symmetry of the differential equation in 1 and 2, the matrix elements of the "*perturbation energy*"[1] H,

[1] With regard to this function H, see what has been said in § 3.

which occur in the secular problems, disappear:

$$\int H\,\alpha^*\,\beta\,d\tau_1\,d\tau_2 = 0,$$
$$\int H\,\beta^*\,\beta\,d\tau_1\,d\tau_2 = 0.$$

So, the secular problem at the intersection of E_α* with E_β is:

$$\begin{vmatrix} \int H\,\alpha^{*2}\,d\tau - E & 0 \\ 0 & \int H\,\beta^2\,d\tau - E \end{vmatrix} = 0.$$

A similar secular problem is obtained for the intersection of E_β* and E_β. From this it follows that the secular *eigenvalues* are the same as the original ones, and that the *eigenfunctions* permeate them unmixed, without forming linear combinations[2].

[2] In a more physical way, one can foresee this immiscibility of β with α* and β*, since β has a rectified twist, α* and β* (i.e. symmetrical He-ground state configuration!) have uneven twist; if the configuration β exists between two atoms, it cannot be α* and β* at the same time. A linear combination of β with α* or β* would be quite incomprehensible.

2. Then α* will have to deal with α. If only the *ionic potential* $-\ \varepsilon^2/R$ were taken into account, E would intersect the zero axis at $R = 2a_0$ (Fig. 2, dotted curve). In this region, however, we are already noticeably inside the charge-cloud (it is much more extensive for H^- than for H), and consequently the ionic attraction will gradually cease there and give way to a repulsion[1].

[1] The presence of an ionic repulsion for shorter distances (on the basis of the quantum mechanical charge distribution) has already been independently established by L. Pauling, [(1927). *Proc. Roy. Soc.* (A), 114, 181] and A. Unsold. [*Zeit. Phys.* (in press).]

It is easy to consider that even if the *exchange terms* are taken into account, the two curves E_α* and E_α in the region of the minimum of E_α still remain distant from each other (Fig. 2, the extended curve E_α* is only quite qualitative). The linear combination of α with α* will still be weak in the region of $R = 2a_0$ - it can be understood as a perturbation of the *eigenfunction* α. It will therefore be justified to calculate E_α without taking into account the linear combination with E_α* and it is precisely for this reason the molecule is characterized as "*predominately homeopolar*". E_α*, for its part, it also has a minimum somewhere; at this point there would be room for a polar molecule. But in so far as this minimum is not an absolute minimum, and in so far as it can pass into the absolute minimum known to us by adiabatic[2] means, it will be rightly said that the polar molecular formation does not represent the stable configuration of the two H-atoms in the molecule.

[2] The word "adiabatic" is not quite appropriate here, since the transfer passes through a point of degeneracy.

Therefore, if the molecules are separated without simultaneous excitation, it is more likely that two neutral atoms will be found. The way in which the two solutions α and α* are linearly combined cannot be foreseen without a more detailed investigation.

Finally, the solution β* (*antisymmetric* in ψ and φ) was undisturbed based on the solution b based on the same considerations as above. But it probably will not come to that. For molecule formation, ß* is out of the question because it cannot be converted into an He-configuration because of its invalidity — it has a *node* between the nuclei.

We would like to believe that the categories of *degeneracies* and *symmetry* relations presented here are typical of a wide range of facts connected with the interactions of atomic systems with each other, especially with the discontinuities of their chemical modes of behaviour.[3]

[3] It will be considered whether the *exchange phenomenon*, which is so decisive for the interaction of atoms, is also noticeable in other branches of physics. We would like to point out two things here. In the case of collision processes, *the exchange of collision and atomic electrons also makes it possible to stimulate quantum leaps between optically non-combining term systems* (e.g. 1^1S — 2^3S for He). *According to the Born theory such transitions are only possible due to the minimal magnetic interaction.* In the interpretation of the law of force of atomic nuclei, which is noticeable in the scattering experiments with α or H particles, one will have to take into account the exchange with nuclear building blocks as an essential influence.

Frenking, G.* (2021). The Chemical Bond – an Entrance Door of Chemistry to the Neighboring Sciences and to Philosophy.

Israel Journal of Chemistry; doi.org/10.1002/ijch.202100070.

Institute of Advanced Synthesis, School of Chemistry and Molecular Engineering, Nanjing Tech University, Nanjing 211816, China and Fachbereich Chemie, Philipps-Universität Marburg, Hans-Meerwein-Strasse 4, D-35043 Marburg, Germany.

* This paper is dedicated to Klaus Ruedenberg.

Chemistry considers the material world essentially as a composition of atoms as well as molecules and solids, which are formed by chemical bonds between the atoms. It is noteworthy that there is no sharp distinction between a chemical bond, as in H_2, and a weak interatomic attraction, such as in He_2, which is most often referred to by the term van-der-Waals interaction. This comes clearly to the fore by the rather vague definition of the chemical bond given by the IUPAC [International Union of Pure and Applied Chemistry], which goes back to a suggestion made by Linus Pauling: "There is a chemical bond between two atoms or groups of atoms in the case that the forces acting between them are such as to lead to the formation of an aggregate with sufficient stability to make it convenient for the chemist to consider it as an independent 'molecular species'." It is a characteristic feature of chemistry that non-precise concepts and models are extremely successful in explaining chemical findings, which is sometimes difficult to accept for orthodox physicists. Chemistry may rightly be called the science of fuzzy concepts, which have been named "unicorns". However, this is not the main topic of the present essay. The focus of this work is on genuine chemical bonding.

The enormous complexity of both animate and inanimate matter is due to the unlimited variation in the linkage of chemical elements, the number of which is relatively small. The number of naturally occurring chemical elements in the periodic table is only 94, but the number of compounds formed and described from them is currently 10^8 with a daily increase of ~40,000. Chemical bonding is at the heart of chemistry, which can be understood as the science of understanding and changing the material world on a molecular scale, and which deals with the phenomenon of chemical bond formation and transformation and their understanding.

It is a curiosity of the history of science and chemistry that a physical explanation for the strong attractive forces between two atoms A and B, which lead to a chemical bond A-B, was unknown until 1927. But at that time, chemistry was already a very successful scientific discipline whose results also formed the basis of an important branch of industry named after it. In a relatively short time during the 19th and 20th century, the chemical industry became a cornerstone of the economic development of many countries and contributed significantly to the prosperity of mankind. One can speculate whether chemistry would have reached its scientific status if its research products were not so immediately commercially viable. It can be said that chemistry is an example of a scientific discipline that can be pursued extremely successfully without understanding its basis - the *chemical bond.*

Physics knows four fundamental forces that can be distinguished for describing the interaction in matter. The *strong* and the *weak* forces are only relevant within atomic nuclei and for processes such as free neutron decay. *Gravitational* forces are far too weak to explain the *strong attraction such as between two hydrogen atoms in* H_2. This leaves electric (Coulombic) forces as the remaining explanation for the chemical bond. But according to classical physics, strong electric attractive forces only exist between particles with opposite charges, not between neutral atoms like in H_2. And why should there be strong electric attraction in H_2 and not in He_2? Despite these obvious problems, Gilbert Lewis presented in 1916 a heuristic model for the chemical bond based on the formation of electron pairs, which he proposed was the essence of the bond. This was a paradoxical suggestion, because according to classical physics electrons repel each other rather than forming a stabilizing electron pair. Lewis wrote: "… a chemical bond is at all times and in all molecules merely a pair of electrons held jointly by two atoms". [Lewis, G. N. (1923). *Valence and the Structure of Atoms and Molecules*. American Chemical Society Monograph Series, New York, p. 78.] The electron-pair model of the bond, which was further developed by Irving Langmuir in 1919–1921[6] and elaborated by Lewis is still the most important model for describing molecular structures and reactivities today, despite its problems and contradictions with classical laws of physics. Lewis recognized the weakness of his model and he speculated: "… electric forces between particles which are very close together do not obey the simple law of inverse squares which holds at greater distances."

In 1927, a paper was published which provided the basis for a physical understanding of the interatomic interactions between two atoms coming from electric forces that lead to a strong attraction and chemical bond in H_2 and no chemical bond in He_2. This year saw the publication of the two young postdocs Walter Heitler and Fritz London [Heitler, W. & London, F. (June, 1927). *Wechselwirkung neutraler Atome und homöopolare Bindung nach der Quantenmechanik.* (Interaction of neutral atoms and homeopolar bonding according to quantum mechanics.) *Zeit. Phys.*, 44, 455–72] from Erwin Schrödinger's research group, in which they described the interatomic interactions in the hydrogen molecule H_2 with the help of the newly developed quantum theory, which was introduced by their mentor and by Werner Heisenberg independently of each other with different mathematical methods.

Heitler and London solved the puzzle of the physical nature of the chemical bond by showing that the electrons must be described not by the charge density ρ(r), which is a particle of classical mechanics in position space r, but by the quantum theoretical wave function Ψ(r1,...,rN), which depends on the 3N spatial coordinates of the number of electrons N. The interference of the wave functions of the electrons obeying the postulates of quantum theory lead to in-phase and out-of-phase combinations that are energetically stabilizing (bonding) or destabilizing (anti-bonding) with respect to the separated atoms. This was a completely new aspect, that the interactions between two electrons can lead to two different discrete results, which depend on the quantum-theoretical state of the wave functions. But the most important and central insight was that the formation of the chemical bond can only be understood if the electrons are described by a wave function and not by the charge.

…

The crucial difference between the two approaches is the occurrence of the *interference term* … $2[\Psi(H_a)\Psi(H_b)]$ …, which shows up only when the *wave function* Ψ is used as basic entity. This is the central aspect of the essay. The key step of the quantum-theoretical approach has a mathematically simple form. It is the solution of the binomial in equation …

$$[\Psi(H_2)]^2 = N^2[\Psi(H_a) \pm \Psi(H_b)]^2 \qquad (8)$$

that leads to equation

$$[\Psi(H_2)]^2 = N^2\{[\Psi(H_a)]^2 + [\Psi(H_b)]^2 \pm 2[\Psi(H_a)\Psi(H_b)]\}. \qquad (9)$$

The calculation of the energy of H_2 using the classical approach yields only a shallow energy minimum with a bond energy of ~10% of the correct value.

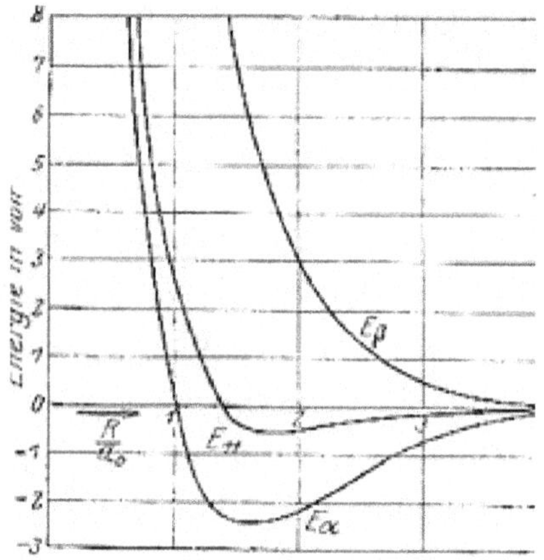

Potential zweier neutraler H-Atome.
(E_α = homöopolare Anziehung,
E_β = elastische Reflexion.)

Figure 1 shows … the original curve of Heitler and London where the classical energy is denoted as E_{11}. The *energy curves* E_α and E_β come from the quantum theoretical approach, which gives two solutions for equation 9 because of the ± sign for the *interference term.* They are related to the *bonding* and *antibonding* combination of the wave functions. Experimental findings are in full accord with the quantum theoretical description of the *covalent bond, where two electrons can occupy the same region of space only if they have opposite spin*; otherwise, they are spatially separated and move out of each other's way due to the Pauli repulsion. *Note that interference between electrons with opposite spin does not necessarily lead to chemical bonding.* A stabilizing electronic interaction *also requires proper spatial symmetry of the orbital.* The *symmetry* aspect of the *wave function* is discussed below.

After the *wave function* Ψ of H_2 including the inference term is constructed from the *wave functions* of the hydrogen atom, the associated *charge density* ρ can be constructed. There is an unambiguous definition Ψ→ρ, *but there is no equally unambiguous definition in the opposite direction.* This is an important finding, which has significant consequences for the understanding of the formation of a chemical bond. *The electronic charge density ρ contains all the information about the chemical bond finally formed, but the bond formation itself can be understood physically only in terms of the wave function Ψ.* Although electrons can be described alternatively as a particle or as a wave function, *it is only the interference of the wave function that provides a physical explanation of chemical bonding.* The conclusion is that *the wave function Ψ*, which is a mathematical function containing imaginary terms for a full consideration of electronic interactions, *is more elementary to chemical bonding than the charge density ρ.* This clearly has implications for the fundamental understanding of chemical processes and molecular matter, but not only for this.

The impact of quantum theory on the understanding of the material world and on the way scientists described physical objects was a revolution for human thinking about the fundamentals of matter, which was not welcomed by everyone. A famous example is Max Planck, whose discovery of the spectral line distribution of thermal blackbody radiation led him propose that the exchange of energy between matter and electromagnetic vacuum takes place in the form of quanta yielding discrete spectra, and it was realized that atomic processes are not continuous but occur in discrete steps. His son later recalled that this insight, which emerged from the analysis of experimental findings, was only reluctantly accepted by Planck because he realized that the seemingly closed view of the material world of classical physics would be destroyed. Albert Einstein made seminal contributions to quantum theory, but he never accepted it as a truly elementary physical theory. The controversial discussions between him and Niels Bohr, one of the founders of quantum theory, are legendary, particularly those at the 1927 Solvay Conference in Brussels. Since then, there has been an ongoing discussion between natural scientists and philosophers about the human understanding of the world in view of the postulates of quantum theory. The quote from Niels Bohr is still valid today: "Those who are not shocked when they first come across quantum theory cannot possibly have understood it".

While the introduction of quantum theory by Heisenberg and Schrödinger led to an intense discussion among physicists and philosophers about the fundamentals of matter and the implications of its quantum theoretical description for the interpretation of physical processes, which continues to this day, the influence of the quantum theoretical insight into chemical bonding in chemistry was, with few exceptions, mostly limited to practical implications. Shortly after the work by Heitler and London in 1927, Erich Hückel explained the particular stability of aromatic compounds with quantum theoretical arguments. Lennard-Jones showed in 1929 that quantum theory easily explains the $X^3\Sigma_g^-$ electronic triplet ground state of O_2, which had been a puzzle for Lewis and which was a failure of his electron-pair model. Friedrich Hund and Robert Mulliken demonstrated that the electronic spectra of diatomic molecules can straightforwardly be explained using a quantum theoretical approach developed by them, which is now known as the MO (*Molecular Orbital*) method. It is an alternative ansatz to the VB (*Valence Bond*) approach that was used by Heitler and London. Mathematically, MO theory rests on the product of sums while VB theory is based on the sum of products. The final results of MO and VB calculations are the same when all terms are considered. A strong impact on the *quantum chemical* description of the *chemical bond* was the work of Linus Pauling, whose book "*The Nature of the Chemical Bond*" provided a bridge between Lewis' heuristic electron pair model and quantum theory for a broad range of chemistry. It was, however, biased towards the VB approach, which obscured the physical understanding of chemical bond and led to misleading conclusions. Other more balanced early textbooks in the field were published by Sidgwick and by Coulson.

The advent of computers and the rapid development of hardware and software in recent decades has made the calculation of the geometries and the electronic structures of molecules and solids a routine procedure in chemical research today. Concerning the interpretation of the numerical results of quantum chemical calculations in terms of bonding models, two types of approaches can be distinguished. One approach analyses the eventually formed electronic structure of the molecule using either wave function-based methods where various types of orbitals are considered, or "real-space" procedures using partitioning techniques of the charge density ρ(r). …

The second approach considers the process of bond formation of a bond A-B between the fragments A and B (or several fragments). …

There is some rivalry between methods that are based on the *analysis of the wave function* $\Psi(\mathbf{r})$ and the *real-space partitioning of the charge density* $\rho(\mathbf{r})$. One advantage of the latter approach is that the *charge density* $\rho(\mathbf{r})$ is subject to routine experimental measurement. The *charge density* can be determined with x-ray crystallography and the results of quantum chemical calculations can directly be compared with experimental values. Furthermore, the *charge density* $\rho(\mathbf{r})$ depends only on three spatial coordinates whereas the *wave function* $\Psi(\mathbf{r}_1,\ldots,\mathbf{r}_N)$ at a given *spin state* depends on the 3 N coordinates of the N electrons. Also, the decomposition of the *total wave function* $\Psi(\mathbf{r}_1,\ldots,\mathbf{r}_N)$ into *one-electron orbitals* $\phi(\mathbf{r})$ is subject to arbitrary decisions depending on what the author considers reasonable. Whatever type of orbitals are used, they are just a model, which plays a crucial role in chemistry. This was poignantly expressed by Michael Dewar who wrote in 1984: "*The only criterion of a model is usefulness, not its "truth"*". A horrifying statement for scientists and nonscientists in search of the truth!

But there is one aspect, which makes the *wave function* $\Psi(\mathbf{r}_1,\ldots,\mathbf{r}_N)$ *clearly superior* to the *charge density* $\rho(\mathbf{r})$ *for the interpretation of the electronic structure of molecules. This is the symmetry and phase of the wave function* $\Psi(\mathbf{r}_1,\ldots,\mathbf{r}_N)$ *and the one-electron orbitals* $\phi(\mathbf{r})$, which carries crucial information about the electronic interactions between the atoms. *Whether two electrons with opposite spin form a chemical bond or repel each other depends on the sign of the orbitals in the overlapping space.*

We skip many technical details here to concentrate on the central point, which is the *sign or phase of the orbital related to the symmetry of the wave function. The sign disappears when the charge density is calculated by squaring the wave function* (Equation 6, $\rho(H_2) = |\Psi|^2$) and *with it the information is lost as to why there is an accumulation or depletion of charge density in the overlapping region of the orbitals*. The resulting *charge density* shows the attractive or repulsive nature of the electronic interaction, but the driving force for the charge formation is not evident without inspection of the interacting fragment orbitals. The *symmetry of the valence orbitals* provides a straightforward explanation why Be_2 does not possess a Be=Be double bond and why B_2 and O_2 have triplet ground states. In fact, *the trend of the E-E bond in diatomic main-group molecules E_2 from E-E single bonds of the alkaline atoms to E≡E triple bonds of the pnictogens up to single bonds E-E for the halogens comes from the symmetry of the MOs which are built from the atomic valence AOs following quantum theoretical rules*. The *frontier orbital model* by Fukui and the *orbital symmetry rules* by Woodward and Hoffmann, which are now widely used in chemistry, make use of the information which is provided by the *symmetry of the wave function* $\Psi(\mathbf{r}_1,\ldots,\mathbf{r}_N)$. …

Thus, while the electronic structure of formed molecules and solids can be analyzed and interpreted using the charge density ρ, that is describing "what is", the *process of forming the chemical bond*, in particular of "why and how it is formed", can only be described in terms of the *interference* (*resonance*) of the wave functions. But what does that mean, "*interference* (*resonance*) of the wave functions"? What is the human understanding of a process that is the chemical basis for material changes in the real world and yet can only be described on the basis of an abstract wave function? It is interesting to read the original description of the phenomenon by Heitler and London, who called it (in German) "*ein charakteristisch quantenmechanisches Schwebungsphänomen, welches nahe verwandt ist mit den von Heisenberg aufgefundenen Resonanzschwebungen*". This may be translated with "*resonance beat vibrational phenomenon*" It is obvious that the authors are trying to express the mathematical insight with simple words that are accessible to the human imagination.

...

There is a sometimes almost religious war between advocates who regard the *electron density* ρ(r) as an object of direct observation and thus as sensually experienceable, with the faithful belief that the *wave functions* $\Psi(r_1,\ldots,\mathbf{r}_N)$ or *orbitals* ϕ(r) only act as a mathematical auxiliary construction, and those of the opposite conviction. I consider this distinction to be untenable. The *electron density* ρ(**r**) is a mathematical quantity derived from measurements, which is just as real/irreal as the *wave function* $\Psi(\mathbf{r}_1,\ldots,\mathbf{r}_N)$. In this context, the fact that the spatial coordinates **r** of the *electron density* ρ(**r**) correspond to the three Cartesian coordinates, while the spatial coordinates **r** of the *wave function* $\Psi(\mathbf{r}_1,\ldots,\mathbf{r}_N)$ represent the 3 N coordinates of Hilbert space, only points to a higher dimensionality of space, such as in classical mechanics of an N-particle system, but not to a change from real human experience to the unreal world. The experimental results may either be used to derive the *electron density* ρ(**r**) or a *wave* $\Psi(\mathbf{r}_1,\ldots,\mathbf{r}_N)$ or *orbitals* ϕ(**r**) in a process of mathematical steps, which are ontically not fundamentally different from each other. The apparently higher accessibility of *electron density* for the human conception of reality can also be called into question by recalling the principally infinite extension of ρ(**r**). In reality, atomic basins are limited by the zero-flux surfaces, which are reasonably defined and yet arbitrary boundaries. Physically, the *electron density* knows no spatial limitation. In total vacuum, an electron at the equator has a finite probability at the north pole. ...

Dirac, P. A. M. (April, 1929). Quantum Mechanics of Many-Electron Systems.

Roy. Soc. Proc., A, 123, 792, 714-33; https://doi.org/10.1098/rspa.1929.0094; also in Underwood, T. G. (2023). *Quantum Electrodynamics - annotated sources*, Volume I, pp. 565-79.

Communicated by R. H. Fowler, F.R.S.

Received March 12, 1929.

St. John's College, Cambridge.

In this paper Dirac introduced the term *exchange energy*. He noted that the general theory of quantum mechanics was now almost complete, and that the imperfections that still remained were in connection with the exact fitting in of the theory with *relativity ideas*, which only gave rise to difficulties when high-speed particles were involved and were therefore *of no importance in the consideration of atomic and molecular structure and ordinary chemical reactions*. The difficulty was only that the exact application of these laws led to equations much too complicated to be soluble. He noted that it was desirable that approximate practical methods of applying quantum mechanics should be developed which could lead to an explanation of the main features of complex atomic systems without too much computation. Current *non-relativistic* quantum theory could not give an explanation of *multiplet structure* without an extraneous assumption of *large forces coupling the spin vectors of the electrons in an atom*. The explanation was provided by *quantum entanglement* through *exchange interaction* arising from electrons being indistinguishable one from another resulted in large *exchange energies* between electrons in different atoms. This accounted for homopolar valency bonds, in which, for each *stationary state* of the atom there was one magnitude of the *total spin vector*. He also noted that developments of the *theory of exchange* made by Heitler & London and Heisenberg made extensive use of *group theory*, which was a theory of certain quantities that did not satisfy the commutative law of multiplication and should thus form a part of quantum mechanics, and then translated the methods and results of *group theory* into the language of *quantum mechanics*. He demonstrated that *exchange interaction* equal to a constant *perturbation energy*, together with *coupling energy* between *spin vectors*, determined *energy* levels; and showed that in the first approximation the *exchange interaction* between the electrons could be replaced *by a coupling between their spins*, the energy of this coupling for each pair of electrons being equal to the scalar product of their *spin vectors* multiplied by a numerical coefficient given by the *exchange energy*.

---

§ 1. *Introduction.*

The general theory of quantum mechanics is now almost complete, the imperfections that still remain being in connection with the exact fitting in of the theory with relativity ideas. These give rise to difficulties only when high-speed particles are involved, and are therefore of no importance in the consideration of atomic and molecular structure and ordinary chemical reactions, in which it is, indeed, usually *sufficiently accurate if one neglects relativity variation of mass with velocity and assumes only Coulomb forces between the various electrons and atomic nuclei.*

> [*Coulomb force*, also called *electrostatic force*, is the attraction or repulsion of particles or objects because of their electric charge.]

The underlying physical laws necessary for the mathematical theory of a large part of physics and the whole of chemistry are thus completely known, and *the difficulty is only that the exact application of these laws leads to equations much too complicated to be soluble.* It therefore becomes desirable that approximate practical methods of applying quantum mechanics should be developed, which can lead to an explanation of the main features of complex atomic systems without too much computation.

Already before the arrival of quantum mechanics there existed a theory of atomic structure, based on *Bohr's ideas of quantized orbits*, which was fairly successful in a wide field. *To get agreement with experiment it was found necessary to introduce the spin of the electron*, giving a doubling in the number of orbits of an electron in an atom. With the help of this spin and *Pauli's exclusion principle, a satisfactory theory of multiplet terms was obtained when one made the additional assumption that the electrons in an atom all set themselves with their spins parallel or antiparallel.*

> [A *multiplet* is the state space for 'internal' degrees of freedom of a particle, that is, degrees of freedom associated to a particle itself, as opposed to 'external' degrees of freedom such as the particle's position in space. Examples of such degrees of freedom are the *spin state* of a particle in quantum mechanics.]

If s denoted the magnitude of the resultant *spin angular momentum*, this s was combined vectorially with the resultant *orbital angular momentum l* to give a multiplet of multiplicity 2s + *1. The fact that one had to make this additional assumption was, however, a serious disadvantage, as no theoretical reasons to support it could be given. It seemed to show that there were large forces coupling the spin vectors of the electrons in an atom*, much larger forces than could be accounted for as due to the interaction of the *magnetic moments* of the electrons. *The position was thus that there was empirical evidence in favor of these large forces, but that their theoretical nature was quite unknown.*

The old orbit theory is now replaced by Hartree's *method of the self-consistent field**, based on quantum mechanics.

* Hartree, D. R. (1928). The Wave Mechanics of an Atom with a Non-Coulomb Central Field. Part I. Theory and Methods. *Proc. Camb. Phil. Soc.*, 24, 89-110; http://dx.doi.org/10.1017/S0305004100011919.

The simplifying feature of the old theory, according to which each electron has its own individual orbit, is retained, but *the orbit is now a quantum-mechanical state of the single electron, represented by a wave function in three dimensions*. The only action of one orbit on another is assumed to be that of a static distribution of electricity, causing a partial screening of the nucleus. A theoretical justification for Hartree's method, showing that its results must be in approximate agreement with those of the exact ***Schrodinger equation*** for the whole system, has been given by Gaunt**.

** Gaunt, J. A. (1928). *Proc. Camb. Phil. Soc.*, 24, 328. It is pointed out by Gaunt that there does not seem to be any theoretical justification for Hartree's method of calculating energies and that its extremely good agreement with observation is probably accidental. The somewhat different method proposed by Gaunt is the one that should be used in connection with the present paper.

The method, however, suffers from the same limitation as the old orbit theory. *It cannot give an explanation of multiplet structure without an extraneous assumption of large forces coupling the spins.*

The solution of this difficulty in the explanation of multiplet structure is provided by the exchange (austausch) interaction of the electrons, which arises owing to the electrons being indistinguishable one from another. Two electrons may change places without our knowing it, and the proper allowance for the possibility of quantum jumps of this nature, which can be made in a treatment of the problem by quantum mechanics, gives rise to the new kind of interaction. The energies involved, the so-called *exchange energies*, are quite large. In fact, it is these *exchange energies* between electrons in different atoms that give rise to *homopolar valency bonds*, as shown by Heitler and London[#].

[#] Heitler, W. & London, F. (June, 1927). Wechselwirkung neutraler Atome und homöopolare Bindung nach der Quantenmechanik. (Interaction of Neutral Atoms and Homopolar Binding in Quantum Mechanics.) *Zeit. Phys.*, 44, 2, 455-72; http://dx.doi.org/10.1007/ BF01397394.

The application of the new *exchange* ideas to the problem of *multiplet structure* has been made by Wigner[§] and Hund[$].

[§] Wigner, E. (1927). Einige Folgerungen aus der Schrödingerschen Theorie für die Termstrukturen. (Some conclusions from Schrödinger's theory for term structures.)

Zeit. Phys., 43, 624-52; https://doi.org/10.1007/BF01397327.

$ Hund, F. (1927). Symmetriecharaktere von Termen bei Systemen mit gleichen Partikeln in der Quantenmechanik. (Symmetry characters of terms in systems with equal particles in quantum mechanics.) *Zeit. Phys.*, 43, 788-804; http://dx.doi.org/10.1007/ BF01397248.

The new theory provides no justification for the assumption that the electrons all set themselves with their spins parallel or antiparallel. In fact, it does not allow any meaning to be given to this assumption, since in quantum mechanics the component of the *spin angular momentum* of an electron in any direction is a q-number with the two *eigenvalues* ± ½ h, so that one cannot *in general* give a meaning to the direction of the *spin* of an electron in a given stationary state. What the new theory shows instead is that *for each stationary state of the atom there is one definite numerical value for s, the magnitude of the total spin vector.* If it were not for this theorem, a measurement of s for the atom in a given stationary state would lead to one or other of a number of possible results, according to a definite probability law. *This theorem forms the basis of the theory of multiplets. It is quite sufficient to replace the previous idea of the electrons all setting themselves parallel or antiparallel*, since it shows that we can take s to be a *quantum number* describing the states of the atom, while s combined vectorially with *1* gives a multiplet of multiplicity 2s + *1*.

Further developments of the *theory of exchange* have been made by Heitler, London and Heisenberg*, containing applications to molecules held together by *homopolar valency bonds* and to *ferromagnetism*.

* See various papers in *Zeit. Phys*, 46-51. An excellent account of the whole theory is also contained in Weyl's book, '*Gruppentheorie und Quantummechanik.*'

The treatment given by these authors makes an extensive use of *group theory* and requires the reader to be well acquainted with this branch of pure mathematics. Now *group theory is just a theory of certain quantities that do not satisfy the commutative law of multiplication, and should thus form a part of quantum mechanics*, which is *the general theory of all quantities that do not satisfy the commutative law of multiplication*. It should therefore be possible to translate the methods and results of *group the*ory into the language of *quantum mechanics* and so *obtain a treatment of the exchange phenomena which does not presuppose any knowledge of groups on the part of the reader. This is the object of the present paper*. The treatment of groups on the lines of *quantum mechanics* has the advantage that it often gives a simple physical meaning to an abstract theorem in the theory of groups, enabling one to remember the theorem more easily and perhaps suggesting a simpler way of proving it. A further advantage of the treatment of the *exchange* phenomena on these lines is that one can avoid doing more work in the theory of groups than is strictly necessary for the physical applications, which results in a considerable shortening in the method.

In §§ 2 and 3 the general theory is given of systems containing a number of similar particles, showing the existence of exclusive sets of states (i.e., sets such that a transition can never take place from a state in one set to a state in another), and giving their main properties. *In § 4 an application is made to electrons*; a proof being obtained of the fundamental theorem in italics above ["*for each stationary state of the atom there is one definite numerical value for s, the magnitude of the total spin vector*"]. The subsequent work is concerned with an approximate calculation of the *energy* levels of the states, the result of this being expressible by the single simple formula (26). *This formula shows that in the first approximation the exchange interaction between the electrons may be replaced by a coupling between their spins, the energy of this coupling for each pair of electrons being equal to the scalar product of their spin vectors multiplied by a numerical coefficient given by the exchange energy*. This form of *coupling energy* is, however, just what was required in the old orbit theory. We obtain in this way a justification for the assumptions of the old theory, in so far as they can be formulated without contradicting the quantum-mechanical description of the *spin*. The formula (26), combined with Hartree's method for determining approximate *wave functions* for the different electrons, should provide a powerful way of dealing with complicated atomic systems.

§ 2. *Permutations as Dynamical Variables.*

We consider a dynamical system composed of similar particles, the rth particle being describable by certain generalized co-ordinates denoted by the single symbol q_r. Thus, a *wave function* representing a state of the system will be a function of the variables q_1, q_2, ... q_n, which may be written

$$\psi\ (q_1, q_2, \ldots q_n) = \psi\ (q)$$

for brevity. Suppose now that P is any permutation of q_1, q_2, ... q_n. This P is an operator which can be applied to any *wave function* ψ (q) to give as result another definite function of the q's, namely

$$P\psi\ (q) = \psi\ (Pq),$$

where Pq denotes the set of q's obtained by applying the permutation P to q_1, q_2, ... q_n. Further P is a linear operator. Now *in quantum mechanics any dynamical variable is a linear operator which can operate on any wave function, and conversely any linear operator that can operate on every wave function may be considered as a dynamical variable*. Thus, P may be considered to be a *dynamical variable*.

The present paper consists in a study of these permutations P as dynamical variables. There are no classical analogues to these variables and hence they give rise to phenomena, e.g., the existence of *exclusive sets of states* and other *exchange phenomena*, which have no classical analogue. There are n! of these variables, one of them, P_1 say, being the identity, which must thus be equal to unity. One can add and multiply these variables and form algebraic functions of them, in exactly the same way in which one can add and multiply and form algebraic functions of the ordinary *co-ordinates* and *momenta*. The product of any two permutations is a third permutation, and hence any function of the permutations is reducible to a linear function of them. Any permutation P has a reciprocal P^{-1} satisfying $PP^{-1} = P^{-1}P = P_1 = 1$.

A permutation P, like any other dynamical variable, can be represented by a matrix. If we take the representation in which the q's are diagonal, P will be represented by a matrix, whose general element may be written

$$(q'_1\, q'_2 \ldots q'_n \mid P \mid q''_1\, q''_2 \ldots q''_n) = (q' \mid P \mid q'')$$

for brevity. This matrix must satisfy

$$\int (q' \mid P \mid q'')\, dq''\, \psi (q'') = P\psi (q') = \psi (Pq'),$$

and hence

$$(q' \mid P \mid q'') = \delta (Pq' - q''). \qquad (1)$$

We are using the notation δ (x), where x is short for a set of variables x_1, x_2, x^3, ..., to denote

$$\delta (x) = \delta (x_1)\, \delta (x_2)\, \delta (x_3) \ldots$$

which vanishes except when each of the x's vanishes. With this notation we have

$$\delta (Pq' - q'') = \delta (q' - P^{-1}q''),$$

since the condition that the left-hand side shall not vanish, which is that the q"'s shall be given by applying the permutation P to the q''s, is the same as the condition that the right-hand side shall not vanish, which is that the q''s shall be given by applying the permutation P^{-1} to the q"'s. Thus, we have an alternative expression for the matrix representing P.

$$(q' \mid P \mid q'') = \delta (q' - P^{-1}q''). \qquad (2)$$

The *conjugate complex* of any dynamical variable is given when one writes –i for i in the matrix representing that variable and also interchanges the rows with the columns. Thus, we find for the conjugate complex of a permutation P, with the help of (2) and (1)

$$[(q' \mid P \mid q'') = \delta (Pq' - q''). \qquad (1)]$$

$$(q' \mid P^* \mid q'') = (q'' \mid P \mid q')^* = \delta (q'' - P^{-1}q') = (q' \mid P^{-1} \mid q'')$$

or $$P^* = P^{-1}.$$

Thus, a permutation is not in general a real variable, its conjugate complex being equal to its reciprocal.

Any permutation of the numbers 1, 2, 3, ..., n may be expressed in the cyclic notation, e.g., for n = 8

$$P_a = (143)\,(27)\,(58)\,(6), \qquad (3)$$

in which each number is to be replaced by the succeeding number in a bracket, unless it is the last in a bracket, when it is to be replaced by the first in that bracket. Thus, P_a changes the numbers 12345678 into 47138625. The type of any permutation is specified by the partition of the number n which is provided by the number of numbers in each of the brackets. Thus, the type of P_a is specified by the partition 8 = 3 + 2 + 2 + 1. Permutations of the same type, i.e., corresponding to the same partition, we shall call similar. (The usual language of *group theory* is to call them *conjugate*.) Thus, for example, P_a in (3) is similar to

$$P_b = (871)\,(35)\,(46)\,(2). \qquad (4)$$

The whole of the n! possible permutations may be divided into sets of similar permutations; each such set being called a class. The permutation $P_1 = 1$ forms a class by itself. Any permutation is similar to its reciprocal.

When two permutations P_a and P_b are similar, either of them P_b may be obtained by making a certain permutation P in the other P_a. Thus, in our example (3), (4) we can take P to be the permutation that changes 14327586 into 87135462, i.e., the permutation

$$P = (18623)\,(475).$$

We then have the algebraic relation between P_b and P_a

$$P_b = PP_aP^{-1}. \qquad (5)$$

To verify this, we observe that the product $P_a\,\psi$ of P_a with any wave function ψ is changed into $P_b\,\psi$ if one applies the permutation P to the P_a in the product but not to the ψ. If we multiply the product by P on the left, we are applying this permutation to both the P_a and the ψ, so that we must insert another factor P^{-1} between the P_a and the ψ, giving us $PP_aP^{-1}\,\psi$ to equate to $P_b\,\psi$.

Equation (5) is the general formula showing when two permutations P_a and P_b are similar. Of course, P is not uniquely determined when P_a and P_b are given, but the existence of any P satisfying (5) is sufficient to show that P_a and P_b are similar.

§ 3. *Permutations as Constants of the Motion.*

We now introduce a Hamiltonian H to describe the motion of the system, so that any *stationary state* of *energy* H' is represented by a *wave function* ψ satisfying

$$H\psi = H'\psi,$$

in which H is regarded as an operator. This Hamiltonian can be an arbitrary function of the dynamical variables provided it is *symmetrical* between all the particles. This symmetry condition requires that an element (q' | H | q'') of the matrix representing H shall be unaltered when one applies any permutation to the q''s and the same permutation to the q'''s, i.e.,

$$(q' \mid H \mid q'') = (Pq' \mid H \mid Pq'') \qquad (6)$$

for arbitrary P.

The fact that H is symmetrical leads at once to the equation

$$PH = HP. \qquad (7)$$

This equation may be verified by a similar argument to that used for equation (5), or alternatively by a direct application of the matrix representatives. Thus from (1)

$$[(q' \mid P \mid q'') = \delta\,(Pq' - q''). \qquad (1)]$$

$$(q' \mid PH \mid q'') = \int \delta\,(Pq' - q''')dq'''\,(q''' \mid H \mid q'') = (Pq' \mid H \mid q'')$$

and from (2)

$$[(q' \mid P \mid q'') = \delta\,(q' - P^{-1}q''). \qquad (2)]$$

$$(q' \mid PH \mid q'') = \int (q' \mid H \mid q''')dq'''\,\delta\,(q''' - P^{-1}q'') = (Pq' \mid H \mid P^{-1}q''),$$

and the two right-hand sides are now equal from (6).

Equation (7)

$$[PH = HP \qquad (7)]$$

shows that *each permutation variable is a constant of the motion*. The P's are still constants when arbitrary perturbations are applied to the system, *provided the perturbation energy to be added to the Hamiltonian is symmetrical. Thus, the constancy of the P's is absolute*.

In dealing with any system in quantum mechanics, when we have found a *constant of the motion* α, we know that if for any *state* α initially has the numerical value α' then it always has this value, so that we can assign different numbers α' to the different *states* and so obtain a classification of the *states*. This procedure is not so straightforward, however, when we have several *constants of the motion* α which do not commute (as is the case with our permutations P), since we cannot assign numerical values for all the α's simultaneously to any *state*. *The existence of constants of the motion α which do not commute is a sign that the system is degenerate*. We must now look for a function β of the α's which has one and the same numerical value β' for all those *states* belonging to one energy level H', so that we can use β for classifying the *energy* levels of the system.

We can express the condition for β by saying that it must be a function of H (a single-valued function is implied) according to the general definition of a function of a variable in quantum mechanics, or that β must commute with every variable that commutes with H, i.e. every *constant of the motion*. If the α's are the only *constants of the motion*, or if they are a set that commute with all other independent *constants of the motion*, our problem reduces to finding a function β of the α's which commutes with all the α's. We can then assign a numerical value β' for β to each *energy* level of the system. If we can find several such functions β, they must all commute with each other, so that we can give them all numerical values simultaneously and obtain a complete classification of the *energy* levels.

An example of this procedure is provided by the study of the *angular momentum* of an isolated system. This *angular momentum* has three components m_x, m_y, m_z, each a constant of the motion, which do not commute. We look for a function of m_x, m_y, m_z which commutes with them all three. We can conveniently take for this function the variable k defined by

$$k(k + h) = m_x^2 + m_y^2 + m_z^2. \qquad (8)$$

For each *energy* level of the system there will now be one definite numerical value k' for k. This constant of the motion k is the only significant one for purposes of classifying the states, as the others merely describe the degeneracy.

We follow this method in dealing with our permutations P. We must find a function χ of the P's such that $P\chi P^{-1} = \chi$ for every P. It is evident that a possible χ is ΣP_c, the sum of all the permutations P_c in a certain class c, i.e. the sum of a set of similar permutations, since $\Sigma P P_c P^{-1}$ must consist of the same permutations summed in a different order. There will be one such χ for each class. Further, there can be no other independent χ, since an arbitrary function of the P's can be expressed as a linear function of them with numerical coefficients and it will not then commute with every P unless the coefficients of similar P's are always the same. We thus obtain all the χ's that can be used for classifying the states. It is convenient to define each χ as an average instead of a sum, thus

$$\chi_c = \Sigma P_c/n_c,$$

where n_c is the number of P's in the class c. An alternative expression for χ is

$$\chi_c = \Sigma_r P_r P_c P_r^{-1}/n!, \qquad (9)$$

the summation being extended over all the n! permutations P_r. For each permutation P there is one χ, $\chi(P)$ say, equal to the average of all permutations similar to P. One of the χ's is $\chi(P_1) = 1$.

The dynamical variables $\chi_1, \chi_2, \ldots \chi_m$ obtained in this way will each have a definite numerical value for every stationary state of the system. Thus, for every permissible set of numerical values $\chi_1', \chi_2', \ldots \chi_m'$ for the χ's there will be a set of *states* of the system. Since the χ's are absolute constants of the motion these sets of *states* will be exclusive, i.e. transitions will never take place from a *state* in one set to a *state* in another.

...

§ 4. *Multiplet Structure.*

The preceding theory of systems composed of similar particles will now be applied to the case when the particles are electrons. The new features which this requires us to take into consideration are *the spin of the electrons* and *Pauli's exclusion principle.*

The three Cartesian co-ordinates x, y, z of the rth electron we denote by the single symbol x_r. The *spin angular momentum* and *magnetic moment* of this electron will be of the form $\frac{1}{2} h \boldsymbol{\sigma}_r$ and $\frac{1}{2} eh/mc \cdot \boldsymbol{\sigma}_r$, where $\boldsymbol{\sigma}_r$ is a vector whose components $\sigma_{rx}, \sigma_{ry}, \sigma_{rz}$ satisfy

$$\sigma_{rx}^2 = 1, \qquad \sigma_{rx}\sigma_{ry} = i\, \sigma_{rz} = -\sigma_{ry}\sigma_{rx}, \qquad (12)$$

with similar relations obtained by cyclic permutation of the suffixes x, y and z. We take x_r and σ_{rz} to be the variables describing the rth electron that appear in the wave function. It is convenient to write the *wave function*

$$\psi\,(x_1\sigma_1 x_2\sigma_2 \ldots x_n\sigma_n) = \psi\,(x\sigma)$$

without the suffixes z attached to the σ's, these suffixes being understood whenever one is dealing with the variables in *wave functions.*

The *exclusion principle* now requires that ψ shall be *antisymmetrical* in the x's and σ's together, i.e., that if any permutation is applied to the x's and also to the σ's, ψ must remain unchanged or change sign according to whether the permutation is an even or an odd one. Thus, permutations applied to the x's and σ's together, produce only trivial effects and *no useful results would be obtained by considering them as dynamical variables.* We can, however, consider permutations P applied to the x's alone and apply our preceding theory to these. Any of these permutations is a *constant of the motion* when *we neglect the forces due to the spins, so that the Hamiltonian does not involve the spin variables* $\boldsymbol{\sigma}$. We can now introduce our χ's as functions of these P's and assert that for any permissible set of numerical values χ' for the χ's there will be one exclusive set of *states. Thus, there exist these exclusive sets of states for systems containing many electrons even when we restrict ourselves to a consideration only of those states that satisfy the exclusion principle.* The exclusiveness of the sets of *states* is now, of course, only approximate, *since the χ's are constants only when we neglect the spin forces*. There will actually be a small probability for a transition taking place from a *state* in one set to a *state* in another.

Since ψ is *antisymmetrical*, the result of any permutation P applied to the x's must equal $\pm$ times the result when the same permutation is applied to the σ's. Thus, if we denote by P^σ a permutation applied to the σ's considered as a dynamical variable, we shall have

$$P_r = \pm P^\sigma{}_r, \tag{13}$$

for each of the n! permutations P_r. Thus, instead of studying the dynamical variables P we can get all the results we want, e.g., the characters χ', by studying the variables P^σ. The P^σ's are much easier to study on account of the fact that the variables σ in the *wave function* have domains consisting each of only the two points 1 and –1, which are the two *eigenvalues* of each σ_z. This fact results in there being fewer characters χ' for the group of permutations of the σ variables than for the group of general permutations, since it prevents a function of the variables $\sigma_1, \sigma_2, \ldots$, from being *antisymmetrical* in more than two of them.

The study of the dynamical variables P^σ is made especially easy by the fact that we can express them as algebraic functions of the dynamical variables $\boldsymbol{\sigma}$. …

…

Hence

$$\chi_{12} = -\tfrac{1}{2}\left[1 + \{4s(s+1) - 3n\}/n(n-1)\right] = -\{n(n-4) + 4s(s+1)\}/2n(n-1). \tag{17}$$

Thus, χ_{12} is expressible as a function of the variable *s* and of *n* the number of electrons. Any of the other χ's could be evaluated on similar lines and would be found to be a function of *s* and *n* only, since there are no other symmetrical functions of all the $\boldsymbol{\sigma}$ variables which could be involved. There is therefore one set of numerical values χ' for the χ's, and thus one exclusive set of *states*, for each eigenvalue *s'* of *s*. The *eigenvalues* of *s* are

$$\tfrac{1}{2}n, \quad \tfrac{1}{2}n - 1, \quad \tfrac{1}{2}n - 2, \ldots$$

the series terminating with ½ or 0.

We obtain in this way a proof of *the fundamental theorem of multiplet structure, that for each stationary state of the atom there is one definite numerical value s' for s*. We obtain further that *the probability of transitions occurring in which s changes is small, of the order of magnitude of the spin forces.*

…

§ 5. *Determination of Energy Levels*

We must now consider the application of perturbation theory for an approximate calculation of the energy levels. We shall take first the general case of a system with n similar particles, discussed in §§ 2 and 3. We shall follow the usual method in the *theory of the perturbations* of the *stationary states* of a degenerate system, according to which, if we label the *states* of the unperturbed system α', α", we obtain the matrix (α' | V | α") representing the *perturbation energy* V *and neglect all those matrix elements α', α" for which the unperturbed states α' and α" have two different energies*. The remaining matrix elements will form a number of small matrices, one referring to each *energy* level of the unperturbed system, and having as the number of its rows and columns the number of independent *states* belonging to this *energy* level. The *eigenvalues* of these matrices will then be, in the first approximation, the changes in the *energy levels* caused by the perturbation.

We suppose that for our unperturbed *states* each of the similar particles has its own "*orbit*", represented by a *wave function* $(q_r \mid \alpha)$ involving only the *co-ordinates* q_r of this one particle. We shall have altogether *n orbits*, one for each particle, which we assume for the present to be all different, and label $\alpha_1, \alpha_2, \ldots \alpha_n$. The *wave function* representing an unperturbed state of the whole system will then be the product

$$(q_1 \mid \alpha_1)(q_2 \mid \alpha_2) \ldots (q_n \mid \alpha_n) = (q \mid \alpha) \qquad (18)$$

say, for brevity. If we apply an arbitrary permutation P_a to the α's, we shall obtain another *wave function*

$$(q_1 \mid \alpha_r)(q_2 \mid \alpha_s) \ldots (q_n \mid \alpha_t) = (q \mid P_a\alpha) \qquad (19)$$

representing another unperturbed state with the same *energy*. There are thus altogether n! unperturbed *states* with this *energy*, if we assume there are no other causes of degeneracy. The matrix elements of V that we must take into consideration are therefore of the type $(P_a\alpha \mid V \mid P_b\alpha)$, where P_a and P_b are two permutations of the α's, and form a matrix with n! rows and columns. *The eigenvalues of this matrix are what we must calculate*.

It is necessary in the present discussion to distinguish between the two kinds of permutations, those of the q's and those of the α's. The essential difference between them can perhaps be seen most clearly in the following way. Let us consider a permutation in the general case, say that consisting of the interchange of 2 and 3. *This may be interpreted either as the interchange of the objects 2 and 3 or as the interchange of the objects in the places 2 and 3, these two operations producing in general quite different results.* The first of these interpretations is the one we have been using throughout §§ 2 and 3, the objects concerned being the q's. A permutation with this interpretation can be applied to an arbitrary function of the q's. A permutation with the second interpretation has a meaning, however, when applied to a function of the q's only if each of the q's has a definite specifiable place in the function. This is not the case for a general function of the s, but it is the case for any of the n! functions of the type (19)

$$[(q_1 \mid \alpha_r)(q_2 \mid \alpha_s) \ldots (q_n \mid \alpha_t) = (q \mid P_a\alpha), \qquad (19)]$$

the place of each q being specified by the α with which it is bracketed. Any permutation applied to the q's in given places now produces the same result as the reciprocal permutation applied to the α's. A permutation of the q's (i.e., one with the first interpretation) since it can be applied to any function of the s, may be regarded as an ordinary dynamical variable. On the other hand, *a permutation of places or of the α's can be considered as a dynamical variable only in a very restricted sense, since it has a meaning only when multiplied into one of the n! wave functions (19) or into some linear combination of them*. We denote such a permutation of the α's, considered as a dynamical variable in this restricted sense, by a symbol P^{α}.

…

… Thus, the whole matrix $(P_a\alpha \mid Y \mid P_b\alpha)$ is equal to the matrix representing $\Sigma\, V_P P^{\alpha}$, where the summation is over all the n! permutations P, and we can put

$$V = \Sigma\, V_P P^{\alpha}. \qquad (24)$$

This formula shows that the *perturbation energy* V is equal to a linear function of the permutation variables P^{α}, with numerical coefficients V_P, which are the *exchange energies*. It is, of course, only an approximate formula, *as it holds only with neglect of those matrix elements of V that refer to two different energy levels of the unperturbed system*. It can, however, be used for the calculation of the *energy* levels in the first approximation, and is very convenient for this purpose as the expression $\Sigma\, V_P P^{\alpha}$ is easily handled. *This expression*, it should be remembered, *is a dynamical variable only in the restricted sense mentioned above*, but this sense is just sufficiently general for equation (24) to be valid *with neglect of those matrix elements of V referring to two different energy levels of the unperturbed system*.

As an example of an application of (24)

$$[V = \Sigma\, V_P P^{\alpha}. \qquad (24)]$$

we shall determine the *average energy* of all those *states* arising from a given state of the unperturbed system that belong to one exclusive set. This requires us to calculate the *average eigenvalue* of V when the χ's have specified numerical values χ'. Now the *average eigenvalue* of P^{α} equals that of $P^{\alpha}{}_a P^{\alpha}(P^{\alpha}{}_a)^{-1}$ for arbitrary $P^{\alpha}{}_a$ and thus equals that of $(n!)^{-1}$ $\Sigma_a\, P^{\alpha}{}_a P^{\alpha}(P^{\alpha}{}_a)^{-1}$, which is $\chi'(P^{\alpha})$ or $\chi'(P)$. Hence the *average eigenvalue* of V is $\Sigma\, V_P\, \chi'(P)$. A similar method could be used for calculating the *average eigenvalue* of any function of V, it being only necessary to replace each P^{α} by $\chi(P)$ to perform the averaging.

…

§ 6. *The Energy Levels in the Case of Electrons.*

We shall now consider the application of the formula (24) to the case of *electrons*. If we assume only Coulomb forces between the electrons, then the perturbation will consist of a number of terms, each involving the coordinates of one or at most two electrons, so that all the *exchange energies* V_P will vanish except those referring to the identical permutation P_1 and to simple interchanges of two *orbits*, $P^{\alpha}{}_{rs}$. Thus (24)

$$[V = \Sigma\, V_P P^{\alpha}. \qquad (24)]$$

reduces to

$$V = V_1 + \Sigma_{r<s}\, V_{rs} P^{\alpha}{}_{rs},$$

V_{rs} being the *exchange energy* of *orbits* r and s. Since the P^{α}'s have exactly the same properties as the P's, we can replace the P^{α}'s in this expression for V by P's without changing its *eigenvalues*. This gives us

$$V = V_1 + \Sigma_{r<s}\, V_{rs} P_{rs}, \qquad (25)$$

where the = sign is now to be interpreted as denoting the equality of the *eigenvalues* of the two sides and not the complete equality of the two sides as dynamical variables or operators.

With, the help of (16)

$$[P_{12} = -\tfrac{1}{2}\, \{1 + (\boldsymbol{\sigma}_1, \boldsymbol{\sigma}_2)\}, \qquad (16)]$$

the result (25) may be put in the more expressive form

$$V = V_1 - \tfrac{1}{2}\, \Sigma_{r<s}\, V_{rs}\, \{1 + (\boldsymbol{\sigma}_1, \boldsymbol{\sigma}_2)\}. \qquad (26)$$

This shows that, *for the purpose of calculating energies, the exchange interaction due to the equivalence of the electrons may be replaced by a constant perturbation energy* $-\tfrac{1}{2}\, \Sigma_{r<s}\, V_{rs}$, *together with a coupling between the spin vectors with energy* $\tfrac{1}{2}\, V_{rs}\, (\boldsymbol{\sigma}_1, \boldsymbol{\sigma}_2)$ *for each pair of electrons r, s*. It is this coupling which may be considered as giving rise, for instance, to the large differences in *energy* between the singlet and triplet terms of helium. The total number of *eigenvalues* of the right-hand side of (26) is a factor 2 occurring for the representation of the *spin vector* of each of the *n* electrons. These 2^n *eigenvalues* will not, in general, all be different, as each one will occur repeated a number of times to give the correct multiplicity of the corresponding term.

When two of the *orbits* of the unperturbed system are the same, say the *orbits* 1 and 2 are the same, the only *eigenvalues* of the right-hand side of (25) or (26)

$$[V = V_1 + \Sigma_{r<s}\, V_{rs} P_{rs}, \qquad (25)$$

$$V = V_1 - \tfrac{1}{2}\, \Sigma_{r<s}\, V_{rs}\, \{1 + (\boldsymbol{\sigma}_1, \boldsymbol{\sigma}_2)\}. \qquad (26)]$$

that will be *eigenvalues* of V are those consistent with the equation $P_{12} = 1$ or $P^{\sigma}{}_{12} = -1$. In this case we have $V_{12} = 0$ and $V_{1r} = V_{2r}$ (r = 3, 4, 5, ...), which results in the right-hand side of (26) being *symmetrical* between $\boldsymbol{\sigma}_1$ and $\boldsymbol{\sigma}_2$. It follows from this that any *eigenfunction* F (σ_{1z}, σ_{2z}, σ_{3z}, ...) of this right-hand side, considered as an operator, must be either *symmetrical* or *antisymmetrical* between σ_{1z} and σ_{2z}. *The condition* $P^{\sigma}_{12} = -1$ *now shows that only the antisymmetrical ones, representing states for which the spins* $\boldsymbol{\sigma}_1$ *and* $\boldsymbol{\sigma}_2$ *are antiparallel, must be taken into account*. The number of *eigenvalues* of (26) that must be used is thus reduced by a factor 4, on account of there being only one *antisymmetrical eigenfunction* for every three *symmetrical* ones. *The case of more than two orbits the same cannot occur with electrons.*

In our theory of the *energies*, we have nowhere had to assume that the *wave functions* ($q \mid \alpha_1$), ($q \mid \alpha_2$), ..., representing the various *orbits* in the unperturbed system, are *orthogonal*, or that they are *eigenfunctions* of any unperturbed Hamiltonian H_0. This enables an important generalization to be made in the application of our results. *It is not necessary that we should be able to split up our Hamiltonian for the whole system H into a Hamiltonian for the unperturbed system* H_0 *and a perturbation energy V, and then use the eigenfunctions of* H_0 *to give our* ($q \mid \alpha_1$), ($q \mid \alpha_2$), ... We can take our ($q \mid \alpha_1$), ($q \mid \alpha_2$) to be any functions giving a good approximation to the *actual distribution of electrons* in the system, and must then throughout the analysis replace V, which now no longer exists, by the whole Hamiltonian H. The *wave functions* supplied by Hartree's theory can thus very conveniently be used. The only mathematical conditions which the ($q \mid \alpha_1$), ($q \mid \alpha_2$), ..., need satisfy is that *in the matrix* ($\alpha' \mid H \mid \alpha''$) *representing H, those matrix elements for which the* α''*'s are not simply a permutation of the* α'*'s must be small.*

As an example of the application of (25)

$$[V = V_1 + \Sigma_{r<s} V_{rs}P_{rs}, \qquad (25)]$$

*Heitler's formula**

* Heitler, W. (1928). Zur Gruppentheorie der homöopolaren chemischen Bindung. (On the group theory of homeopolar chemical bonding.) *Zeit. Phys.*, 47, 835-58; https://doi.org/10.1007/BF01328643, equation (33).

for the interaction of two atoms A and B, each with valency electrons, will be deduced. The fundamental theorem of multiplet structure shows that there will be a *quantum number s* describing the magnitude of the resultant *spin* of the electrons in both atoms. This same theorem shows that, provided the interaction between the atoms is small compared with the *exchange energies* within either of them, the whole *energy* of the system will depend very largely on the magnitudes s_A and s_B of the resultant *spins* for the two atoms separately, so that for each *energy* level of the whole system there must be definite numerical values for s_A and s_B, which will thus be two more *quantum numbers* describing states of the whole system. For *valency electrons* the resultant *spin vector* has its maximum possible value (we can if we like in this case speak of the *electron spins* all being parallel) and hence

$$s_A = s_B = \tfrac{1}{2} n \qquad (27)$$

Again, if ς is the *valency* of the *homopolar bond* uniting the two atoms (i.e., $\varsigma = 1, 2, \ldots,$ for a single, double, ... bond)

$$s = n - \varsigma \qquad (28)$$

We now apply our formula (25)

$$[V = V_1 + \Sigma_{r<s} V_{rs}P_{rs}, \qquad (25)]$$

taking the summation only over pairs of *orbits r, s*, of which one is in each atom, since we want the *interaction energy* between the two atoms. This gives

$$V = V_1 + \Sigma_{AB} V_{rs} P_{rs}.$$

As a rough approximation we may take all the *exchange energies* V_{rs} between two *orbits*, one in each atom, to be equal. Calling these *exchange energies* V_Q, we get

$$V = V_1 + V_Q \Sigma_{AB} P_{rs}.$$

We must now evaluate $\Sigma_{AB} P_{rs}$, summed over all pairs, one in each atom, which we can best do by first summing over all possible pairs and then subtracting the two sums for pairs both in atom A and both in atom B respectively. Thus

$$\begin{aligned}\Sigma_{AB} P_{rs} &= \Sigma P_{rs} - \Sigma_{AA} P_{rs} - \Sigma_{BB} P_{rs} \\ &= -\tfrac{1}{4}\{2n(2n-4) + 4s(s+1)\} + \tfrac{1}{4}\{n(n-4) + 4s_A(s_A+1)\} \\ &\quad + \tfrac{1}{4}\{n\{n-4) + 4s_B(s_B+1)\}\end{aligned}$$

by a three-fold application of (17)

$$[\chi_{12} = -\tfrac{1}{2}[1 + \{4s(s+1) - 3n\}/n(n-1)] = -\{n(n-4) + 4s(s+1)\}/2n(n-1), \qquad (17)]$$

in the first case to a system of 2n electrons.

This reduces to

$$\begin{aligned}\Sigma_{AB} P_{rs} &= -\tfrac{1}{4}\{2n^2 + 4s(s+1) - 4s_A(s_A+1) - 4s_B(s_B+1)\} \\ &= \varsigma - (n-\varsigma)^2.\end{aligned}$$

Thus

$$V = V_1 + V_Q\{\varsigma - (n-\varsigma)^2\},$$

which is Heitler's result.

PART III Susceptibility of materials to a magnetic field.

While heuristic explanations based on classical physics can be formulated, *diamagnetism, paramagnetism and ferromagnetism can be fully explained only using quantum theory*. A successful model was developed in 1927, by Walter Heitler and Fritz London. The Heitler-London considerations can be generalized to the Heisenberg model of magnetism. In 1828, Heisenberg published a paper in which he applied the Heitler-London calculation to show that two conditions are necessary for the appearance of *ferromagnetism*: the crystal lattice must be a type such that *any atom has at least 8 neighbors*; and the *principal quantum number* of the electrons that are responsible for magnetism must be $n \geq 3$.

Diamagnetism

Diamagnetism appears in all materials and *is the tendency of a material to oppose an applied magnetic field*, and therefore, *to be repelled* by a magnetic field. However, in a material with *paramagnetic* properties (that is, with *a tendency to enhance an external magnetic field*), *the paramagnetic behavior dominates*. Thus, despite its universal occurrence, *diamagnetic behavior is observed only in a purely diamagnetic material. In a diamagnetic material, there are no unpaired electrons*, so the *intrinsic ('spin') magnetic moments* cannot produce any bulk effect. In these cases, *the magnetization arises from the electrons' orbital motions*, which can be understood classically as follows.

When a *material* is put in a *magnetic field, the electrons circling the nucleus will experience*, in addition to their Coulomb attraction to the nucleus, *a Lorentz force from the magnetic field.*

> [The *Lorentz force* (or electromagnetic force) is the combination of electric and magnetic force on a point charge due to electromagnetic fields. A *particle of charge* q moving with a *velocity* **v** in an *electric field* **E** and a *magnetic field* **B** experiences a force (in SI units) of
>
> $$\mathbf{F} = q(\mathbf{E} + \mathbf{v} \times \mathbf{B})$$
>
> This states that the *electromagnetic force* on a *charge* q is a combination of (1) a force in the direction of the *electric field* **E** (proportional to the magnitude of the field and the quantity of charge), and (2) a force at right angles to both *the magnetic field* **B** and the *velocity* **v** of the charge (proportional to the magnitude of the field, the charge, and the velocity).]

Depending on which direction the electron is orbiting, this force may increase the centripetal force on the electrons, pulling them in towards the nucleus, or it may decrease the force, pulling them away from the nucleus. *This effect systematically increases the orbital magnetic moments that were aligned opposite the field and decreases the ones aligned parallel to the field* (in accordance with *Lenz's law*).

> [*Lenz's law* states that the *direction* of the *electric current* induced in a conductor by a changing *magnetic field* is such that *the magnetic field created by the induced current opposes changes in the initial magnetic field.* It is named after physicist Heinrich Lenz, who formulated it in 1834. It is a qualitative law that specifies the direction of induced current, but states nothing about its magnitude. *Lenz's law* may be seen as analogous to *Newton's third law* in classical mechanics.]

This results in a small bulk magnetic moment, with an opposite direction to the applied field. This description is meant only as a heuristic; the *Bohr–Van Leeuwen theorem* shows that *diamagnetism* is impossible according to classical physics, and that *a proper understanding requires a quantum-mechanical description.*

All *materials* undergo this *orbital* response. However, *in paramagnetic and ferromagnetic substances, the diamagnetic effect is overwhelmed by the much stronger effects caused by the unpaired electrons.*

Paramagnetism

In a paramagnetic material there are unpaired electrons; i.e., atomic or molecular orbitals with exactly one electron in them. While *paired electrons* are required by the *Pauli exclusion principle* to have their *intrinsic ('spin') magnetic moments* pointing in opposite directions, causing their magnetic fields to cancel out, an *unpaired electron* is free to align its *magnetic moment* in any direction. *When an external magnetic field is applied, these magnetic moments will tend to align themselves in the same direction as the applied field, thus reinforcing it.*

Ferromagnetism

A *magnetic field* contains energy, and physical systems move toward configurations with lower energy. When a *diamagnetic material* is placed in a *magnetic field*, a *magnetic dipole* tends to align itself in opposed polarity to that field, thereby lowering the net field strength. When a *ferromagnetic material* is placed within a *magnetic field*, the *magnetic dipoles* align to the applied field, thus expanding the domain walls of the magnetic domains.

A *ferromagnet*, like a paramagnetic substance, has *unpaired electrons*. However, in addition to the electrons' *intrinsic ('spin') magnetic moment's* tendency to be parallel to an applied field, *there is also in these materials a tendency for these magnetic moments to orient parallel to each other to maintain a lowered-energy state*. Thus, *even in the absence of an applied field, the magnetic moments of the electrons in the material spontaneously line up parallel to one another*.

Every *ferromagnetic* substance has its own individual temperature, called the Curie temperature, or Curie point [named for Pierre Curie], above which it loses its ferromagnetic properties. This is because the thermal tendency to disorder overwhelms the energy-lowering due to ferromagnetic order. *Ferromagnetism* only occurs in a few substances; common ones are *iron, nickel, cobalt*, their alloys, and some alloys of rare-earth metals.

The *magnetic moments of atoms* in a *ferromagnetic* material cause them to behave something like tiny permanent magnets. They stick together and align themselves into small regions of more or less uniform alignment called *magnetic domains* or *Weiss domains*. *Magnetic domains* can be observed with a magnetic force microscope to reveal magnetic domain boundaries.

When a domain contains too many molecules, it becomes unstable and divides into two domains aligned in opposite directions so that they stick together more stably. *When exposed to a magnetic field, the domain boundaries move, so that the domains aligned with the magnetic field grow and dominate* the structure. When the magnetizing field is removed, the domains may not return to an unmagnetized state. This results in the *ferromagnetic* material being magnetized, forming a *permanent magnet*.

When magnetized strongly enough that the prevailing domain overruns all others to result in only one single domain, the material is magnetically saturated. When a magnetized ferromagnetic material is heated to the Curie point temperature, the molecules are agitated to the point that the magnetic domains lose the organization, and the magnetic properties they cause cease. When the material is cooled, this domain alignment structure spontaneously returns, in a manner roughly analogous to how a liquid can freeze into a crystalline solid.

Antiferromagnetism

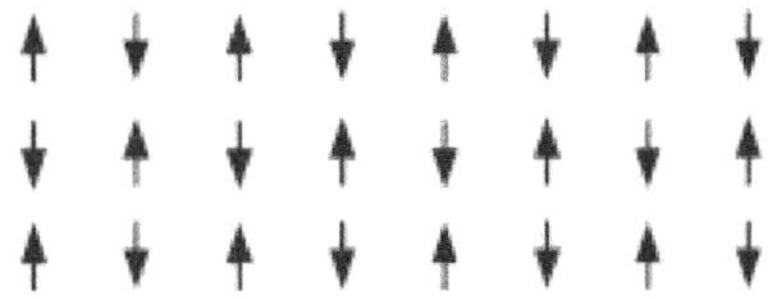

Antiferromagnetic ordering.

In an *antiferromagnet*, unlike a ferromagnet, there is a tendency for the *intrinsic magnetic moments* of neighboring *valence electrons* to point in *opposite* directions. *When all atoms are arranged in a substance so that each neighbor is anti-parallel, the substance is antiferromagnetic*. *Antiferromagnet*s have a zero net magnetic moment because adjacent opposite moment cancels out, meaning that no field is produced by them. *Antiferromagnets* are less common compared to the other types of behaviors and are mostly observed at low temperatures. In varying temperatures, antiferromagnets can be seen to exhibit diamagnetic and ferromagnetic properties.

In some materials, neighboring electrons prefer to point in opposite directions, but there is no geometrical arrangement in which *each* pair of neighbors is anti-aligned. This is called a canted antiferromagnet or spin ice and is an example of geometrical frustration.

Ferrimagnetism

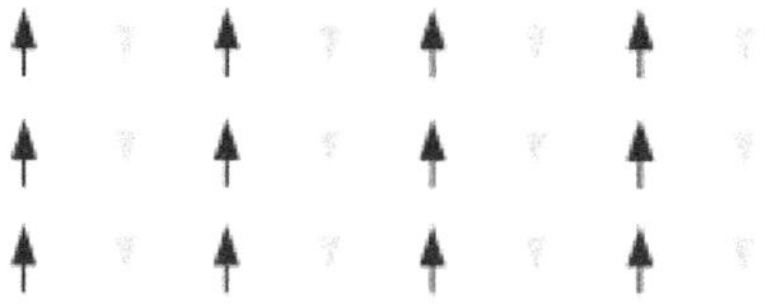

Ferrimagnetic ordering.

Like ferromagnetism, *ferrimagnets* retain their magnetization in the absence of a field. However, *like antiferromagnets, neighboring pairs of electron spins tend to point in opposite directions*. These two properties are not contradictory, because in the optimal geometrical arrangement, there is more *magnetic moment* from the sublattice of electrons that point in one direction, than from the sublattice that points in the opposite direction.

Most *ferrites* are *ferrimagnetic*. The first discovered magnetic substance, *magnetite, is a ferrite* and was originally believed to be a ferromagnet; Louis Néel disproved this, however, after discovering ferrimagnetism.

The metals iron, cobalt and nickel, and some rare earths, are *ferromagnetic*. Most of the other metals, are *diamagnetic* (e.g. sodium, aluminum, and magnesium) or *antiferromagnetic* (e.g. manganese). Diatomic gases are also almost exclusively *diamagnetic*, and not *paramagnetic*. However, the oxygen molecule, because of the involvement of π-orbitals, is an exception.

Feynman, R. P. (1963). The Magnetism of Matter.

The Feynman Lectures on Physics, Volume II, Chapter 34. Basic Books, New York, 34-7 to 34-8. https://www.feynmanlectures.caltech.edu/II_34.html.

...

"***34–7 Angular momentum in quantum mechanics***

We have already given you a relation between the *magnetic moment* and the *angular momentum*. ... But what do the magnetic moment and the angular momentum *mean* in quantum mechanics? *In quantum mechanics it turns out to be best to define things like magnetic moments in terms of the other concepts such as energy*, in order to make sure that one knows what it means. Now, it is easy to define a magnetic moment in terms of energy, because the energy of a moment in a magnetic field is, in the classical theory, **μ·B**. Therefore, *the following definition has been taken in quantum mechanics: If we calculate the energy of a system in a magnetic field and we find that it is proportional to the field strength (for small field), the coefficient is called the component of magnetic moment in the direction of the field.* (We don't have to get so elegant for our work now; we can still think of the magnetic moment in the ordinary, to some extent classical, sense.)

Now we would like to discuss the idea of *angular momentum in quantum mechanics*—or rather, the characteristics of what, in quantum mechanics, is called angular momentum. You see, when you go to new kinds of laws, you can't just assume that each word is going to mean exactly the same thing. You may think, say, "Oh, I know what angular momentum is. It's that thing that is changed by a torque." But what's a torque? In quantum mechanics we have to have new definitions of old quantities. It would, therefore, be legally best to call it by some other name such as "quantangular momentum," or something like that, because it is the angular momentum as defined in quantum mechanics. But if we can find a quantity in quantum mechanics which is identical to our old idea of angular momentum when the system becomes large enough, there is no use in inventing an extra word. We might as well just call it angular momentum. With that understanding, this odd thing that we are about to describe *is* angular momentum. It is the thing which in a large system we recognize as angular momentum in classical mechanics.

Here Feynman associates spin with an atom, and ignores the existence of electrons.

First, *we take a system in which angular momentum is conserved*, such as an *atom* all by itself in empty space. Now such a thing (like the earth spinning on its axis) could, in the ordinary sense, be spinning around any axis one wished to choose. And *for a given spin, there could be many different "states," all of the same energy, each "state" corresponding to a particular direction of the axis of the angular momentum.* So, in the classical theory, with a given angular momentum, there is an infinite number of possible states, all of the same energy.

It turns out *in quantum mechanics*, however, that several strange things happen. First, *the number of states in which such a system can exist is limited—there is only a finite number.* If the system is small, the finite number is very small, and if the system is large, the finite number gets very, very large. Second, *we cannot describe a "state" by giving the direction of its angular momentum, but only by giving the component of the angular momentum along some direction*—say in the z-direction. Classically, an object with a given total angular momentum J could have, for its z-component, any value from +J to −J. But *quantum-mechanically, the z-component of angular momentum can have only certain discrete values.* Any given system—a particular atom, or a nucleus, or anything—with a given energy, has a characteristic number j, and its z-component of angular momentum can only be one of the following set of values:

$$\begin{matrix} j\hbar \\ (j-1)\hbar \\ (j-2)\hbar \\ \vdots \\ -(j-2)\hbar \\ -(j-1)\hbar \\ -j\hbar \end{matrix} \qquad (34.23)$$

The largest z-component is j times ħ; the next smaller is one unit of ħ less, and so on down to −jħ. The number j is called "the spin of the system." (Some people call it the "total angular momentum quantum number"; but *we'll call it the "spin.")*

You may be worried that what we are saying can only be true for some "special" z-axis. But that is not so. For a system whose spin is j, the component of angular momentum along *any* axis can have only one of the values in (34.23). Although it is quite mysterious, we ask you just to accept it for the moment. We will come back and discuss the point later. You may at least be pleased to hear that the z-component goes from some number to minus the *same* number, so that we at least don't have to decide which is the plus direction of the z-axis. (Certainly, if we said that it went from +j to minus a different amount, that would be infinitely mysterious, because we wouldn't have been able to define the z-axis, pointing the other way.)

Now if the z-component of angular momentum must go down by integers from +j to −j, then j must be an integer. No! Not quite; twice j must be an integer. It is only the *difference* between +j and −j that must be an integer. *So, in general, the spin j is either an integer or a half-integer, depending on whether 2j is even or odd.* Take, for instance, a nucleus like lithium, which has a spin of three-halves, j = 3/2. Then the angular momentum around the z-axis, in units of ħ, is one of the following:

$$\begin{array}{c} +3/2 \\ +1/2 \\ -1/2 \\ -3/2. \end{array}$$

There are four possible states, each of the same energy, if the nucleus is in empty space with no external fields. If we have a system whose spin is two, then the z-component of angular momentum has only the values, in units of ħ,

$$\begin{array}{c} 2 \\ 1 \\ 0 \\ -1 \\ -2. \end{array}$$

If you count how many states there are for a given j, there are (2j + 1) possibilities. In other words, *if you tell me the energy and also the spin j, it turns out that there are exactly (2j + 1) states with that energy, each state corresponding to one of the different possible values of the z-component of the angular momentum.*

We would like to add one other fact. If you pick out any *atom* of known j at random and measure the z-component of the angular momentum, then you may get any one of the possible values, *and each of the values is equally likely.* All of the states are in fact single states, and each is just as good as any other. Each one has the same "weight" in the world. (We are assuming that nothing has been done to sort out a special sample.) This fact has, incidentally, a simple classical analog. If you ask the same question classically: What is the likelihood of a particular z-component of angular momentum if you take a random sample of systems, all with the same total angular momentum? —the answer is that all values from the maximum to the minimum are equally likely. (You can easily work that out.) The classical result corresponds to the equal probability of the (2j + 1) possibilities in quantum mechanics.

From what we have so far, we can get another interesting and somewhat surprising conclusion. In certain classical calculations the quantity that appears in the final result is the *square* of the magnitude of the angular momentum **J**—in other words, **J·J**. It turns out that it is often possible to *guess* at the correct quantum-mechanical formula by using the classical calculation and the following simple rule: Replace $J^2 = \mathbf{J \cdot J}$ by $j(j + 1)\hbar^2$. This rule is commonly used, and usually gives the correct result, but *not* always. We can give the following argument to show why you might expect this rule to work.

The scalar product $\mathbf{J \cdot J}$ can be written as

$$\mathbf{J \cdot J} = J_x^2 + J_y^2 + J_z^2.$$

Since it is a scalar, it should be the same for any orientation of the spin. Suppose we pick samples of any given atomic system at random and make measurements of J_x^2, or J_y^2, or J_z^2, the *average value* should be the same for each. (There is no special distinction for any one of the directions.) Therefore, the average of $\mathbf{J \cdot J}$ is just equal to three times the average of any component squared, say of J_z^2;

$$\langle \mathbf{J \cdot J} \rangle_{av} = 3\langle J_z^2 \rangle_{av}.$$

But since $\mathbf{J \cdot J}$ is the same for all orientations, its average is, of course, just its constant value; we have

$$\mathbf{J \cdot J} = 3\langle J_z^2 \rangle_{Av}. \tag{34.24}$$

If we now say that we will use the same equation for quantum mechanics, we can easily find $\langle J_z^2 \rangle_{av}$. We just have to take the sum of the $(2j + 1)$ possible values of J_z^2, and divide by the total number;

$$\langle J_z^2 \rangle_{Av} = \{j^2 + (j - 1)^2 + \cdots + (-j + 1)^2 + (-j)^2\}/\{2j + 1\}\ \hbar^2. \qquad (34.25).$$

For a system with a spin of 3/2, it goes like this:

$$\langle J_z^2 \rangle_{av} = \{(3/2)^2 + (1/2)^2 + (-1/2)^2 + (-3/2)^2\}/4\ \hbar^2 = 5/4\ \hbar^2.$$

We conclude that

$$\mathbf{J \cdot J} = 3\langle J_z^2 \rangle_{av} = 3\ 5/4\ \hbar^2 = 3/2\ (3/2 + 1)\hbar^2.$$

We will leave it for you to show that Eq. (34.25), together with Eq. (34.24), gives the general result

$$\mathbf{J \cdot J} = j(d + 1)\hbar^2. \tag{34.26}$$

Although we would think classically that the largest possible value of the z-component of $\mathbf{J}$ is just the magnitude of $\mathbf{J}$—namely, $\sqrt{(\mathbf{J \cdot J})}$ − *quantum mechanically the maximum of* J_z

is always a little less than that, because $j\hbar$ is always less than $\sqrt{j(j+1)}\hbar$. The angular momentum is never "completely along the z-direction."

34–8 The magnetic energy of atoms

Now we want to talk again about the *magnetic moment*. We have said that in quantum mechanics the *magnetic moment* of a particular atomic system can be written in terms of the angular momentum by Eq. (34.6);

$$\mu = -g\,(q_e/2m)\,\mathbf{J}, \qquad (34.27),$$

where $-q_e$ and m are the charge and mass of the electron.

An atomic magnet placed in an external magnetic field will have an *extra magnetic energy* which depends on the component of its *magnetic moment* along the field direction. We know that

$$U_{mag} = -\,\boldsymbol{\mu}\cdot\mathbf{B}. \qquad (34.28).$$

Choosing our z-axis along the direction of **B**,

$$U_{mag} = -\,\mu_z B. \qquad (34.29).$$

Using Eq. (34.27), we have that

$$U_{mag} = g\,(q_e/2m)\,J_z B.$$

Quantum mechanics says that J_z can have only certain values: $j\hbar$, $(j-1)\hbar$, ..., $-j\hbar$. Therefore, the *magnetic energy of an atomic system* is not arbitrary; it can have only certain values. Its maximum value, for instance, is

$$g\,(q_e/2m)\,\hbar j B.$$

The quantity $q_e\hbar/2m$ is usually given the name "the *Bohr magneton*" and written μ_B:

$$\mu_B = q_e\hbar/2m.$$

The possible values of the *magnetic energy* are

$$U_{mag} = g\mu_B B\, J_z/\hbar,$$

where $J_z/\hbar$ takes on the possible values j, (j−1), (j−2), ..., (−j + 1), −j.

> In other words, *the energy of an atomic system is changed when it is put in a magnetic field by an amount that is proportional to the field, and proportional to J_z. We say that the energy of an atomic system is "split into 2j + 1 levels" by a magnetic field.* For instance, an *atom* whose energy is U_0 outside a magnetic field and whose j is 3/2, will have four possible energies when placed in a field. We can show these energies by an energy-level diagram like that drawn in Fig. Sec. 34-5. Any particular *atom* can have only one of the four possible energies in any given field B. *That is what quantum mechanics says about the behavior of an atomic system in a magnetic field.*

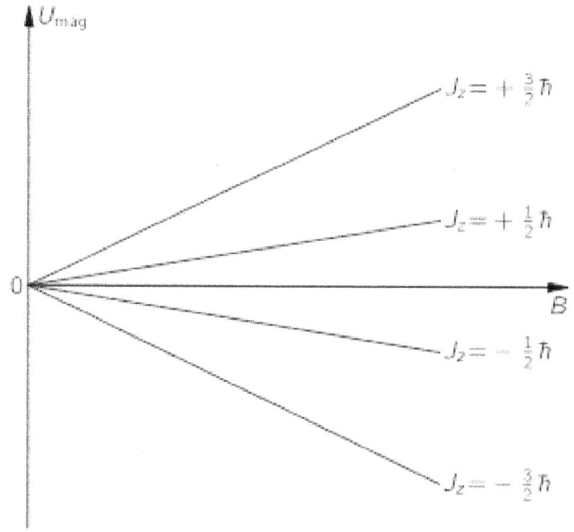

Figs. 34–5. The possible magnetic energies of an atomic system with a spin of 3/2 in a magnetic field **B**.

Here Feynman switches to associating the spin with an electron.

> The simplest "atomic" system is a *single electron.* The spin of an electron is 1/2, so there are two possible states: $J_z = \hbar/2$ and $J_z = -\hbar/2$. *For an electron at rest (no orbital motion),* the spin magnetic moment has a g-value of 2, so the magnetic energy can be either $\pm \mu_B B$. The possible energies in a magnetic field are shown in Fig. 34-6. Speaking loosely, we say that *the electron either has its spin "up" (along the field) or "down" (opposite the field).*

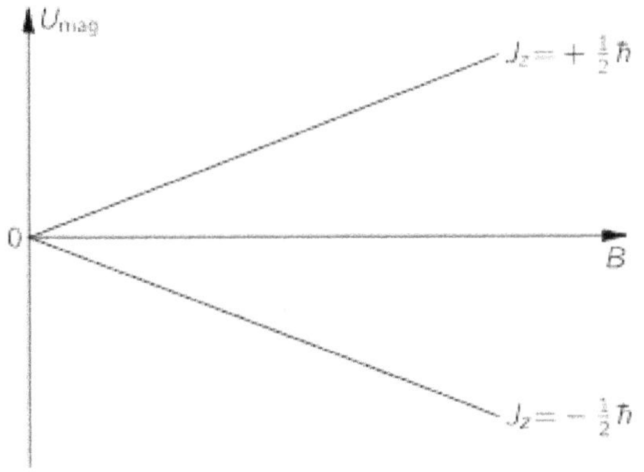

Figs. 34–6. The two possible energy states of an electron in a magnetic field **B**.

For systems with higher spins, there are more states. We can think that the *spin* is "up" or "down" or cocked at some "angle" in between, depending on the value of J_z.

We will use these quantum mechanical results to discuss the magnetic properties of materials in the next chapter."

Werner Karl Heisenberg (December 5, 1901–February 1, 1976).

Heisenberg was a German theoretical physicist and one of the key pioneers of quantum mechanics. He published his seminal work in 1925 in a breakthrough paper. In the subsequent series of papers with Max Born and Pascual Jordan, during the same year, this matrix formulation of quantum mechanics was substantially elaborated. He is known for the uncertainty principle, which he published in 1927. Heisenberg was awarded the 1932 Nobel Prize in Physics "for the creation of quantum mechanics".

Heisenberg also made important contributions to the theories of the hydrodynamics of turbulent flows, the atomic nucleus, ferromagnetism, cosmic rays, and subatomic particles. He was a principal scientist in the German nuclear weapons program during World War II. He was also instrumental in planning the first West German nuclear reactor at Karlsruhe, together with a research reactor in Munich, in 1957.

Heisenberg was born in Würzburg, Germany, to Kaspar Ernst August Heisenberg, and his wife, Annie Wecklein. His father was a secondary school teacher of classical languages who became Germany's only ordentlicher Professor (ordinarius professor) of medieval and modern Greek studies in the university system.

In his youth Heisenberg was a member and Scoutleader of the Neupfadfinder, a German Scout association and part of the German Youth Movement.

From 1920 to 1923, he studied physics and mathematics at the Ludwig Maximilian University of Munich under Arnold Sommerfeld and Wilhelm Wien; and at the Georg-August University of Göttingen with Max Born and James Franck and mathematics with David Hilbert. In June 1922, Sommerfeld took Heisenberg to Göttingen to attend the Bohr Festival, because Sommerfeld knew of Heisenberg's interest in Niels Bohr's theories on atomic physics. At the event, Bohr was a guest lecturer and gave a series of comprehensive lectures on quantum atomic physics and Heisenberg met Bohr for the first time.

Heisenberg's doctoral thesis, the topic of which was suggested by Sommerfeld, was on turbulence; the thesis discussed both the stability of laminar flow and the nature of turbulent flow. The problem of stability was investigated by the use of the Orr–Sommerfeld equation, a fourth order linear differential equation for small disturbances from laminar flow. He received his doctorate in 1923.

At Göttingen, under Born, he completed his habilitation in 1924 with a Habilitationsschrift (habilitation thesis) on the anomalous Zeeman effect.

From 1924 to 1927, Heisenberg was a Privatdozent at Göttingen, meaning he was qualified to teach and examine independently, without having a chair. From September 17, 1924 to May 1, 1925, under an International Education Board Rockefeller Foundation fellowship,

Heisenberg went to do research with Niels Bohr, director of the Institute of Theoretical Physics at the University of Copenhagen.

In Copenhagen, Heisenberg and Hans Kramers collaborated on a paper on dispersion, or the scattering from atoms of radiation whose wavelength is larger than the atoms. They showed that the successful formula Kramers had developed earlier could not be based on Bohr orbits, because the *transition frequencies* are based on level spacings which are not constant. The frequencies which occur in the Fourier transform of sharp classical orbits, by contrast, are equally spaced. But these results could be explained by a semi-classical virtual state model: the incoming radiation excites the valence, or outer, electron to a virtual state from which it decays. In a subsequent paper Heisenberg showed that this virtual oscillator model could also explain the polarization of fluorescent radiation.

These two successes, and the continuing failure of the Bohr–Sommerfeld model to explain the outstanding problem of the anomalous Zeeman effect, led Heisenberg to use the virtual oscillator model to try to calculate *spectral frequencies*. The method proved too difficult to immediately apply to realistic problems, so Heisenberg turned to a simpler example, the anharmonic oscillator.

The dipole oscillator consists of a simple harmonic oscillator, which is thought of as a charged particle on a spring, perturbed by an external force, like an external charge. The motion of the oscillating charge can be expressed as a Fourier series in the frequency of the oscillator. Heisenberg solved for the quantum behavior by two different methods. First, he treated the system with the virtual oscillator method, calculating the transitions between the levels that would be produced by the external source. He then solved the same problem by treating the anharmonic potential term as a perturbation to the harmonic oscillator and using the perturbation methods that he and Born had developed. Both methods led to the same results for the first and the very complicated second order correction terms. This suggested that behind the very complicated calculations lay a consistent scheme. Heisenberg returned to Göttingen and, with Max Born and Pascual Jordan over a period of about six months, developed the matrix mechanics formulation of quantum mechanics.

In his 1925 paper, which assumes that the reader is familiar with Kramers-Heisenberg transition probability calculations, Heisenberg set out to *try to construct a theory of quantum mechanics in which only relationships among observable quantities occur*. In place of assigning to the electron a point in space as a function of time he *assigned to the electron an emitted radiation;* where the observables are the *energies W(n) of the (Bohr) stationary states*, together with the associated (Einstein-Bohr) *frequencies v* and *amplitudes* which characterize the radiation emitted in the transition between the stationary states. Recognizing that quantum theory describes transitions between two stationary states he substituted two variables in place of one in the classical theory. He justified this replacement by an appeal to *Bohr's correspondence principle* and the *Pauli doctrine* that quantum mechanics must be limited to observables. In order to calculate the *energy* of a harmonic oscillator, in which the *amplitudes* do not combine in the same way as the classical harmonics, but rather in accordance with the *Ritz combination principle,* instead of reinterpreting x(t) as a *sum* over transition components, he represented the position by the *set* of *transition components*, thereby introducing non-commutative multiplication of matrices by physical reasoning, based on the *correspondence principle*, despite the fact that he was not then familiar with the mathematical theory of matrices.

After addressing what he referred to as the kinematic of the quantum theory, Heisenberg turned to mechanical problems aiming at the determination of the *amplitudes, frequencies* and *energies in order to construct the line spectrum of an atom from the given force on the electron*. He achieved this by translating the old quantum condition that fixes the properties of the states to a new condition that fixes the properties of the transitions between states, replacing the differential in the equation for the *phase* integral by a difference, resulting in an equation that has a simple quantum theoretical connection to the *Kramer's dispersion theory*.

On July 9, Heisenberg gave Born his paper to review and submit for publication. Heisenberg's seminal paper was published in September 1925. [Heisenberg, W. (July, 1925). Über quantentheoretische Umdeutung kinematischer und mechanischer Beziehungen. (On the Quantum-Theoretical Re-interpretation of Kinematic and Mechanical Relations.) *Zeit. Physik*, 33, 879-93.] This is the first paper in the famous trilogy which launched the matrix mechanics formulation of quantum mechanics.

When Born read the paper, he recognized the formulation as one which could be transcribed and extended to the systematic language of matrices, which he had learned from his study under Jakob Rosanes at Breslau University. Up until this time, matrices were seldom used by physicists; they were considered to belong to the realm of pure mathematics. Gustav Mie had used them in a paper on electrodynamics in 1912; and Born had used them in his work on the lattice theory of crystals in 1921. While matrices were used in these cases, the algebra of matrices with their multiplication did not enter the picture as they did in the matrix formulation of quantum mechanics.

Born, with the help of his assistant and former student Pascual Jordan, began immediately to make the transcription and extension, and they submitted their results for publication; the paper was received for publication just 60 days after Heisenberg's paper. [Born, M. & Jordan, P. (December, 1925). Zur Quantenmechanik. (On Quantum Mechanics.) *Zeit. Phys.*, 34, 858-88.] A follow-on paper was submitted for publication before the end of the year by all three authors. [Born, M., Heisenberg, W. & Jordan, P. (August, 1926). Zur Quantenmechanik II. (On Quantum Mechanics II.) *Zeit. Phys.*, 35, 557-615.]

On May 1, 1926, Heisenberg began his appointment as a university lecturer and assistant to Bohr in Copenhagen. It was in Copenhagen, in 1927, that Heisenberg developed his *uncertainty principle*, while working on the mathematical foundations of quantum mechanics. On February 23, Heisenberg wrote a letter to fellow physicist Wolfgang Pauli, in which he first described his new principle. In his paper on the principle, Heisenberg used the word "Ungenauigkeit" (imprecision), not "uncertainty", to describe it.

In 1928, Heisenberg was appointed ordentlicher Professor (professor ordinarius) of theoretical physics and head of the department of physics at the University of Leipzig; he gave his inaugural lecture there on 1 February 1928. In his first paper published from Leipzig, Heisenberg used the *Pauli exclusion principle* to solve the mystery of ferromagnetism.

In 1928, the British mathematical physicist Paul Dirac had derived his relativistic wave equation of quantum mechanics, which implied the existence of positive electrons, later to be named positrons. In early 1929, Heisenberg and Pauli submitted the first of two papers laying the foundation for *relativistic quantum field theory*. Also in 1929, Heisenberg went on a lecture tour of China, Japan, India, and the United States. In the spring of 1929, he was a visiting lecturer at the University of Chicago, where he lectured on quantum mechanics.

In 1932, from a cloud chamber photograph of cosmic rays, the American physicist Carl David Anderson identified a track as having been made by a positron. In mid-1933, Heisenberg presented his theory of the positron. His thinking on Dirac's theory and further development of the theory were set forth in two papers. The first, [Heisenberg, W. (March, 1934). Bemerkungen zur Diracschen Theorie des Positrons. (Remarks on the Dirac theory of positron.) *Zeit. Phys*., 90, 3-4, 209-31] was published in 1934, and the second, Heisenberg, W., Euler, H. (1936). Folgerungen aus der Diracschen Theorie des Positrons. *Zeit. Phys.,* 98, 714-32; https://doi.org/10.1007/BF01343663], was published in 1936.

In these papers Heisenberg was the first to reinterpret the Dirac equation as a "classical" field equation for any point particle of spin ħ/2, itself subject to quantization conditions involving anti-commutators. Thus, reinterpreting it as a quantum field equation accurately describing electrons, Heisenberg put matter on the same footing as electromagnetism: as being described by *relativistic quantum field equations* which allowed the possibility of particle creation and destruction. (Hermann Weyl had already described this in a 1929 letter to Albert Einstein.)

In 1928, Albert Einstein nominated Heisenberg, Born, and Jordan for the Nobel Prize in Physics. The announcement of the Nobel Prize in Physics for 1932 was delayed until November 1933. It was at that time that it was announced Heisenberg had won the Prize for 1932 "for the creation of quantum mechanics, the application of which has, inter alia, led to the discovery of the allotropic forms of hydrogen".

Heisenberg enjoyed classical music and was an accomplished pianist. His interest in music led to meeting his future wife. In January 1937, Heisenberg met Elisabeth Schumacher (1914–1998) at a private music recital. Elisabeth was the daughter of a well-known Berlin economics professor, and her brother was the economist E. F. Schumacher, author of Small Is Beautiful. Heisenberg married her on April 29. Fraternal twins Maria and Wolfgang were born in January 1938, whereupon Wolfgang Pauli congratulated Heisenberg on his "pair creation"—a word play on a process from elementary particle physics, pair production. They had five more children over the next 12 years: Barbara, Christine, Jochen, Martin and Verena. In 1936 he bought a summer home for his family in Urfeld am Walchensee, in southern Germany.

Heisenberg was involved in the German nuclear weapons program, known as *Uranverein*, which was formed on September 1, 1939, the day World War II began, The Kaiser-Wilhelm Institut für Physik (KWIP, Kaiser Wilhelm Institute for Physics) in Berlin-Dahlem, was placed under the authority of the Heereswaffenamt (HWA, Army Ordnance Office), and the military control of the nuclear research commenced. In February 1942, at a scientific conference called by the Army Weapons Office, Heisenberg presented a lecture to Reichs officials on energy acquisition from nuclear fission entitled *Die theoretischen Grundlagen für die Energiegewinning aus der Uranspaltung* (The theoretical basis for energy generation from uranium fission). He lectured on the enormous energy potential of nuclear fission, stating that 250 million electron volts could be released through the fission of an atomic nucleus. Heisenberg stressed that pure U-235 had to be obtained to achieve a chain reaction; and explored various ways of obtaining isotope $^{235}_{92}U$ in its pure form, including uranium enrichment and an alternative layered method of normal uranium and a moderator in a machine.

In April 1942, Reichs Minister Rust decided to move the nuclear project from the Physics Institute to the Reichs Research Council; returning the Physics Institute to the Kaiser Wilhelm Society, and naming Heisenberg as Director at the Institute. With this appointment, Heisenberg obtained his first professorship. Heisenberg still also had his department of physics at the University of Leipzig.

In February 1943, Heisenberg was appointed to the Chair for Theoretical Physics at the Friedrich-Wilhelms-Universität (today, the Humboldt-Universität zu Berlin). In April, his election to the Preußische Akademie der Wissenschaften (Prussian Academy of Sciences) was approved. That same month, he moved his family to their retreat in Urfeld as Allied bombing increased in Berlin.

The Alsos Mission, an Allied effort to determine if the Germans had an atomic bomb program and to exploit German atomic related facilities, research, material resources, and scientific personnel for the benefit of the US, generally moved into areas which had just come under control of the Allied military forces, but sometimes they operated in areas still under control by German forces. The Kaiser-Wilhelm-Institut für Physik (KWIP, Kaiser Wilhelm Institute for Physics) had been bombed so it had mostly been moved in 1943 and 1944 to Hechingen and its neighboring town of Haigerloch, on the edge of the Black Forest, which eventually became included in the French occupation zone. This allowed the American task force of the Alsos Mission to take into custody a large number of German scientists associated with nuclear research. In January 1945, Heisenberg, with most of the rest of his staff, moved from the Kaiser-Wilhelm Institut für Physik to the facilities in the Black Forest.

On 30 March, 1945, the Alsos Mission reached Heidelberg, where important scientists were captured. Their interrogation revealed that Otto Hahn was at his laboratory in Tailfingen, while Heisenberg and Max von Laue were at Heisenberg's laboratory in Hechingen, and that the experimental natural uranium reactor that Heisenberg's team had built in Berlin had been moved to Haigerloch. Thereafter, the main focus of the Alsos Mission was on these nuclear facilities in the Württemberg area. Heisenberg was captured and arrested in Urfeld, on May 3, 1945, in an alpine operation in territory still under control by German forces.

Germany surrendered on May 7. Heisenberg would not see his family again for eight months, as he was moved across France and Belgium and flown to England on July 3, 1945. Nine prominent German scientists, including Heisenberg, who were members of the *Uranverein* were captured and incarcerated at Farm Hall in England. The facility had been a safe house of the British foreign intelligence MI6. During their detention, their conversations were recorded. Conversations thought to be of intelligence value were transcribed and translated into English. The Farm Hall transcripts reveal that Heisenberg, along with other physicists interned at Farm Hall including Otto Hahn and Carl Friedrich von Weizsäcker, were glad the Allies had won the war. Heisenberg told other scientists that he had never contemplated a bomb, only an atomic pile to produce energy.

On 3 January 1946, the Operation Epsilon detainees were transported to Alswede in Germany. Heisenberg settled in Göttingen, which was in the British zone of Allied-occupied Germany. Following the Kaiser Wilhelm Society's obliteration by the Allied Control Council, and the establishment of the Max Planck Society in the British zone, the Kaiser Wilhelm Institute for Physics was renamed, and Heisenberg became the director of the Max Planck Institute for Physics.

In 1951, Heisenberg agreed to become the scientific representative of the Federal Republic of Germany at the UNESCO conference, with the aim of establishing a European laboratory for nuclear physics. Heisenberg's aim was to build a large particle accelerator, drawing on the resources and technical skills of scientists across the Western Bloc. On July 1, 1953 Heisenberg signed the convention that established CERN on behalf of the Federal Republic of Germany. Although he was asked to become CERN's founding scientific director, he declined. Instead, he was appointed chair of CERN's science policy committee and went on to determine the scientific program at CERN.

In 1958, the Max Planck Institute for Physics was moved to Munich and renamed Max Planck Institute for Physics und Astrophysics, of which Heisenberg was a co-director, and then sole director until he resigned his directorship on December 31, 1970.

> [Heisenberg gave a joint lecture with Dirac at the old Cavendish Laboratory on May 22, 1963, which the author attended. [Underwood, T. G. (1962-3). *Cambridge University lecture notebook 6*.]

Heisenberg died age 74 of kidney cancer at his home, on February 1, 1976. Heisenberg is buried in Munich Waldfriedhof.

Heisenberg, W. (September, 1928). Zur Theory of Ferromagnetismus. (On the theory of ferromagnetism.)

Zeit. Phys., 49, 619–36 (in German); https://doi.org/10.1007/BF01328601; translation by D. H. Delphenich; https://neo-classical-physics.info/uploads/3/0/6/5/3065888/ Heisenberg _-_on_the_theory_of_ferromagnetism.pdf.

Received May 20, 1928.

Leipzig, Institut für theoretische Physik der Universität.

With 1 Figure.

In another brilliant paper, Heisenberg noted that empirical results exhibit *ferromagnetism* as an entirely similar state of affairs to what was previously observed in the spectrum of the helium atom; and it seemed to follow from the levels in the helium atoms that *a powerful interaction prevailed between the spin directions of two electrons* that led to the splitting of the level structure into systems of singlets and triplets. He also noted that this was closely related to explaining *ferromagnetic* phenomena as being implied by the *exchange phenomenon* (resulting from local *quantum entanglement*). He proceeded to show that the *Coulomb interaction*, together with the *Pauli principle*, succeeded in evoking the same effects as the molecular field that Weiss postulated, noting that it was only in recent times that mathematical methods were developed for the treatment of such a complicated problem in the important investigations of Wigner, Hund, and Heitler and London. As a first approximation, Heisenberg assumed that the lattice separations were very large, and that every electron belonged to its own atom, and *applied Heitler-London's calculations to the case of 2n electrons in a state of interaction*, finding 2n electrons in 2n positionally different quantum cells. Due to their smallness, he was able to leave the magnetic interactions outside of consideration, and showed that *the spin moments of all electrons become partly parallel and partly anti-parallel as a result of the exchange processes*. By adding the fundamental Pauli principle to this, viz., that the *eigenfunctions* of the total system should be *anti-symmetric* in all electrons, he showed that an entirely well-defined *total magnetic moment* belonged to each level value of the perturbed system, and there were (2n)! levels in the unperturbed system (*ignoring the Pauli principle and spin*). He then showed that a *statistical treatment of ferromagnetism was possible when all energy values had been calculated.* Heisenberg concluded that *an atom in a lattice can only be exchanged with its "neighbors"*; exchanges with atoms that lie further away that the "neighboring atoms" could then be neglected. The *number of "neighbors" of an atom* is, e.g., 1 in a molecular lattice of diatomic molecules, 2 in a linear chain, 4 in a quadratic surface lattice, 6 in a simple cubic lattice, 8 *in a cubic, space-centered lattice*, and 12 in a cubic, face-centered lattice. By assuming a distribution of energy values about the mean had the approximate form of a Gaussian error curve, Heisenberg showed that small or negative values of the constant β [= zJ_0/kT)] resulted in *paramagnetism*; and that *ferromagnetism was only possible for lattice types for which an atom had at least eight neighbors*, which was the case for Fe, Co, Ni, whose lattices were all cubic, some of which were space-centered ($z = 8$) and some of which were face-centered ($z = 12$). He concluded that two conditions were necessary for the appearance of *ferromagnetism*: the *crystal lattice* must be a type such that *any atom has at least 8 neighbors*; and the *principal quantum number* of the electrons that are responsible for magnetism must be $n \geq 3$.

Abstract.

Weiss's molecular forces will be attributed to a *quantum-mechanical exchange phenomenon*, and indeed, it will be treated as the *exchange process* that was successfully enlisted in recent times by Heitler and London in order to interpret homopolar valence forces.

Introduction.

Ferromagnetic phenomena have been interpreted in a formally satisfying way by the well-known Weiss theory*.

* Weiss, P. (1907). *Journ. de phys.*, 4, 6, 661 and (1908). *Phys. Zeit.*, 9, 358.

That theory is based upon the assumption that every atom in a crystal experiences a directed force from the remaining atoms of the lattice that should be proportional to the number of already-directed atoms. By contrast, the origin of these atomic fields was unknown, and several obstacles stood in the way of any interpretation of the Weiss forces on the basis of classical theory: magnetic interactions between atoms are always a few orders of magnitude smaller than the atomic fields that follow from ferromagnetic experiments. Indeed, electric interactions lead to the correct order of magnitude; however, one would rather expect that the electrical interactions of two atoms would be proportional to the square of the cosine of their mutual angle of inclination, rather than the cosine, which contradicts the assumptions of Weiss's theory. Other complications were discussed more thoroughly by Lenz**,

** Lenz, W. (1920). *Phys. Zeit.*, 21, 613.

and Ising***

*** Ising, E. (1925). *Zeit. Phys.*, 31, 253.

who succeeded in showing that the assumption of directed, sufficiently large forces between any two neighboring atoms of a chain did not suffice to generate *ferromagnetism*.

The ferromagnetic complex of questions has entered a new arena with the Uhlenbeck-Goudsmit theory of *spin electrons*. In particular, it follows from the known factor g = 2 in the Einstein-de Haas effect (which was, in fact, measured for ferromagnetic substances) that in a ferromagnetic crystal *only the magnetic eigenmoment of the electrons is oriented*, but not, by any means, the atoms. Thus, the possibility of interpreting the Weiss forces as electrical interactions, independent of the relative spin directions of the electrons, goes away, since we know that such forces do not exist. Furthermore, by applying Pauli-Fermi-Dirac statistics, Pauli†

† Pauli, W. (1927). Über Gasentartung und Paramagnetismus. (On Gas Degeneration and Paramagnetism.) *Zeit. Phys.*, 41, 81-102; https://doi.org/10.1007/BF01391920.

has been able to show that *paramagnetism or diamagnetism will always result from neglecting the interaction of the electrons in a metal.*

§ 1. ***A model for the foundations of the theory***. The basic idea of the theory that we seek here is this: Empirical results exhibit *ferromagnetism* as an entirely similar state of affairs to what was previously observed in the spectrum of the helium atom. At the time, *it seemed to follow from the levels in the helium atoms that a powerful interaction prevailed between the spin directions of two electrons that led to the splitting of the level structure into systems of singlets and triplets*. At the time, this difficulty could be resolved by verifying that the apparently large interaction would emerge indirectly from a *resonance* or *exchange phenomenon* that would be characteristic of all quantum-mechanical systems of identical particles. *This is also closely related to explaining ferromagnetic phenomena as being implied by this exchange phenomenon.* We will attempt to show that the *Coulomb interaction*, together with the *Pauli principle*, succeeds in evoking the same effects as the molecular field that Weiss postulated. It was only in recent times that mathematical methods were developed for the treatment of such a complicated problem in the important investigations of Wigner*, Hund**, Heitler and London***.

* Wigner, E. (1927). *Zeit. Phys.* 40, 883; (1927). Einige Folgerungen aus der Schrödingerschen Theorie für die Termstrukturen. (Some conclusions from Schrödinger's theory for term structures.) *Ibid.*, 43, 624-52; https://doi.org/10.1007/BF01397327.

** Hund, F. (1927). Symmetriecharaktere von Termen bei Systemen mit gleichen Partikeln in der Quantenmechanik. (Symmetry characters of terms in systems with equal particles in quantum mechanics.) *Zeit. Phys.*, 43, 788-804; http://dx.doi.org/10.1007/ BF01397248.

*** Heitler, W. & London, F. (June, 1927). Wechselwirkung neutraler Atome und homöopolare Bindung nach der Quantenmechanik. (Interaction of Neutral Atoms and Homopolar Binding in Quantum Mechanics.) *Zeit. Phys.*, 44, 2, 455-72, cited as I in what follows [See above]; Heitler, W. (1927). *Ibid.*, 46, 47 (cited as II); (1928). *ibid.*, 47, 835 (cited as III); London, F. (1928). *Ibid.*, 46, 455.

Before I go into the actual calculations, I would like to give a brief overview of the methods of approximation that can be applied in the treatment of *electronic motions in metals*.

Method I. From Pauli (*loc. cit.*) and Sommerfeld****, in the first approximation, *electrons can be assumed to be completely free*.

**** Sommerfeld, A. (1928). *Zeit. Phys*., 47, 1; cf., also Houston, W. V. (1928). *Ibid*., 47, 33, and Eckart, C. (1928). *Ibid*., 47, 38.

In the second approximation, one might, perhaps, add in the *interactions with the lattice points* [Houston†].

† Houston, W. V. (1928). *Zeit. Phys*., 48, 449.

The interaction of electrons with each other is neglected completely.

Method II. As a first approximation, *one calculates the motion of an electron in a force field* (that, by no means, needs to be small) *that is periodic* (in three directions). In the next approximation, *one might perhaps consider the perturbations that arise from the deviations from periodicity in the lattice. The treatment of the interaction of electrons with each other encounters the same difficulties here as it does in method I.*

Method III. In the first approximation, *one thinks of the lattice separations as being very large and assumes that every electron thus belongs to its own atom*. In the next approximation, *one considers the exchange of electrons that move in the unperturbed system with equal energies at different points*, which was first considered by Heitler and London (*loc. cit*. I). States in which more electrons are found in comparison to the number that is found in one atom in the unperturbed system will not be considered in this approximation.

The difference between these three methods becomes clearer when we explain it by another example, namely, the *hydrogen molecule*, which was treated rigorously by Heitler and London (*loc. cit*. In method I, *the electrons were, once more, first treated as free*, which would naturally not yield any suitable starting solution for the calculation. In method II, *one starts with the solutions of the two-center problem* [cf., Hund*].

* Hund, F. (1927). *Zeit. Phys*., 40, 742.

A level that describes electron 1 as being in a 1s state around nucleus *a* and electron 2 as being a 1s state around nucleus b in the limiting case of infinite nuclear separations would split into four levels (1 to 4) that might be characterized by the table:

	Nucleus a	*Nucleus b*
1	1	2
2	2	1
3	1,2	–
4	–	1,2

The interaction of the two electrons will first be considered in higher-order approximations. Method II will be directly identical to the method that was employed by Heitler and London. Only levels 1 and 2 included in an unperturbed system. It will be assumed that levels 3 and 4 lead to substantially higher-lying energy values. The diversity of levels in unperturbed systems will then be more meager for method III than it is for methods I or II.

There is, indeed, no argument, a priori, for preferring any of the three approximation procedures over the other ones. Method I will be most closely applicable to metals of very large conductivity, while method III is most applicable to metals of very feeble conductivity. Method II is in the middle between these two limiting cases.

I have based the following calculations upon method III, since only it can permit a quantitative treatment of the electron interactions.

§ 2. ***The distribution of the level values***. The following calculations define a simple generalization of the Heitler-London investigations (*loc. cit.* I) to the case of 2n electrons in a state of interaction (the number of electrons is now assumed to be even, upon purely formal grounds). One will then find 2n electrons in 2n different (indeed, they are not different energetically, but positionally) quantum cells.

We shall first assume only that the quantum numbers of the electrons in their atoms are the same for all atoms. Other stationary states of the unperturbed system will not be considered, since it will be assumed that they would lead to much high energy values.

One is then dealing with the determination of the energy values of the stationary states of the total system, which will belong to the state that was described above when the Coulomb interaction of the charges in an atom with the charges of any other atom is considered to be a perturbation. Due to the great computational complications that have appeared up to now, it will only be possible for us to attempt the perturbation calculations up to the first approximation. Whether this first approximation will actually be successful for the cases that nature presents must remain undecided. We take the eigenfunctions of the unperturbed system to be, say, products of the Schrodinger eigenfunctions of the *hydrogen atom*, or better yet, *the eigenfunctions that correspond to the rest of the atoms considered*, just like in the cited paper of Heitler and London; it is entirely superfluous to repeat those Ansätze here explicitly.

[An *ansatz* is the establishment of the starting equation(s), the theorem(s), or the value(s) describing a mathematical or physical problem or solution. It typically provides an initial estimate or framework to the solution of a mathematical problem, and can also take into consideration the boundary conditions (in fact, an ansatz is sometimes thought of as a "trial answer" and an important technique in solving differential equations).]

These eigenfunctions are certainly not orthogonal, but the deviation from the usual treatment that is required first differs in the terms of order two, so we can apply the usual method of treating things in the case of orthogonal eigenfunctions. *The electrons of an atom can be exchanged with those of any other atom as a result of perturbations.* As long as one overlooks the perturbing terms of order two, only simple transpositions between two neighboring atoms will occur. If one chooses the simplest case to be the one in which any atom in the unperturbed system possesses one valence electron then the "exchange terms" for the perturbing energy will reduce to the expressions that were given by Heitler and London:

$$J_{(kl)} = 1/2 \int \psi_k^\kappa \psi_k^\lambda \psi_l^\kappa \psi_l^\lambda \, (2\varepsilon^2/r_{kl} + 2\varepsilon^2/r_{\kappa\lambda} - \varepsilon^2/r_{\kappa k} - \varepsilon^2/r_{\kappa l} - \varepsilon^2/r_{\lambda l}) \, d\tau_k d\tau_l. \qquad (1)$$

Here, k and *l* mean the numbers of *two electrons*, while κ and λ mean the numbers of the *remaining atoms to which k and l belong in the unperturbed state.* A very important constant that enters into the perturbation calculations is the purely "static" interaction:

$$J_E = \int d\tau_1 d\tau_2 \dots d\tau_{2n} \, (\psi_1^{\,1})^2 (\psi_2^{\,2})^2 \dots (\psi_{2n}^{\,2n})^2 \left[\sum_{\substack{k,l \\ k>l}} \varepsilon^2/r_{kl} + \sum_{\substack{\kappa,\lambda \\ \kappa>\lambda}} \varepsilon^2/r_{\kappa\lambda} - \sum_{\substack{k,\lambda \\ k<\lambda}} \varepsilon^2/r_{k\lambda}\right] \qquad (2)$$

Due to their smallness, we can leave the magnetic interactions completely outside of consideration. Nevertheless, *the spin moments of all electrons will become partly parallel and partly anti-parallel as a result of the exchange processes.* If one adds the fundamental Pauli principle to this, viz., that the eigenfunctions of the total system should be *anti-symmetric* in all electrons, then an entirely well-defined *total magnetic moment* will belong to each level value of the perturbed system that will be characterized by the rotational moment $s\hbar/2\pi$ of the system. In all, there will be (2n)! levels in the unperturbed system (if one ignores the Pauli principle and spin). *A statistical treatment of ferromagnetism will be possible when all energy values that belong to a given value of s have been calculated.* This problem is generally not soluble in this form, since 2n is a very large number. We can only hope to obtain a general insight into the distribution of the eigenvalues for a given s. In what follows, we will calculate the *number* of levels, the center of mass of the energy (thus, the mean value of energy for a given s), and the *mean-square variance* of the energy about that mean value. We shall then make the generally somewhat arbitrary assumption that, in the first approximation, the energy values are distributed around the mean in a Gaussian error curve, such that the breadth of the error curve is calculated from the mean-square variance.

From the investigations of Wigner, Hund, and Heitler (*loc. cit.*) and the assumption of the Pauli principle, every value s of the total *spin* moment belongs to one system of levels ("σ") that are characterized by the well-defined partitioning of 2n into summands:

$$2n = \underbrace{2 + 2 + \ldots + 2}_{(n-s)\ times} + \underbrace{1 + 1 + \ldots + 1}_{2s\ times}. \tag{3}$$

The partition of the "reciprocal" system is then called simply:

$$2n = (n - s) + (n + s). \tag{4}$$

Heitler (*loc. cit.* II) has given the following formula for the mean value – i.e., the "center of mass of energy" – of the system σ:

$$E_\sigma = 1/f_\sigma \sum_P \chi_\sigma^P J_P. \tag{5}$$

In this, χ_σ^P means the group character that belongs to the permutation P, and $f_\sigma = \chi_\sigma^E$ is the *number* of levels in the system. The energy of the unperturbed levels is omitted, as an additive constant. We further calculate the mean-square variance ΔE^2 of the energy about the value E_σ: The energy value is given by the square root of an equation of degree f_σ that one obtains when one sets the following determinant equal to zero:

$$\ldots \tag{6}$$

... It next yields for the reciprocal system of levels:

$$\chi^E_{n-s,n+s} = (2n)!\,(2s + 1)/\{(n - s)!(n + s + 1)!\} \tag{13}$$
$$\chi^{(12)}_{n-s,n+s} = \ldots$$
$$\chi^{(123)}_{n-s,n+s} = \ldots$$
$$\chi^{(12)(34)}_{n-s,n+s} = \ldots$$

The characters of the system of levels that actually is present differ from the characters of their reciprocals that are employed here only by their signs. Indeed, the character of the reciprocal system is equal to (equal and opposite to, resp.) that of the system itself when the permutation P arises from an even (odd, resp.) number of transpositions.

Up to this point, everything is true in complete generality, with no relationship to any special assumptions that we might make about the crystal lattice or the atomic structure of the ferromagnetic substance.

In order to be able to calculate, we must now specialize our assumptions somewhat further. It follows from formula (1) and the calculations of Heitler and London that $J_{(12)}$ decreases exponentially with increasing distance. *For the most part, one can exchange an atom in a lattice only with its "neighbors"*; exchanges with atoms that lie further away that the "neighboring atoms" will then be neglected. The *number of "neighbors" of an atom* [z] is, e.g., 1 in a molecular lattice of diatomic molecules, 2 in a linear chain, 4 in a quadratic surface lattice, 6 in a simple cubic lattice, 8 in a cubic, space-centered lattice, and 12 in a cubic, face-centered lattice.

We shall make only the assumption that all non-vanishing exchange terms J_P should be equal (we call that value J_0). That must be case when the remaining atoms are non-magnetic; i.e., centrally-symmetric. We then now calculate E_σ and ΔE_σ^2 for a lattice in which every atom has z neighbors. ...

...

§ 3. ***Statistics: connection with Weiss's formulas***. The following arguments will be founded upon the aforementioned, generally somewhat arbitrary, assumption that the distribution of energy values about the mean has the approximate form of a Gaussian error curve. ...

...

For small or negative values of the constant β [= zJ_0/kT)], one will get *paramagnetism*. *Ferromagnetism* enters in when the tangent of the curve II for y = 0 subtends a smaller angle with the x-axis than the tangent of I; the influence of the cubic terms is first ignored in this. The *condition for ferromagnetism* then reads:

$$\beta(1 - \beta/z) \geq 2. \tag{25}$$

This condition can be fulfilled only for high values of z. The maximal value on the left-hand side of (25) is ($\beta_{max} = z/2$) z/2(1 – 1/2), and it follows that:

$$z \geq 8. \tag{26}$$

Ferromagnetism is then possible only for lattice types for which an atom has at least eight neighbors. That is the case for Fe, Co, Ni, whose lattices are all cubic, some of which are space-centered (z = 8) and some of which are face-centered (z = 12). ...

...

One must improve the provisional theory that was attempted here by calculating the higher variances of the mean ΔE^3, ΔE^4, etc. and correspondingly construct improved distribution curves for the values of the terms. Corresponding higher powers of β would appear on the left-hand side of equation (25) in this improved theory; the left-hand side of (25) is thus actually a transcendental function of β. ... Such a more precise examination of the distribution curve would also most likely displace the limiting value (26) of z. However, nothing in our results would change very much qualitatively.

...

§ 4. ***Magnitudes and signs of the "molecular field"***. The constant β must have order of magnitude 1, in order for *ferromagnetism* to be possible; one must then have $J_0 \sim kT$, where T will assume values on the order of 10^3 degrees for Fe, Co, Ni. *It follows that* *$J_0 \sim 10^{-13}$ erg ~ 1/100 the energy of the hydrogen ground state. That is just the order of the energy contribution that one would expect for the exchange term* of the form (1) when the atoms lie close to each other. If the atomic separations become larger, then the exchange terms will decay exponentially. *That is the basis for the fact that iron or nickel salt solutions are never ferromagnetic.*

The question of the sign of J_0 is much more difficult to answer. In their theory of the homopolar bond, *Heitler and London make that assumption that J_0 is negative in complete generality, which would exclude ferromagnetism.* For the special case in which the electrons are found to be unperturbed in the 1s state, it follows, in fact, from general theorems that the energy values must lie in a way that would correspond to negative values of J_0. *Such an argument is, in turn, applicable only to electrons in the 1s state, and one can show that J_0 will generally be positive for high principal quantum numbers.* One must then deal with the expression:

$$J_0 = 1/2 \int \psi_k^\kappa \psi_k^\lambda \psi_l^\kappa \psi_l^\lambda \left(2\varepsilon^2/r_{kl} + 2\varepsilon^2/r_{\kappa\lambda} - \varepsilon^2/r_{\kappa k} - \varepsilon^2/r_{\kappa l} - \varepsilon^2/r_{\lambda l}\right) d\tau_k d\tau_l. \qquad (1)$$

in which κ and λ are the indices for the atomic nuclei, while k and l are the those of the electrons. Initially, ψ will be a hydrogen eigenfunction, but later on, it will be shown that the argument is just as valid for other central fields in the vicinity of the nucleus. *One can then say with certainty that J_0 will be positive for very small values of $r_{\kappa\lambda}$, since the term $1/r_{\kappa\lambda}$ will then outweigh all of the other ones.* However, that result does not need to have any physical meaning, since for very small values of $r_{\kappa\lambda}$, even the entire approximation becomes illusory (cf., the case of 1s terms!). One then comes to the values of J_0 for very large $r_{\kappa\lambda}$. When J_0 is positive there, one must assume that it remains positive for all values of $r_{\kappa\lambda}$, in general. We thus investigate further how a charge distribution of density $\psi_k^\kappa \psi_k^\lambda$ appears at large distances $r_{\kappa\lambda}$, first for perhaps the higher s terms.

The Schrödinger functions contain an e-function as the most important term, and $\psi_k^\kappa \psi_k^\lambda$ thus contains the factor $e^{-(r_{k\kappa} + r_{k\lambda})/a_0 n}$ (a_0 = Bohr hydrogen radius, n = principal quantum number; thus, no confusion with the electron number 2n should be created). *If one drops the remaining factors then the density will be constant on confocal ellipsoids of rotation around the two nuclei.* For increasing distance between the nuclei, the charge ellipsoid degenerates into a cylinder around the connecting line between the nuclei. (This happens for both values of the principal quantum number.) Furthermore, the e-function appears multiplied by a polynomial in $r_{\kappa\kappa}$ ($r_{\kappa\lambda}$, resp.) of degree n – 1. The zero locus of this polynomial lies entirely in the neighborhood of the nucleus; at greater distances from it, it will suffice to replace the polynomial with its highest power r^{n-1}. The behavior of the central force at distances of order a_0 from the nucleus is entirely inessential when only $r_{\kappa\lambda}$ is sufficiently large. The density distribution of the charge over the length of the aforementioned cylinder is therefore non-uniform, but otherwise approximately proportional to $r_{k\kappa}^{n-1} r_{k\lambda}^{n-1}$. For small values of n, this distribution is still quite uniform and one can easily see that the negative terms in J_0 can substantially predominate. For increasing n, by contrast, the density distribution assumes an ever steeper maximum at the midpoint between the two nuclei. In the limit of very large values of n, the mean value of the terms of type $1/r_{\kappa\kappa}$, when taken over the density distribution that was given above, tends to the value $2/r_{\kappa\lambda}$:

$$1/r_{k\kappa} = 1/r_{k\lambda} = 1/r_{l\kappa} = 1/r_{l\lambda} \to 2/r_{\kappa\lambda}.$$

By contrast, the term with $1/r_{kl}$ – viz., the "*self-potential*" of the density distribution – increases beyond all limits with increasing n. J_0 is then certainly positive *for sufficiently high principal quantum numbers*. One can easily show that nothing will change in this result when one carries out the calculations for p, d, ..., or any other higher state. The limiting value of n for which J_0 can become positive for the first time is difficult to determine exactly. A rough calculation yields n = 3. This limiting value will possibly depend upon values of the remaining quantum numbers. The fact that, e.g., the oxygen molecule empirically possesses a *magnetic moment* of $2 \cdot 1/2\ h/2\pi$ in the ground state seems to show that J_0 can already be positive for n = 2. On the other hand, it can follow from the many-times-observed critical temperatures (e.g., for γ-iron) that there are many times J_0 that can also be negative for higher principal quantum numbers.

Concluding remarks. The calculations that were described here lead to two conditions for the appearance of *ferromagnetism*:

1. The crystal lattice must be a type such that any atom has at least 8 neighbors.
2. The principal quantum number of the electrons that are responsible for magnetism must be $n \geq 3$.

Both conditions together do not reach far enough to single out Fe, Co, Ni from all other materials; however, Fe, Co, Ni do satisfy the conditions. It was certainly also to be expected that the theory that was contrived here can meanwhile serve as only a qualitative schema in which *ferromagnetic* phenomena will perhaps be classified later. The theory admits an extension for the case of several exchanges per atom; an incisive study of the $J_{(kl)}$ values, as well as the distribution curve of the term values, will be requisite. I hope to be able to go into these questions, as well as a thorough comparison of the theory with the experimental results later.

John Hasbrouck Van Vleck (March 13, 1899–October 27, 1980).

Van Vleck was an American physicist and mathematician. He was co-awarded the Nobel Prize in Physics in 1977, for his contributions to the understanding of the behavior of electronic magnetism in solids.

Van Vleck was born to mathematician Edward Burr Van Vleck and Hester L. Raymond in Middletown, Connecticut. His father was an assistant professor at Wesleyan University, where his grandfather, astronomer John Monroe Van Vleck, was also a professor. He grew up in Madison, Wisconsin, and received an A.B. degree from the University of Wisconsin in 1920, before earning his Ph.D. at Harvard University in 1922 under the supervision of Edwin C. Kemble. He joined the University of Minnesota as an assistant professor in 1923, then moved to the University of Wisconsin.

Van Vleck met Abigail Pearson, a student at University of Minnesota, during his professorship there, and married her on June 10, 1927. He and his wife Abigail were important art collectors, particularly in the medium of Japanese woodblock prints (principally Ukiyo-e), known as Van Vleck Collection. It was inherited from his father Edward Burr Van Vleck. They donated the collection to the Chazen Museum of Art in Madison, Wisconsin in 1980s.

Van Vleck established the fundamentals of the *quantum mechanical theory of magnetism*, crystal field theory and ligand field theory (chemical bonding in metal complexes). He is regarded as the Father of Modern Magnetism. In 1932, he published his 400-page magnum opus on the non-relativistic quantum theory of electric and magnetic susceptibilities in materials [Van Vleck J. H. (1932). *The Theory of Electric and Magnetic Susceptibilities*. Oxford at Clarendon Press, Oxford. See below.] Forty-five years later, he was awarded the Nobel Prize in Physics 1977, along with Philip W. Anderson and Sir Nevill Mott, "for his contributions to the understanding of the behavior of electrons in magnetic solids".

During World War II, Van Vleck worked on radar at the MIT Radiation Lab. He was half time at the Radiation Lab and half time on the staff at Harvard. He showed that at about 1.25-centimeter wavelength water molecules in the atmosphere would lead to troublesome absorption and that at 0.5-centimeter wavelength there would be a similar absorption by oxygen molecules. This was to have important consequences not just for military (and civil) radar systems but later for the new science of radioastronomy.

Van Vleck participated in the Manhattan Project. In June 1942, J. Robert Oppenheimer held a summer study for confirming the concept and feasibility of a nuclear weapon at the University of California, Berkeley. Eight theoretical scientists, including Van Vleck, attended it. From July to September, the theoretical study group examined and developed the principles of atomic bomb design.

Van Vleck's theoretical work led to the establishment of the Los Alamos Nuclear Weapons Laboratory. He also served on the Los Alamos Review committee in 1943. The committee's important contribution was a reduction in the size of the firing gun for the Little Boy atomic bomb, a concept that eliminated additional design weight and sped up production of the bomb for its eventual release over Hiroshima. However, it was not employed for the Fat Man bomb at Nagasaki, which relied on implosion of a plutonium shell to reach critical mass.

From 1951, Van Vleck was Hollis Professor of Mathematics and Natural Philosophy at Harvard. He concurrently held the first deanship of Harvard's Division of Engineering and Applied Physics until 1957. In 1961/62 he was George Eastman Visiting Professor at University of Oxford and held a professorship at Balliol College.

Van Vleck died in Cambridge, Massachusetts, aged 81. He was buried at Forest Hill Cemetery.

Van Vleck, J. H. (1932). The Theory of Electric and Magnetic Susceptibilities.

Clarendon Press, Oxford; https://dn790000.ca.archive.org/0/items/theoryofelectric031070mbp/theoryofelectric031070mbp.pdf.

Professor of Theoretical Physics, University of Wisconsin.

Extracts are provided from John Van Vleck's 400-page book, published in 1932, which provides an extremely comprehensive explanation of the *non-relativistic* quantum theory of the *electric and magnetic susceptibities of materials*, including the quantum theory of *diamagnetism* and *paramagnetism*, and detailed explanations of the *exchange effect* and Heisenberg's *theory of ferromagnetism.* As Van Vleck noted in the Preface, "the analysis of experimental *magnetic susceptibilities* cannot be attempted until the quantum chapters, since *the numerical values of magnetic susceptibilities are inextricably connected with the quantization of angular momentum*". Van Vleck also noted that *the theory of the electron spin may be presented in two ways,* viz. by means of what he calls a *semi-mechanical model* (the Uhlenbeck-Goudsmit model) or by means of *Dirac's 'quantum theory of the electron'.* In the *semi-mechanical model*, matrix expressions for the *spin angular momentum* are written down by analogy with the *orbital angular momentum* matrices, with certain postulates regarding the occurrence of a half-quantum of *spin* per electron and *the ratio of spin magnetic moment to spin angular momentum*, in order to bring the theory into line with experiments. He noted that the interaction of the *spin* with *external magnetic fields* was handled perfectly well by the *semi-mechanical model*, but Einstein's *special theory of relativity* is inconsistent with this. He noted, on the other hand, that *the extension of Dirac's relativistic theory to many electron systems was at that time in a rather unsettled state* resulting in a nonvanishing probability of the mass of the electron changing sign, *an obvious absurdity*. Consequently, Van Vleck decided to present the quantitative aspects of the *spin* entirely with the aid of the older *semi-mechanical model*. In considering *diamagnetism*, [the property of materials that are repelled by a magnetic field; which creates an induced magnetic field in them in the opposite direction] Van Vleck supposed that the atoms were in *singlet S states* [in which all electrons in the S orbital are paired], as otherwise there was an overwhelming *paramagnetism* [a form of magnetism whereby some materials are weakly attracted by an externally applied magnetic field, and form internal induced magnetic fields in the direction of the applied magnetic field]. In such states there cannot be even an instantaneous *magnetic moment* in the absence of external fields. In other cases, Van Vleck distinguished different cases, including where the *inter-atomic forces* were so small that the magnetism could be calculated by treating the atoms of the solid to be free (exemplified by *rare earth salts*); cases where *the orbital and spin magnetic effects were both largely destroyed*, resulting in feeble *paramagnetism* (which most elements exhibit in the solid state), or even *diamagnetism*; and solids in which *the Heisenberg exchange forces tended to align the spins parallel and so create ferromagnetism* (as for iron, nickel, and cobalt). Van Vleck noted that the *exchange forces had the effect of introducing a very strong coupling between the spins of paramagnetic atoms or ions. Diamagnetic* atoms or ions have no resultant spin and so do not give rise to any *exchange forces* tending to orient the *spins* of other atoms. Moreover, these *exchange forces* became of subordinate importance in media in which the density of *paramagnetic* atoms or ions was low because the great majority of the atoms are *diamagnetic*, and in most salts involving the iron group which are consequently only *paramagnetic*. Van Vleck then provided a detailed account of how *entanglement* between *electrons* with the same *quantum spin states* creates an *exchange effect* which results in a strong coupling between their *spins*. He concluded by explaining how Heisenberg applied this in his theory of *ferromagnetism*.

PREFACE

The new quantum mechanics is perhaps most noted for its triumphs in the field of spectroscopy, but it's less heralded successes in the theory of *electric and magnetic susceptibilities* must be regarded as one of its great achievements. At the same time the accomplishments of classical mechanics in this field must not be overlooked, and so the first four chapters are devoted to purely classical theory. Most of the comparison with experiment regarding *dielectric constants* is included in one of these (Chap. III). This can be done without making the comparison obsolete because the new quantum mechanics has restored the validity of many classical theorems violated in the old quantum theory. On the other hand, *the analysis of experimental magnetic susceptibilities cannot be attempted until the quantum chapters, since the numerical values of magnetic susceptibilities are inextricably connected with the quantization of angular momentum*. At the outset 1 intended to include only gaseous media, but the number of *paramagnetic* gases is so very limited that any treatment of magnetism not applicable to solids would be rather unfruitful. Therefore, salts of the *rare earth* and *iron* groups are examined in considerable detail. A theory is developed to explain why, as conjectured by Stoner, *inter-atomic forces obliterate the contribution of the orbital angular momentum to the magnetic moment in the iron group*. Chapter XII includes the aspects of *ferromagnetism* so far amenable to the Heisenberg theory, which has at last divested the Weiss molecular field of its mystery. This means that here the discussion is centered on the thermal behavior of the saturation, rather than on hysteresis and retentivity. As far as practicable, I have striven throughout the volume to avoid duplication of the existing literature, especially Debye's *Polar Molecules* and Stoner's *Magnetism and Atomic Structure*.

In the preface to a book on theoretical physics it is customary for the author to express the laudable but, alas, usually unwarranted hope that the volume will prove simultaneously rigorous to mathematical readers and intelligible to the non-mathematical, at least provided the latter omit the particular sections where the density of equations is excessive. At any rate this has been the aim of the present volume, and I hope that it has not fallen too far short. A detailed knowledge of quantum mechanics or of spectroscopic nomenclature has not been presupposed only an elementary acquaintance with the Schrodinger wave equation. The necessary perturbation theory and theorems of spectroscopic stability are developed in Chapter VI. Here I have tried to correlate and intermingle the use of *wave functions* and of *matrices*, rather than relying exclusively on the one or the other, as is too often done. It is hoped that this chapter may be helpful as a presentation of the perturbation machinery of quantum mechanics, quite irrespective of the magnetic applications.

I am much indebted to the Guggenheim Memorial Foundation for a travelling fellowship which enabled me to visit many European institutes for theoretical physics. I wish to take this occasion to thank the staffs of these institutes for their cordiality and helpful discussions. The list is rather extensive Cambridge, Leipzig, Munich, Gottingen, Berlin, Zurich, Copenhagen, Leiden, Utrecht, Groningen, Bristol, Paris. I am also indebted to the University of Wisconsin for extension of leave which permitted me to attend the sixth Solvay Congress, devoted to magnetism, and to Professors W. Weaver and J. W. Williams of this university for valuable criticisms on Chapters I and III respectively. I also wish to thank Miss A. Frank and Mr. R. Serber for assistance in some of the computations and in proof-reading.

J. H. V. V.

DEPARTMENT OF PHYSICS,
UNIVERSITY OF WISCONSIN,
June, 1931.

CONTENTS

I. CLASSICAL FOUNDATIONS

I. CLASSICAL FOUNDATIONS

1. *The Macroscopic versus Microscopic Field Equations*

The conventional Maxwell equations [where **E** is the *electric field*, **D** is the *electric displacement field*, **B** is the *magnetic field*, **H** is the *magnetic H-field*, and **J** and **i** are the *electric current*] are

$$\text{curl}\ \mathbf{E} = -\ 1/c\ \partial\mathbf{B}/\partial t \qquad (1)$$

[$\nabla \times \mathbf{E} = -\ \partial\mathbf{B}/\partial t$ The Maxwell–Faraday version of
Heaviside formulation Faraday's Law of Induction]

$$\text{curl } \mathbf{H} = -1/c\,(4\pi\mathbf{i} + \partial\mathbf{D}/\partial t) \qquad (1)$$

[$\nabla \times \mathbf{B} = -\mu_0(\mathbf{J} + \varepsilon_0\,\partial\mathbf{E}/\partial t) = -\mu_0\varepsilon_0\,\partial\mathbf{E}/\partial t$ — The Maxwell-Ampère Law with Maxwell's addition]
Heaviside formulation

together with

$$\text{div } \mathbf{D} = -4\pi\rho \qquad (2)$$

[$\nabla . \mathbf{E} = \rho/\varepsilon_0 = 0$ — Gauss's Law
Heaviside formulation]

$$\text{div } \mathbf{B} = 0. \qquad (2)$$

[$\nabla . \mathbf{B} = 0$ — Gauss's Law for Magnetism
Heaviside formulation.]

We shall term these the '*macroscopic field equations*' as they do not aim to take direct cognizance of the atomicity of matter or electricity. Throughout the volume all expressions printed in bold-face type are vectors. Between the four field vectors there exist the so-called *constitutive relations*

$$\mathbf{D}/\mathbf{E} = \varepsilon, \qquad \mathbf{B}/\mathbf{H} = \mu, \qquad (3)$$

which may be regarded as defining the *dielectric constant* ε and the *permeability* μ. The ratios ε and μ are, *except for ferromagnetic media*, independent of the field strength for sufficiently small fields, and in general we must have such an independence, or at least a known dependence on the field strength, before Eqs. (1), (2) become unambiguous enough to be useful. We suppose throughout the volume that *the medium is isotropic*; *in crystalline media directional effects make it necessary to use six dielectric constants or permeabilities instead of one, and D, B cease in general to be parallel to E, H respectively as presupposed by (3)*.

Of course, ε and μ depend on many factors, notably on the temperature, density, chemical constitution, and frequency, as well as on the field strength if great. *The theoretical description of their modes of dependence is the main aim of the present volume*. This description is accomplished by means of the *molecular theory of matter*, and especially by means of *the dynamics governing the electrons within each atom or molecule*. The dawn of the twentieth century brought to light *the electrical origin of matter*, unknown to Maxwell when he developed his macroscopic equations in 1861-73. This electrical origin implied that by probing down to sub-atomic distances *it should be possible to formulate the equations of electrodynamics in terms of charges in vacua without the introduction of ponderable dielectric and magnetic media*. H. A. Lorentz[2] therefore proposed and studied what we shall term the *'microscopic field equations'*

[2] Cf., for instance, Lorentz, H. A. (1916). *The Theory of Electrons*. Leipzig. His original papers were published considerably earlier in the *Proceedings of the Amsterdam Academy*.

curl $\mathbf{e} = -1/c\ \partial\mathbf{h}/\partial t$, curl $\mathbf{h} = -1/c\ (4\pi\rho'\mathbf{v} + \partial\mathbf{D}/\partial t)$ (4)

[curl $\mathbf{E} = -1/c\ \partial\mathbf{B}/\partial t$, curl $\mathbf{H} = -1/c\ (4\pi i + \partial\mathbf{D}/\partial t)$ (1)]
[$\nabla \times \mathbf{E} = -\partial\mathbf{B}/\partial t$ $\nabla \times \mathbf{B} = -\mu_0 (\mathbf{J} + \varepsilon_0\, \partial\mathbf{E}/\partial t) = -\mu_0\varepsilon_0\, \partial\mathbf{E}/\partial t$
Heaviside formulation]

div $\mathbf{e} = -4\pi\rho'$, div $\mathbf{h} = 0$ (5)

[div $\mathbf{D} = -4\pi\rho$, div $\mathbf{B} = 0$ (2)]
[$\nabla \cdot \mathbf{E} = \rho/\varepsilon_0$ $\nabla \cdot \mathbf{B} = 0$
Heaviside formulation]

which are similar in structure to the macroscopic equations in vacuo (where, of course, **B** = **H**, **D** = **E**), except that instead of the *ponderable current density* **i** Lorentz introduced the *convection current density* ρ'**v** due to motion of the *charge density* ρ' with the vector *velocity* **v**. The microscopic *fields* **e**, **h** and *charge* ρ' are not the same as the macroscopic fields or charge, and have therefore been printed in small letters or else designated by a prime. *Eqs. (4), (5) are more fundamental than (1), (2), (3), as (4), (5) are supposed to hold at every point either inside or outside the molecule*, whereas (1), (2), (3) are essentially *statistical* in nature, and the expressions **E**, **D**, **H**, **B**, p which they involve must be correlated in some way with averages of microscopic fields and charges over a large number of molecules. How this correlation is achieved will be discussed in the two following sections.

...

VI. QUANTUM-MECHANICAL FOUNDATIONS

38. *The Electron Spin*

The writer begins this section with considerable trepidation, as *the theory of the spin is neither particularly simple nor particularly rigorous*. The concept that an *electron* has an internal degree of freedom about which it is free to spin has been extraordinarily fruitful in clarifying the analysis of spectra. This idea is due primarily to Uhlenbeck and Goudsmit [30]

[30] Uhlenbeck, G. E. & Goudsmit, S. (November, 1925). Ersetzung der Hypothese vom unmechanischen Zwang durch eine Forderung bezuglich des inneren Verhaltens jedes einzelnen Elektrons. (Replacement of the hypothesis of unmechanical coercion by a requirement regarding the internal behavior of each individual electron.) *Naturw.*, 13, 47, 953-4; (February 20, 1926). Spinning Electrons and the Structure of Spectra. *Nature*, 117, 264-5 [See above].

although the *spin* has been proposed in other connections at earlier dates by Compton, Kennard[31], and others.

[31] Compton, A. H. (August, 1921). The Magnetic Electron. *Journ. Frankl. Inst.*, 192, 2, 145-55 [See above]; Kennard, E. H. (1922). *Phys. Rev.* 19, 420 (abstract). Kennard's note is often overlooked; in it the spin was proposed explicitly in connection with the gyromagnetic anomaly.

The theory of the electron spin may be presented in two ways, viz. by means of what we shall call a semi-mechanical model or by means of Dirac's 'quantum theory of the electron'.

In the *semi-mechanical model*, matrix expressions for the *spin angular momentum* are written down by analogy with the *orbital angular momentum* matrices, with certain postulates regarding the occurrence of a half-quantum of *spin* per electron which will be explained below. It is further assumed that *the ratio of spin magnetic moment* M_s *to spin angular momentum* P_s has twice the classical value – e/2mc for *the ratio of orbital magnetic moment to orbital angular momentum*, so that

$$M_s/P_s = -\, e/mc. \qquad (73)$$

The assumption (73) is made to explain the fact that in experiments on rotation by magnetization (the Einstein-Richardson-de-Haas effect) as well as on the converse magnetization by rotation (Barnett effect), the ratio of magnetic moment to angular momentum has approximately the value (73) instead of the classical orbital value[32].

[32] For description of these gyromagnetic experiments, and references, see Stoner, *Magnetism and Atomic Structure*, p. 184.

The anomalous ratio (73) for the *spin* is also required by the *anomalous Zeeman effect*, as will be seen more fully in § 42. Lande[33]

[33] Lande, E. (1923). *Zeit. Phys.*, 15, 189; or Back and Lande, *Zeemaneffekt und Multiplettstruktur der Spektrallinien*, pp. 43, 79.

found that his celebrated g-formula could be explained, except for certain characteristic modifications resulting from the new quantum mechanics not understood prior to 1926, by assuming that the atom contained a rather mysterious 'atom-core' (*Atomrumpf*) whose *ratio of magnetic moment to angular momentum* has the value (73). This mystical 'atom-core' now turns out in reality to be the *spin*.

Besides the arbitrary character of its postulates, the *semi-mechanical model* has the drawback that it is able to describe only to a first approximation (i.e. through terms of the order $1/c^2$) the internal magnetic forces of the atom. That is to say, *it does not furnish an adequate dynamics of the interaction of the spins with each other and with orbital forces*. Practically, this is not a serious handicap, as the terms of higher order $1/c^4$ are entirely too small to be of any consequence in the optical region, although they are large enough to be observable in the case of X-ray doublets in heavy atoms. The interaction of the *spin* with external magnetic fields, which is our particular concern, is handled perfectly well by the *semi-mechanical model*. However, an approximate theory of internal magnetic forces is never as satisfying logically as an exact theory, and because these forces are only approximately described, *the Hamiltonian function used in the semi-mechanical model does not behave properly under a Lorentz transformation*, and so *does not meet the requirements of the special theory of relativity*.

It is this need of relativity invariance which led Dirac to the discovery of his brilliant 'quantum theory of the electron'.[34]

[34] Dirac, P. A. M. (February, 1928). The Quantum Theory of the Electron. *Roy. Soc. Proc., A*, 117, 778, 610–24; (March, 1928). The Quantum Theory of the Electron. Part II. *Loc. cit*., 118, 779, 351-61; or *The Principles of Quantum Mechanics*, Chap. XIII. The explicit calculation of the *susceptibility* of an *atom* with one valence *electron* by means of Dirac's four simultaneous equations is given by Sommerfeld in the report of the 1930 Solvay Congress. The results are the same as with the *semi-mechanical model* except for terms too small to be observable.

In the case of a system with one electron, he boldly replaced the single second-order Schrodinger wave equation by four simultaneous first-order *wave equations*, involving the use of four *wave functions*. In a system with f electrons there would be 4f wave functions, but *the extension of Dirac's theory to many electron systems is at present in a rather unsettled state*, and this is one reason we do not incorporate it in the present volume. Previously to Dirac, Pauli had shown that the existence of two *wave functions* per *electron*, and of two corresponding simultaneous second-order equations, was necessary in order to interpret in wave language the *spin* matrices of the *semi-mechanical model*. One wave function corresponds in a certain sense to the alignment of *spin* parallel to the axis of quantization, and the other to it anti-parallel. *Four wave functions are twice too many*, and in order to vest them with a physical interpretation it seems necessary to interpret certain states as representing an electron of negative mass. If Dirac's quartet of wave functions were separable into two non-combining pairs, i.e. into pairs such that integrals of the form (14) always vanish if the two wave functions belong to different pairs, the difficulty would not be so serious. Actually, the two pairs of wave functions do 'combine', so that in the ordinary quantum-mechanical interpretation of wave functions there is a nonvanishing probability of the mass of the electron changing sign, *an obvious absurdity. This difficulty is probably the most serious flaw in the logical framework of present-day quantum mechanics*[35], and very likely will not be cleared up until the long-awaited theory is evolved which explains the differences in mass of the electron and proton.

> [35] Dirac, P. A. M. [(January, 1930). A theory of electrons and protons. *Roy. Soc. Proc., A*, 126, 801, 360-65; https://doi.org/10.1098/rspa.1930.0013] has made the bold but interesting suggestion that the states with negative mass may be nearly 'all full', as the Pauli exclusion principle allows each state to occur only once. What we interpret as ordinary electric neutrality is then really a maximum, infinite charge density of electrons with negative mass, and a proton is a vacancy or ' hole ' in the infinity of negative states. This idea, however, encounters many serious difficulties, and its ultimate significance is uncertain.

However, Dirac's theory is marvelously successful in explaining all *spin* phenomena. After setting up his four first-order equations, Dirac magically extracts all the properties of the *spin*, such as the anomalous ratio (73). His equations have the necessary *relativity invariance*, and give the internal magnetic interactions exactly rather than approximately. They yield *spin* doublets of exactly the same width as Sommerfeld's *relativity* doublets in the old quantum theory, thus yielding one of the most amazing fortuitous coincidences in the history of physics. The previous *semi-mechanical model* gave this coincidence only to terms of the order $1/c^2$ inclusive.

Van Vleck opts for a *non-relativistic* theory.

To many readers it will doubtless appear a step backwards that *we shall dismiss Dirac's theory* after this cursory qualitative discussion, and present the quantitative aspects of the spin entirely with the aid of the older *semi-mechanical model*. However, besides the difficulty of the physical interpretation of the superfluous pair of wave functions, Dirac's theory, with its four simultaneous equations, has necessarily a certain amount of mathematical complexity, and the *semi-mechanical model* is easier to visualize more '*anschaulich*' as the Germans say. This property makes results on susceptibilities easier to remember and interpret, and perhaps less liable to computational errors if the *semi-mechanical model* is used. There is no loss of rigor, as it can be shown that Dirac's theory yields the same matrices for the *spin energy* in an external magnetic field as the previous Uhlenbeck-Goudsmit model. We can thus regard Dirac's theory as the most refined way of deriving the matrix elements of the *spin, which in the semi-mechanical model are taken as sheer postulates*. Our omission of derivation of the spin matrix elements by Dirac's method is in accord with our policy of not attempting to solve dynamical problems exactly, but only to show how the perturbed energy can be approximately found once the matrix elements of the perturbative *potential* are known. One reason that we use the *semi-mechanical model* is that while Dirac's quantum theory of the electron is discussed in most recent texts on quantum mechanics, Heisenberg and Jordan's very compact and elegant treatment of the anomalous Zeeman effect by means of the pure-matrix theory is too generally ignored.

We shall present the *semi-mechanical model* in the pure matrix language, without giving the allied wave functions, as the latter do not help in setting up the appropriate secular equations (35). *The first attempt at finding wave functions associated with the spin was made by Darwin*[36].

> [36] Darwin, C. G. (June, 1927). The Zeeman Effect and Spherical Harmonics. *Proc. Roy. Soc.*, 115A, 770, 1.

In natural analogy with *orbital* motions, he supposed that there was an azimuthal rotational coordinate ϕ_s associated with precession of the spin axis. The *wave function* would then contain a factor $e^{im_s\phi_s}$ where m_s is a quantum number specifying the axial component of *spin angular momentum*. Unfortunately, *this function then does not have the necessary property of single-valuedness*, as for a single *electron* m_s has the values ± ½ instead of being an integer, and $e^{1/2i(\phi_s + 2\pi)} \neq e^{1/2i\phi_s}$. Because of this difficulty we speak of the *Uhlenbeck-Goudsmit model* as '*semi-mechanical*' rather than '*mechanical*'. As a matter of fact, Darwin ingeniously found that *spin* matrix elements could be calculated by means of the fundamental quadrature (14) even with multiple-valued wave functions, but this appears a little fortuitous. Pauli[37] later showed that the difficulty of multiple-valuedness could be overcome by taking the arguments of the *wave functions* to be the axial component s_z of *spin angular momentum* instead of a *rotational coordinate*.

> [37] Pauli, W. (September, 1927). Zur Quantenmechanik des magnetischen Elektrons. (On the quantum mechanics of magnetic electrons.) *Zeit. Phys.*, 43, 601-23; https://doi.org/10.1007/BF01397326.

The *Dirac-Jordan transformation theory* indeed permits us to use any set of coordinates and momenta as arguments of the *wave function*, which is a special case of a 'probability amplitude'. Now s_z has only the two discrete characteristic values ± ½(h/2π) whereas ϕ_s assumes a continuous range of values. A function whose argument only assumes two values is equivalent to a pair of functions, so *Pauli's scheme involves two wave functions per electron*. For definition of the operators corresponding to *spin angular momenta*, which cannot be expressed as differentiations, and for modification of the fundamental quadrature (14) to include summation over the discrete spin characteristic values as well as integration over the continuous orbital coordinates, the reader is referred to Pauli's paper[37] and closely allied work by Darwin[38].

> [38] Darwin, C. G. (September, 1927). The Electron as a Vector Wave. *Roy. Soc. Proc.*, A, 116, 773, 227-53; https://royalsocietypublishing.org/doi/pdf/10.1098/ rspa.1927.0134.

The treatment of the anomalous Zeeman effect either by the Pauli operators or by Darwin's multiple-valued wave functions is, of course, only superficially different from that with matrices (§ 42). All methods inevitably lead to the same secular equation.

...

40. *Russell -Saunders Coupling, Spectroscopic Notation, &c.*

In the previous section we have neglected internal forces, but actually there are powerful forces of this character tending to couple together the various angular momentum vectors of the atom. The simplest assumption is that the *energy of interaction* between any two vectors is proportional to the cosine of the angle included between them or, what is equivalent, to their *scalar product*. The Hamiltonian function will then contain terms of the form

$$a_{ik}\mathbf{l}_i \cdot \mathbf{s}_k, \qquad b_{ik}\mathbf{l}_i \cdot \mathbf{l}_k, \qquad c_{ik}\mathbf{s}_i \cdot \mathbf{s}_k, \qquad (i,k = 1, \ldots, \eta), \qquad (79)$$

where η is the *number of electrons in the atom.*

Throughout the rest of the volume *all expressions in bold-face type are to be construed as vector matrices*, i.e. vectors of which each component is a matrix rather than an ordinary number. The proportionality constants a_{ik}, b_{ik}, c_{ik} will in general be functions of all quantum numbers (such as, for instance, n, l) other than those quantizing the relative orientations of the vectors involved. The expression $a_{ik}\mathbf{l}_i \cdot \mathbf{s}_k$, for instance, means the *energy associated with the force which the orbital angular momentum of the ith exerts on the spin of the kth electron.*

Ordinarily it turns out that $|a_{ii}| > |a_{ik}|$, $k \neq i$, meaning that *the orbital angular momentum of a given electron interacts more strongly with its own than with other spins*. It can be shown that [43]

$$a_{ii} = h^2 Z_{eff} \varepsilon^2 / 8\pi^2 m^2 c^2 \, \overline{1/r^3}, \qquad (80)$$

where r is the *distance of the electron from the nucleus*, provided the given electron is not so highly perturbed by other electrons but that it may be regarded as moving in a Coulomb field.

[43] Thomas, L. H. (April, 1926). The Motion of the Spinning Electron. *Nature*, 117, 2945, 514; https://doi.org/10.1038/117514a0; (1927). *Phil. Mag.*, 3, 1; Frenkel, J. (1926). *Zeit. Phys.*, 37, 243. Formula (80), which is of course also yielded by Dirac's 'Quantum Theory of the Electron' with appropriate approximations, differs by a factor 2 from what one would expect from elementary, over-simplified calculations.

Although the validity of the cosine law (70) is only approximate, and can be justified theoretically only with the aid of many simplifying assumptions, the departures from (79) need not cause us concern, as the general different types of quantization which we delineate by means of (79) are significant even when (79) is not strictly applicable. Actually, the coupling of the *l* vectors often departs widely from the cosine form given in the second term of (79). Also, our arguments are not affected by the fact, to be discussed in § 76, that the constants c_{ik} coupling the various *spin* vectors with each other *are due primarily not to magnetic forces but to the Heisenberg exchange effect*.

...

VIII THE DIELECTRIC CONSTANTS AND DIAMAGNETIC SUSCEPTIBILITIES OF ATOMS AND MONATOMIC IONS

...

49. *The Diamagnetism of Atoms, especially Hydrogen and Helium*

In considering *diamagnetism*, we may suppose the atoms in *singlet S states*, as otherwise there is an overwhelming *paramagnetism*.

[*Diamagnetism* is the property of materials that are *repelled* by a magnetic field; an applied magnetic field creates an induced magnetic field in them *in the opposite direction*, causing a repulsive force.

Paramagnetism is a form of magnetism whereby some materials are *weakly attracted* by an externally applied magnetic field, and form internal, induced magnetic fields *in the direction of the applied magnetic field*.

In quantum mechanics, a *singlet state* refers to a system in which *all electrons are paired*. The term 'singlet' originally meant a linked set of particles whose net angular momentum is zero, that is, whose overall spin quantum number $s = 0$. As a result, there is only one spectral line of a singlet state.

The S refers to the *orbital* of the electron in the atom. S orbital, P orbital, D orbital, and F orbital refer to orbitals with angular momentum quantum number $\ell = 0, 1, 2,$ and 3 respectively. These names, together with the value of n, are used to describe the electron configurations of atoms]

In such states the *paramagnetic* terms in the Hamiltonian function, which were given in Eq. (97), Chap. VI, and which yield a perturbative potential proportional to the first power of H, *disappear completely*. This is so inasmuch as in atoms the squares of the orbital and spin angular momenta are respectively L(L + 1) and S(S + 1), and consequently in the ^1S states, which have S = L = 0, there cannot be even an *instantaneous magnetic moment* in the absence of external fields. It may be cautioned that *molecules* have such a moment even in $^1\Sigma$ states, and for them the following formula (2) must be modified, as will be done in § 69. In ^1S atoms, there remains only the *diamagnetic* term in the perturbative *potential*, which is proportional to H^2, and the *magnetic moment* is entirely an induced one coming from the Larmor precession. The resulting change in energy due to this term was seen in Eq. (105), Chap. VI, to be $(e^2/8mc^2)\ \Sigma\ (x^2 + y^2)^{-}H^2$; and furthermore, it was shown in § 35 that on averaging over the different spacial orientations one may replace $x^2 + y^2$ by $2/3\ r^2$ because of spectroscopic stability. This, of course, assumes that the *Boltzmann factor* is sensibly the same for the different allowed spacial orientations, which it surely is in gases, and also in solids as long as the energy of orientation in the solid's intermolecular field is small compared to kT. If we suppose that the atoms are all in the same stationary state except for spacial orientation, as is usually the case because the first excited states involve energy increments large compared to kT, it is unnecessary to average over different electronic states weighted in accordance with the Boltzmann factor. The *susceptibility* $L(-\partial W/\partial H)/H$ per gramme mol is then

$$\chi_{mol} = -\ e^2L/6mc^2\ \Sigma\ r^2 \qquad (2)$$

where r^2 is the *time-average value*, i.e. the diagonal matrix element for the state under consideration. Eq. (2) is exactly the same as the Pauli form of Langevin's formula in classical theory, already given in (2), Chap. IV. Thus, again the new mechanics restores a classical formula.

Eq. (2) is valid regardless of whether or not the atom is hydrogenic. For hydrogen-like atoms, we may, however, proceed farther and use the formula for the mean value of r^2 given in (107), Chap. VI, and then

$$\chi_{mol} = \ldots = \ldots\ . \qquad (3)$$

The normal state of atomic hydrogen has n = 1, *l* = 0, Z = 1, and thus its *molar diamagnetic susceptibility* is $-\ 2.37 \times 10^{-6}$. This value cannot, of course, be tested directly because of the difficulty of dissociating molecular hydrogen, also because monatomic hydrogen has a ^2S normal state and hence would be highly *paramagnetic* because of the spin. Instead, we have only Pascal's[11] indirect value -2.93×10^{-6}, obtained by applying the additivity method to diamagnetic organic compounds containing hydrogen.

[11] A. Pascal, numerous references listed in *Jahrb. d. Rad. und Elektr.*, 17, 184 (1920); cf. also Weiss, *J. de Physique*, 1, 185 (1930).

Exact agreement cannot be expected, as we have seen in § 21, on refractivities, that the analysis of compounds by assumed additivity rules does not necessarily furnish true atomic properties. The error, however, is probably not so great as to permit the discrepancy by a factor about 3 ½ which there was between his value and that -0.79×10^{-6} furnished by the old quantum theory (p. 210). Thus, Pascal's result must be regarded as distinct evidence favoring the new mechanics in preference to the old.

Direct Calculation of r^2 from the Wave Functions for Helium. Turning now to non-hydrogenic atoms, the theoretical calculation of *diamagnetic* susceptibilities is much easier than of the electric, *as in the diamagnetic case it is only necessary to know the unperturbed wave function of the normal state*. Once this is known, the requisite mean value needed for (2) is given by the simple quadrature

$$\int \ldots \int |\psi_n|^2 \, \Sigma \, r^2 \, d\upsilon$$

the integration of course being over the coordinate space of all the electrons. On the other hand, to make calculations of *electric susceptibilities* such as were quoted on p. 205 *one must know the effect of the perturbing electric field on the wave function*. This is because the perturbing *potential* was linear in the field rather than quadratic as in the *magnetic* case; and so to obtain the *energy* to the second power of the *field strength*, as needed for *susceptibilities*, it was necessary to find a second rather than a first approximation to the effect of the perturbation, which demands knowledge of the *wave functions* to the 1st rather than 0th approximation in E.

The requisite quadrature (4) has been performed for *neutral helium* by Slater[12], using a *wave function* which he shows to be a good approximation to the three-body problem of the normal state of helium.

[12] Slater, J. C. (1928). *Phys. Rev.*, 32, 340.

He thus finds $\chi_{mol} = -1.85 \times 10^{-6}$, in gratifying accord with Hector and Wills' experimental value -1.88×10^{-6}. The discrepancy is less than the experimental error, as well as less than the amount of uncertainty in our knowledge of the helium wave functions. The quadrature for helium has also been evaluated independently by Stoner[13] with Hartree's wave functions. He finds $\chi_{mol} = -1.90 \times 10^{-6}$, likewise in exceedingly good agreement with experiment.

[13] Stoner, E. C. (1929). *Proc. Leeds Phil. Soc.*, 1, 484.

...

IX THE PARAMAGNETISM OF FREE ATOMS AND RARE EARTH IONS

53. ***Adaptation of Proof of Langevin-Debye Formula given in § 46***

The general proof of the Langevin-Debye formula was given in Chapter VII explicitly only for electric polarization, but can be applied to the magnetic case merely by substituting everywhere the *magnetic moment vector* **m** = (**L** + 2**S**)he/4πmc for the *electric moment vector* **p** used in § 44 and § 46. There are, however, two points which require comment.

In the first place, besides the *paramagnetic* part (**L** + 2**S**)Heh/4πmc (§ 42) of the Hamiltonian function, there is ever present the *diamagnetic* term proportional to $\Sigma(x^2 + y^2)$ which has been discussed in § 43 and § 49, and which has no analogue in the electric case. Therefore, to all formulae for the *susceptibility* calculated by the methods of Chapter VII, we must add the expression for the *diamagnetic susceptibility* given in Eq. (2), Chap. VIII.

The second point is the following. The *magnetic moment* in general consists of two parts, viz. the '*orbital*' and '*spin*' portions. In the different 'normal states' (cf. p. 187) these two parts may be inclined to each other at different angles. This will be the case if the normal states embrace a *spin multiplet* whose components are separated by intervals small compared to kT, as these different components correspond to different relative alinements of L and S ranging in atomic spectra from the '*anti-parallel*' alinement J = |L – S| to the parallel one J = L + S (cf. § 40). Because of this flexibility in the coupling of L and S we cannot in general suppose that the resultant magnetic moment is 'permanent', i.e. the same for all normal states, and so we cannot always effect the simplification made in passing from Eq. (22) to (24) in Chapter VII. Instead, we must use the more general expression (25), Chap. VII, for the contribution of the low-frequency elements, which does not require the hypothesis of a permanent moment.

If, then, we make only the fundamental assumption that the *moment matrix* involves only elements whose frequencies are either small or large compared to kT/h, the analysis in § 46 shows that the formula for the *susceptibility* is

$$\chi = N\bar{\bar{\mu}}^2/3kT + N\bar{\bar{\alpha}}, \qquad (1)$$

where $\mu^{=2}$ is defined as in Eq. (25), Chap. VII, and is thus the time average of the square of the low-frequency part μ of the *magnetic moment* vector, this average itself being averaged over the various normal states weighted in accordance with the *Boltzmann factor*

...

In practice the *diamagnetic* correction given in the second term of (2) is relatively small if the material is really *paramagnetic*, as usually molar *paramagnetic* and *diamagnetic susceptibilities* are respectively of the orders 10^{-4}-10^{-2} and 10^{-6}-10^{-5}. Consequently, we shall henceforth omit writing the *diamagnetic* term except when we explicitly consider *diamagnetism* in 69 and 81. Of course allowance for *diamagnetism* ought to be made in the most refined calculations of *paramagnetic moments* from observed *susceptibilities*. Most of the experimental measurements of the *susceptibilities* of *paramagnetic* salts which we quote in the balance of the volume, also the *'effective Bohr magneton numbers'* deduced therefrom, are corrected for the *diamagnetism* of the *anion* but not that of the *cation*. In other words, the quoted susceptibilities are the *measured susceptibilities* of the compound augmented by the absolute magnitude of the *diamagnetism* of the non-paramagnetic ingredient (*anion*), but not that of the paramagnetic ingredient itself (*cation*). One reason why the correction for the *diamagnetism* of the cation is usually omitted in the literature is that the *diamagnetism* of *rare earth ions* is rather hard to estimate quantitatively[2].

[2] It must be cautioned that even though the diamagnetic correction is inconsequential for the given atom or ion itself, it can well be exceedingly important in solutions, since ordinary solvents are *diamagnetic*, or in salts of high '*magnetic dilution*', where the *diamagnetic* atoms or ions greatly outnumber the *paramagnetic*. In these cases the total *diamagnetism* can clearly be appreciable compared to the *paramagnetism*.

...

X. THE PARA- AND DIAMAGNETISM OF FREE MOLECULES

69. *The Diamagnetism of Molecules*

The fact that most gases are *diamagnetic* shows that ordinarily the first term of (11) is the greater in magnitude. ...

...

Quantitative calculation of the two parts of (11) has been attempted only for the hydrogen molecule. Even here, direct evaluation of the sum over the excited states n' would be excessively difficult, ...

...

XI THE PARAMAGNETISM OF SOLIDS, ESPECIALLY SALTS OF THE IRON GROUP

71. *Delineation of Various Cases*

We shall stress primarily only the new quantum developments rather than the innumerable classical theories of magnetization in solids. As the present and following chapters are a digression from our intent to study only rarefied media, and as the quantum theory of *magnetism of solids* has so far achieved success more in the bold qualitative outlines of the phenomena rather than quantitative detail, we shall not document the experimental measurements quite as completely as in the two preceding chapters. A whole volume would be required to digest the copious experimental work on the iron family alone.

Different solids can exhibit *susceptibilities* of entirely different natures, and it may be well to outline in advance the various cases which can occur and in what materials they are commonly found.

(a) Instances where the *inter-atomic forces are so small that the magnetism can be calculated by treating the atoms of the solid to be as free as in an ideal gas*. The criterion for this is that the work required to orient an atom against the inter-atomic forces be small compared to kT. This case is exemplified remarkably well by *rare earth salts*, which have consequently been discussed at length in Chapter IX on *free atoms* and *ions*. As noted to the writer by Professor Bohr, the extraordinary freeness of the 4f orbits is revealed not only by the magnetism but also by the sharpness of the spectral lines from rare earth salts. This can only mean that *the 4f wave functions of the various rare earth atoms project out very little from the interiors of their respective atoms* and so 'overlap' other atoms only very slightly even in the solid state.

(b) Solids or solutions in which *inter-atomic forces quench the orbital angular momentum but leave the spin free*. This is what probably occurs in most salts of the iron group, as we shall see in § 72.

(c) Solids *in which there is such strong internal magnetic coupling*, i.e. such wide multiplets, that irrespective of the Heisenberg exchange effect *the inter-atomic forces of necessity quench the spin angular momentum when they do the orbital*. It is hard to distinguish experimentally between this case (c) and (e), (f) below, but case (c) is possibly sometimes realized in some salts of the platinum and palladium groups (§ 75).

(d) Solids in which *the Heisenberg exchange forces tend to align the spins parallel and so create ferromagnetism*. This is, of course, the case of iron, nickel, cobalt, also a few alloys which are ferromagnetic.

(e) Solids in which *these forces have the opposite sign from that in (d) and so tend to aline the spins antiparallel and destroy magnetism.*

(f) Materials in which *the spin angular momenta compensate each other because of the restrictions imposed by the Pauli exclusion principle rather than because of the exchange effect.*

In cases (e) and (f) any *orbital angular momentum* is ordinarily quenched as in (6). Hence in (c), (e), and (f) *the orbital and spin magnetic effects are both largely destroyed*, so that these cases all give feeble *paramagnetism*, or even *diamagnetism*. One of these cases must be the commonest of all, as most elements (distinct as from salts) exhibit only a feeble *paramagnetism*, if any, in the solid state. Cases (e) and (f), which will be discussed in § 80, are more probable than (c).

We throughout use the term '*quenched* ' when the constancy of angular momentum is so completely destroyed by inter-atomic forces as to blot out most of the *paramagnetism* which would be found in the ideal gas state. The distinction between the various cases is, of course, usually not a hard and fast one. Besides (a) - (f) there is also the trivial case of solids composed exclusively of atoms which are in ^1S states when free and which are hence without appreciable magnetism.

The Heisenberg '*exchange*' or *Austausch* forces[1] play a very important role in the magnetism of solids, especially in ferromagnetism.

> [1] Heisenberg, W. (June, 1926). Mehrkörperproblem und Resonanz in der Quantenmechanik. (Multibody Problem and Resonance in Quantum Mechanics.) *Zeit. Phys.*, 38, 411-426; https://doi.org/10.1007/BF01397160; (September, 1928). Zur Theory of Ferromagnetismus. (On the theory of ferromagnetism.) *Ibid.*, 49, 619–36 [See above.].

As far as the present chapter is concerned, it will be sufficient to say that *the exchange forces have the effect of introducing a very strong coupling between the spins of paramagnetic atoms or ions. Diamagnetic* atoms or ions have no resultant spin and so do not give rise to any exchange forces tending to orient the spins of other atoms. The mathematical basis for these statements will be given in Chapter XII. The important thing for present purposes is that the *exchange forces* become of subordinate importance in media of considerable 'magnetic dilution', i.e. media in which the density of *paramagnetic* atoms or ions is low because the great majority of the atoms are *diamagnetic*. Such media are the primary concern of the present chapter, and so it seems best to defer until Chapter XII the detailed description of the nature and workings of the Heisenberg *exchange effect*.

72. *Salts and Solutions Involving the Iron Group*

Pure solid elements of the iron group have high magnetic concentrations and large exchange effects, leading to the ferromagnetic phenomena to be discussed in the next chapter.

> [*Ferromagnetism* is a property of certain materials (such as iron) that results in a significant, observable *magnetic permeability*, and in many cases, a significant *magnetic coercivity*, allowing the material to form a *permanent magnet*. *Ferromagnetic materials* are noticeably *attracted* to a magnet, which is a consequence of their substantial *magnetic permeability*.]

On the other hand, *most salts involving ions of the iron group are only paramagnetic*, except possibly at extremely low temperatures. In these salts the magnetic dilution is usually sufficient to warrant neglect of the exchange forces. This is perhaps obvious only if the salt is in solution, or is highly hydrated in the solid state. However, it is found that in true salts (not oxides) the *susceptibility* is usually affected comparatively little (not over 10 per cent, in many cases) by whether or not water molecules are present to increase the magnetic dilution.

...

XII HEISENBERG'S THEORY OF FERROMAGNETISM. FURTHER TOPICS IN SOLIDS

76. *The Heisenberg Exchange Effect*

An outstanding characteristic feature of the new quantum mechanics is the so-called '*Austausch*' or *exchange effect*, first discovered by Heisenberg[1].

> [1] Heisenberg, W. (June, 1926). Mehrkörperproblem und Resonanz in der Quantenmechanik. (Multibody Problem and Resonance in Quantum Mechanics.) *Zeit. Phys.*, 38, 411-426; https://doi.org/10.1007/BF01397160. The same effect was also discovered almost simultaneously by Dirac, P. A. M. (October, 1926). On the Theory of Quantum Mechanics. *Roy. Soc. Proc., A*, 112, 762, 661-77. [See above.]

It is concerned with the degeneracy associated with the possibility of two electrons trading places, and is best explained by considering first *a system with only two electrons and with neglect of spin*. First suppose that the electrons do not influence each other and that they are subject to fields derived from similar *potential* functions, so that the ***Schrodinger wave equation*** is

$$\nabla_1^2 \Psi + \nabla_2^2 \Psi + 8\pi^2 m/h^2 \, [W - V(x_1, y_1, z_1) - V(x_2, y_2, z_2)] \, \Psi. \qquad (1)$$

A solution of this equation is

$$\Psi_I = \Psi_k (x_1, y_1, z_1)\, \Psi_m (x_2, y_2, z_2), \qquad W = W_k + W_m, \qquad (2)$$

where Ψ_k, Ψ_m are solutions of the ***Schrodinger equation*** for a single electron subject to a potential V, as *in the absence of interaction* it is, of course, possible to consider each electron separately rather than together as in (1). We shall suppose the *wave functions* Ψ_k, Ψ_m are real, orthogonal, and normalized to unity[2].

[2] The restriction to real solutions involves no essential loss of generality for our purposes, and avoids the necessity of introducing complex coefficients in equations such as (4) or for distinguishing between J_{12} and J_{21}. The requirement of orthogonality is usually not met in the important case that k, m relate to different atoms, but the resulting error is not great if the wave functions of the different atoms do not overlap too much (cf. Heitler, W. & London, F. (June, 1927). Wechselwirkung neutraler Atome und homöopolare Bindung nach der Quantenmechanik. (Interaction of neutral atoms and homeopolar bonding according to quantum mechanics.) *Zeit. Phys.*, 44, 455–72. [See above.]

The physical interpretation of solution (2) is that electron 1 is in the state k and electron 2 in the state m (not necessarily states belonging to the same atom). This solution is, however, not the only one belonging to the energy $W_k + W_m$. An alternative solution is clearly

$$\Psi_{II} = \Psi_k (x_2, y_2, z_2)\, \Psi_m (x_1, y_1, z_1), \qquad W = W_k + W_m, \qquad (3)$$

in which the electrons have traded places as compared to (2). More generally, any linear combination of (2) and (3) is a solution.

[***Quantum superposition*** is a fundamental principle of quantum mechanics that ***states that linear combinations of solutions to the Schrödinger equation are also solutions of the Schrödinger equation.*** This follows from the fact that the Schrödinger equation is a linear differential equation in time and position. More precisely, the state of a system is given by a linear combination of all the eigenfunctions of the Schrödinger equation governing that system.]

The question now arises as to what is the proper combination to use when the degeneracy of interchange is removed by adding to the potential energy in (1) a potential energy V_{12} of interaction between the two electrons, which we may suppose *symmetrical* in the coordinates x_1, y_1, z_1 and x_2, y_2, z_2. Most readers doubtless already know that the answer is the '*symmetric*' and '*antisymmetric*' combinations

$$\Psi_{sym} = 1/2^{1/2}\ (\Psi_I + \Psi_{II}), \qquad \Psi_{ant} = 1/2^{1/2}\ (\Psi_I - \Psi_{II}), \qquad (4)$$

One way of proving (4) is to note that (4) diagonalizes the energy as far as the *exchange degeneracy* is concerned, since one can easily show that the fundamental quadrature (14), Chap. VI, vanishes if $\psi_{n'}$, ψ_n are respectively *symmetrical* and *antisymmetrical* or vice versa, and if f is a *symmetrical* function, such as V_{12}. ...

...

The *Pauli exclusion principle* demands that one use only *antisymmetric wave functions*[3].

> [3] The interpretation of the *exclusion principle* in terms of the *symmetry* of the *wave functions* appears to have first been given by Heisenberg and by Dirac, *loc. sit.* [1].

The symmetry properties, however, are profoundly modified by inclusion of the spin. If we neglect the *'magnetic'* coupling between the *spin* and *orbital angular momenta*, the *wave functions* are the product of the *orbital* and *spin* ones. Therefore, *when the orbital wave function is symmetrical, the spin one must be antisymmetrical and vice versa.* Now it can be shown that in a two-electron problem *the spin wave function is symmetrical when the spin quantum number S is 1 and is antisymmetrical when it is 0.*[4]

> [4] Cf, for instance, Dirac, *The Principles of Quantum Mechanics*, p. 214, or Sommerfeld, *Wellenmechanischer Erganzungsband*, p.274.

In other words, the triplet and singlet spectra (e.g. ortho- and parhelium) are respectively *antisymmetrical* and *symmetrical* in the *orbital* part of the *wave function.* Let $\mathbf{s}_1$ and $\mathbf{s}_2$ be the *spin angular momentum vectors* of the two *electrons*, measured in multiples of the quantum unit $h/2\pi$ as in previous chapters. The *characteristic values* [eigenvalues] of the matrix $(\mathbf{s}_1 + \mathbf{s}_2)^2$, the square of the magnitude of their resultant, are $S(S + 1)$, with S equal to 0 or 1. (By the *characteristic values* of a matrix are meant its diagonal elements after it is converted into a diagonal matrix by a proper canonical transformation. Cf. § 35.) Now $\mathbf{s}_1^2$ and $\mathbf{s}_2^2$ are invariably diagonal matrices whose diagonal elements are all $\frac{1}{2}(\frac{1}{2} + 1) = \frac{3}{4}$, as the spin quantum number for one electron is invariably ½; in other words, $\mathbf{s}_1^2$ and $\mathbf{s}_2^2$ are 'c-numbers' in Dirac's terminology. As ... it now follows that the *characteristic values* of the scalar product $\mathbf{s}_1 \cdot \mathbf{s}_2$ are ... $-\frac{3}{4}$ and ... $\frac{1}{4}$ corresponding respectively to $S = 0$ and $S = 1$, or to the symmetric and antisymmetric *orbital* solutions. The *characteristic values* of the *potential energy* V_{12} of interaction between the electrons are seen from (8) to be $K_{12} + J_{12}$ and $K_{12} - J_{12}$ respectively for the *symmetrical* and *antisymmetrical orbital* solutions, as the remaining terms in the Hamiltonian function have the *characteristic value* W_0 independent of the symmetry. Thus, the matrix V_{12} has the *characteristic value* $K_{12} + J_{12}$ when $\mathbf{s}_1 \cdot \mathbf{s}_2$ has the *characteristic value* $-\frac{3}{4}$, and $K_{12} - J_{12}$ when the latter has $+\frac{1}{4}$. In other words, $V_{12} - K_{12} + \frac{1}{2} J_{12} + 2J_{12}\, \mathbf{s}_1 \cdot \mathbf{s}_2$ has the *characteristic values* zero. Now a matrix whose *characteristic values* arc all zero is identically zero regardless of the system of representation, as any canonical transformation applied to a null matrix clearly still gives only a null matrix. Consequently, we have the matrix equality

$$V_{12} = K_{12} - ½\, J_{12} - 2J_{12}\, \mathbf{s}_1 \cdot \mathbf{s}_2, \qquad (9)$$

which applies regardless of whether the matrices in question have been transformed to diagonal form.

Eq. (9) shows that *the two electrons behave as though there were a strong coupling between their two spins which apart from an additive constant is proportional to the scalar product of these spin angular momenta, or to the cosine of the angle between the two spin vectors*. The latter is precisely the dependence of angle found in one term[5] of the mutual potential energy of two dipoles, so that the *exchange effect* has a partial semblance to *a very powerful magnetic coupling between the spins*.

[5] The mutual *potential energy* of two dipoles is $\mu_1 \cdot \mu_2/r^3 - 3(\mu_1 \cdot \mathbf{r})(\mu_2 \cdot \mathbf{r})/r^5$. Thus, only the first term is of the type form (10) [below].

This is not at all the same as saying that actually there is a real magnetic coupling of such magnitude, as the actual magnetic forces are so very weak that we have neglected them entirely in the present connection. The semblance of large direct coupling between the spins is only because the exclusion principle requires one type of orbital solution when the spins are parallel and another when they are antiparallel. Nevertheless, the interpretation (9), due to Dirac, of the *exchange effect* as formally equivalent to coupling between spins is exceedingly useful, as it enables us to picture and also to follow quantitatively the workings of the *exchange effect* by means of the vector model. The large par-ortho energy separations were shrouded in mystery before the new mechanics, as they require the constant of proportionality J_{12} in the coupling (9) to be fairly large. This trouble now disappears, as J_{12} is an *exchange integral* rather than a small magnetic factor.

Let us now pass to systems with more than two electrons, say a crystal composed of N atoms each having Z electrons. *The exchange degeneracy now becomes exceedingly complicated.* It is, in fact, (NZ)! fold rather than twofold as above, since *in order to treat the interatomic forces such as interest us for magnetism in solids it is necessary to consider the permutations of electrons not necessarily in the same atom of the crystal*. Even the problem of the Z!-fold *exchange degeneracy* for a single atom is complicated. *Regardless of the number of electrons, the Pauli exclusion principle requires that the wave functions still be antisymmetric in any two electrons if both the spin and orbital coordinates be interchanged*, but they will no longer in general be *symmetrical* or *antisymmetrical* in the orbital and spin parts considered separately. (The latter characteristic is peculiar to systems with only two electrons.) Eq. (9) shows that this is equivalent to saying that the spins of two electrons taken at random in the crystal (or even in the same atom) will not in general be parallel or anti-parallel, a result which seems quite obvious. The proper linear combinations of the (NZ)! original wave functions are usually deduced by rather involved group theory. We owe to Dirac[6] and Slater[7]

[6] Dirac, P. A. M. (April, 1929). Quantum Mechanics of Many-Electron Systems. *Roy. Soc. Proc., A*, 123, 792, 714-33 [See above], or *The Principles of Quantum Mechanics*, Chap. XI.

[7] Another method of avoiding group theory has boon given by Slater, [(1930). *Phys. Rev*., 35, 509]. Slater's method could doubtless be used to obtain the mean values (22-3) which we calculate in § 78. In fact, it is used by Bloch [(1930). *Zeit. Phys*., 57, 545] and Pauli [Report of the 1930 Solvay Congress] to obtain the *mean energy* (22), or its equivalent, but they do not give the more difficult computation of the *mean square energy* (23). Dirac's and Slater's procedures resemble each other in that their strength arises from recognizing at the outset that the *exclusion principle* severely restricts the symmetry character. Slater's method is very powerful for computing purposes when spacial degeneracy in the orbital motion must be considered, but does not give quite as much kinematical insight as Dirac's.

the elucidation that this is not really necessary because the *exclusion principle* limits so severely the allowable '*characters*' in the group theory. Dirac points out that the important results can instead all be obtained in an elementary way from the fact that Eq. (9) shows that any two *electrons* k, l in the crystal can be considered as having their *spins* coupled together by a *potential* of the form

$$-2J_{kl}\,\mathbf{s}_k\,.\,\mathbf{s}_l \qquad (10)$$

where the *coupling constant* or *exchange integral* J_{kl} will depend on the states assumed to be occupied by these two electrons, k and l, before allowing for the permutations. We here drop the first two right-hand terms of (9) as they do not depend on the orientations of the *spin*, and are of no interest for our problems in *magnetism*. These terms should, of course, be added when one requires absolute, as distinct from relative, *energies*. When there are more than two *electrons*, solution of the *exchange degeneracy* does not transform the matrix (10) into diagonal form, but only the expression

$$-2\,\Sigma\,J_{kl}\,\mathbf{s}_k\,.\,\mathbf{s}_l \qquad (11)$$

which is the *total exchange energy* of the *crystal* except for the additive term $-\frac{1}{2}\,\Sigma\,J_{kl}$ which we have dropped. The summation is over all the ½ NZ(NZ – 1) pairs of *electrons* in the *crystal*. The fact that individual terms in the sum (11) are not diagonalized does not impair the kinematical representation (10) of the *exchange effect*, as we have already mentioned that the validity of (9), which is basic to (10), (11), is invariant of the system of representation. We shall, for instance, show that use of (11) yields the mean values *employed by Heisenberg in his theory of magnetism.*

It is clearly to be understood that (11) is only an approximation, in that it embodies only the '*exchange*' secular problem connected with the *interaction* between the various members of a family of (NZ)! states having the same original *energy*, and neglects the *interaction* with the infinity of *states* with other unperturbed *energies*. An analogous approximation in the *two-electron problem* was made in (4)-(9). In other words, we use (32) rather than (15) of Chapter VI, i.e. we seek to express the perturbed *wave function* as a linear combination of a finite number of unperturbed *wave functions*, whereas an infinity is required for a complete development. This means that by solving the secular problem connected with (11) the *energy* is obtained only to a first approximation in a parameter λ proportional to the *coupling forces* between *electrons*.

This not only suffices to give all the essential qualitative features of the *exchange effect*, but is often a fair quantitative approximation in the case of *inter-atomic forces*, our primary interest. In the latter case the related integrals such as (6) and (7) are usually small since the *wave functions* of different *atoms* overlap but little. The most important thing for *magnetism*, however, is merely the fact that *the exchange effect, though entirely orbital in nature, is, because of the exclusion principle, very sensitive to the way the spin is aligned, and is formally equivalent to 'cosine coupling' between the spin magnets of the various atoms.*

A very vital point is that the alignment of the *spin* of a given *atom* having a non-vanishing *spin* is not influenced by the interaction with *atoms* which have closed shells of electrons and are thus in ^1S states. It is not correct to say that the *exchange effects* disappear entirely between a pair of *atoms* if at least one of them is in a ^1S state, as there is in any case the additive *exchange term* $-\frac{1}{2}\Sigma J_{12}$ which we have dropped in going from (9) to (10). This term, however, does not involve the *spin* and so is not of significance for our magnetic work. The significant part of the *exchange energy* for us does, however, vanish if one of the atoms is in an S state. To prove this, consider the *interaction* of a given *electron* k of one *atom* with a closed shell of r similar *electrons* ($l = 1, \ldots, r$) in another atom. According to (10) the part of the *exchange energy* depending on alinement of the *spin* is
$-2J_{kl}\,\mathbf{s}_k \cdot \Sigma_l\,\mathbf{s}_l$. This vanishes, as $\Sigma_l\,\mathbf{s}_l$ is zero for a closed shell. In other words, for our purposes (viz. neglecting terms which have no aligning effect on the *spin*), the exchange forces can be considered as existing only between the *paramagnetic* atoms or ions of a solid. These forces will thus be subordinate if the material has a high *'magnetic dilution'*, i.e. consists primarily of *diamagnetic* rather than *paramagnetic* atoms. Hence, *exchange effects* have played only a subordinate role in the preceding chapter.

77. *Heisenberg's Theory of Ferromagnetism*[9]

[9] Heisenberg, W. (September, 1928). Zur Theory of Ferromagnetismus. (On the theory of ferromagnetism.) *Zeit. Phys.*, 49, 619–36. [See above.]

The explanation of *ferromagnetism* has long been a conundrum. The early work of Ewing, subsequently amplified by Honda and others, showed that many of the phenomena of hysteresis and of magnetization in crystals could be described by assuming a large *potential energy* between adjacent molecular magnets. Also, Weiss, in his well-known theory, showed that many properties of *ferromagnetic* media, especially the thermal ones, could be imputed to a *local field* of the form H + qM. The portion qM proportional to the *intensity of magnetization* M is called the *'molecular field'*. The great difficulty, however, has been that tremendously large values must be assumed for the constant q, of the order 10^4, quite different from the value 4π/3 calculated under ordinary electromagnetic assumptions (§ 5). The magnetic forces between molecules are clearly too feeble to account for such enormous values of q, or for the large amount of *interaction* between molecular magnets in the Ewing theory. *Classical electrostatic forces lead to interactions of the right order of magnitude, but do not give the desired linearity of the Weiss molecular field in M or, what is more or less equivalent, the right dependence of the Ewing interaction energy on the angle between the elementary magnets*[12].

[12] For further discussion of this and related points and of the magnitude required for q see p. 703 of Debye's article in *Handbuch der Radiologie*, vol. vi.

This dilemma has been beautifully solved by the quantum-mechanical exchange forces described in 76. These forces are *electrostatic*, but *because of the constraints imposed by the Pauli exclusion principle are formally equivalent to a tremendously large coupling between spins*. In fact, reference to Eq. (10) or (11) shows that this coupling is proportional to the cosine of the angle between two *spins*, just as in the classical theories of Weiss and Ewing. Even without the following further analysis the empirical successes of these theories are thus already qualitatively understandable. *A crystal is nothing but a large molecule*. Hence, if we neglect the usually subordinate, purely magnetic coupling between spin and orbital angular momenta, the *total spin of the entire crystal is conserved*, like that of an ordinary *free molecule*, and its square has the characteristic values S'(S' + 1), where S' is a whole or half-integer according as *the number of electrons in the entire crystal is even or odd*. Also, if we continue to neglect the magnetic forces in comparison with the larger electrostatic ones, the energy of the crystal will not depend on the orientation (as distinct from the absolute value of) its resultant *spin*. The truth of these propositions can be seen by invoking the formal similarity of a crystal to the arbitrary polyatomic molecule. …

At this point it is perhaps well to say a word on notation. We employ primes, as in S', M'_S, &c., to distinguish *'crystalline quantum numbers'* and other expressions which relate to the entire *crystal*, regarded as one big *molecule*. Quantum numbers written in capital letters without primes, such as S, M_S, refer to a complete single *atom*, while those written in small letters, such as s, m_s, are, as usual, for a single *electron* within the *atom*.

If a *magnetic field* is applied along the z direction, the z component of the crystal's *spin* assumes a quantized value M'_S. Let us suppose that the *crystal* is composed of n identical *atoms* each having a given *spin* S. The maximum value of S' is then nS. The number of *states* $\Omega(M'_S)$ of the crystal having a given M'_S is best obtained by imagining a field so strong as to break down inter-atomic coupling and give each *atom* individual spacial quantization of *spin* and *orbit*. That ordinary laboratory fields are not adequate to do this is immaterial since we are merely counting the number of terms. Each atom, then, has a spacial *spin quantum number* M_S, and Ω is clearly the number of different combinations of the M_S consistent with the condition $M'_S = \Sigma M_S$. In case S = ½, the expression for Ω takes the simple form

$$\Omega(M'_S) = n!/\{(\tfrac{1}{2} n + M'_S)!\ (\tfrac{1}{2} n - M'_S)!\}, \qquad (12)$$

as here ½ n + M'_S *atoms* must have M_S = + ½, and ½ n – M'_S must have M_S = – ½; hence Ω is merely the number of permutations of n things between two classes. …

…

Heisenberg's calculations appear to assume that the *atoms* are in *S states*, but this is not really the case, as it is only necessary to suppose that the *orbital angular momentum* is quenched after the fashion explained in § 73. He also assumes that a given *atom* has an appreciable *exchange coupling* only with adjacent *atoms*, and possesses z such neighbors equidistant from it. Thus z = 2 for a linear chain, 4 for a quadratic surface grating, 6 for a simple cubic grating, 8 for body-centered cubic, and 12 for face-centered cubic. Let us further suppose that the *valence electrons*, or *electrons* not in closed shells, are in similar *states*. The part of the Hamiltonian function which involves *interatomic spin coupling*, and which we shall denote by $\mathcal{H}'$, is then

$$\mathcal{H}' = -2J\,\Sigma_{\text{neighbors}}\,\mathbf{S}_i \cdot \mathbf{S}_j. \qquad (16)$$

Here J is the *exchange integral* (7) between two *valence electrons* of adjacent atoms, and the summation extends over all neighboring pairs of *atoms*. The result (16) can be seen from (9) or (11), since $\Sigma\,\mathbf{s}_k \cdot \mathbf{s}_i = \Sigma\,\mathbf{s}_k \cdot \mathbf{s}_l = \mathbf{S}_i \cdot \mathbf{S}_j$ if k and *l* refer to different atoms i and j and if we sum over the *valence electrons* of both atoms. Closed shells contribute nothing to (11) or (16), as explained at the end of § 76, while *exchange effects* between *electrons* of the same *atom* merely give an additive constant to the *energy* as far as we are concerned, since we may suppose the inter-atomic forces not large enough to destroy the quantization S of the *spin* of each individual *atom*.

The fundamental problem of Heisenberg's *theory of ferromagnetism* is to calculate the characteristic values of (16) and hence the *energy states* belonging to various resultant *spins* of the crystal. Before explaining the mathematical details of how this is done, or rather circumvented, it will perhaps be illuminating to consider qualitatively three limiting cases.

1. $J/kT \gg 1$. Here *the exchange coupling is so exceedingly great* that the state $S' = nS$ of maximum crystalline spin has much less *energy* than all other states of less S' and hence is the only normal state. By regarding the whole crystal as a single molecule of spin nS, its susceptibility is seen to be $(2nS\beta/H)B_{nS}(2nS\beta H/kT)$, where B_{nS} is the Brillouin function defined in § 61. As the number n of *atoms* is very great, virtually any field is sufficient to make $n\beta H/kT \gg 1$, and so $B_{nS} = 1$, thus giving the full *saturation magnetization* $2nS\beta$. The crystal is then, so to speak, infinitely *ferromagnetic*. In fact, it would possess a *magnetic moment* even without an external *field*. This difficulty is, of course, avoided by supposing that our crystal is really a micro-crystal and that the macro-crystal is composed of a large number of micro-crystals, whose spins have random orientations and hence compensate each other without an external aligning field.

2. $|J/kT| \ll 1$. Here *the inter-atomic exchange coupling is negligible*, and the *susceptibility* will be $\chi = 4NS(S+1)\beta^2/3kT$, disregarding the here negligible saturation effects. This is the case which arises in *paramagnetic* iron salts (Chap. XI). We may here remark that the derivation of the Langevin-Debye formula given in Chapter VII can still be applied if the unit of structure is the (micro-) crystal instead of individual atom. We showed in Eq. (6), § 54, that the *orbit* and *spin* made the same contribution to the *susceptibility* as though both were entirely free, provided only their *interaction energy* is small compared to kT. Similarly, one can show that the *susceptibility* is the same as that coming from the individual *atomic spins*, considered separately, provided only the inter-atomic *exchange couplings* are small compared to kT.

3. $J/kT \ll -1$. Here *J is negative and the energy will be lowest when as many spins as possible are anti-parallel*, and the normal states are those of least S'. The inter-atomic coupling thus here erases practically all the *paramagnetism*, as will be discussed more fully in § 80.

Even *ferromagnetic* bodies conform to case 1 only asymptotically at $T = 0$. In such bodies the state $S' = nS$ of maximum spin for the crystal does, to be sure, represent the least energy, as when the spins are all parallel, each term in the summation (11) has its minimum characteristic value (assuming $J > 0$). However, whereas there is only one state with the maximum spin nS, there are by (13)-(15) n – 1 states of spin S' = nS – 1, approximately $n^2/2$ states of spin nS – 2, &c. This increasing number of states as S' is diminished from its maximum is important for two reasons. First of all, strength of numbers will partially offset the smaller Boltzmann factors for *spins* less than the maximum. In other words, the probability of the crystal being in *some* state having a given S'< nS may be appreciable even though the probability of its being in one particular designated state of this S' may be negligible. Secondly, all the states with a given S'< nS do not have the same *energy*, and a few favored ones may have quite low energies, even though they can never lie as deep as the state S' = nS. Whereas the infinitely *ferromagnetic* case 1 is thus too much of an idealization, it may nevertheless well be that most of the crystals have very large resultant *spins*. This seems to be the characteristic of *ferromagnetic materials*. A field of ordinary magnitude is then not able to produce the true saturation magnetization 2nSβ, as this would require that the field be able to convert the crystal into the state S' = nS, and only enormous fields can have an appreciable effect on the distribution of S' as distinct from M'_S. …

…

We must now seek to make these ideas more quantitative. …

…

81. ***The Diamagnetism of Free Electrons in Quantum Mechanics***

Landau[54] has discovered the very remarkable fact that *the orbital motions of free electrons give a diamagnetic contribution in quantum mechanics*, whereas we saw in § 26 that classically they were without such an effect. This difference is a little hard to explain intuitively, but arises from the fact that the boundary *electrons* have different quantized velocities than those which do not touch the walls of the vessel, and so the *magnetic moments* of these two types of *electrons* do not compensate each other as in classical theory. (Classically, both types have the Maxwell-Boltzmann distribution of velocities.) …

…

The Nobel Prize in Physics 1977, Press release, October 11, 1977.

The Royal Swedish Academy of Sciences has decided to award the 1977 Nobel Prize for physics to be shared equally between Dr. Philip W. Anderson, Bell Telephone Laboratories, Murray Hill, New Jersey, USA, Professor Sir Nevill F. Mott, Cambridge University, Cambridge, England and Professor John H. Van Vleck, Harvard University, Cambridge, Massachusetts, USA, for their fundamental theoretical investigations of the electronic structure of magnetic and disordered systems.

The three prize-winners are theoreticians within the field of solid-state physics – the branch of physics which lies behind essential parts of the current technical development, particularly in electronics. All three have added many new basic concepts to the theory, which have made it possible to understand new experimental results. The distance between fundamental results in basic research and technical applications is as a rule comparatively short in this field. As an example, one can mention that Van Vleck's ideas have played a central role for the development of the laser, whereas the technical development of amorphous materials like glass, which is now going on, would be unthinkable without Mott's and Anderson's contributions to the fundamental theory.

Van Vleck has been called "the father of modern magnetism". He has developed methods which make it possible to understand *how a foreign ion or atom behaves in a crystal*. At first the electrons of such a perturbing ion feel the influence of the electric field – the crystal field – which is generated by the atomic nuclei and the electrons of the host crystal. Through its electrons, the perturbing ion can also enter into chemical bonding with its environment which is usually called the *ligands*. Van Vleck was the first to develop the *crystal field theory* as well as the *ligand field theory* to describe such phenomena in greater detail. These quantum chemistry methods have now almost become routine tools, particularly within inorganic chemistry with important extensions to molecular biology, medicine and geology.

Another important part of Van Vleck's work deals with the Jahn-Teller effect, which is associated with an interaction between the *electrons* and the positions and motions of the *atomic nuclei*. A perturbing atom in a crystal can sometimes replace a host atom without essential changes in the surrounding lattice. Under certain circumstances the electronic structure of the perturbing atom is so incompatible with the symmetry of the environment that it leads to a local distortion of the lattice. This so-called Jahn-Teller effect was predicted in the 1930's but, only during the last decade, has one essentially through Van Vleck's work succeeded in understanding this phenomenon in greater detail and in realizing its experimental importance.

Van Vleck was the first to point out the essential importance of electron correlation – the *interaction* between the motions of the electrons – for the appearance of *local magnetic moments*, i. e. "mini-magnets" in materials. His former student P. W. Anderson, has further developed these ideas and succeeded in explaining how *local magnetic moments* can occur in metals, as for instance copper and silver, which in pure form are not magnetic at all. These phenomena can be quite complicated – the strength of the "mini-magnets" can, for instance, change abruptly when the concentration of the perturbing ion varies only a few percent. In a simple *quantum mechanical model*, Anderson has caught all the aspects which seem to be of decisive importance for understanding what happens in such situations

Mott and Anderson have separately given essential contributions to our knowledge of disordered systems. *In crystallic materials, the atoms form regular lattices*, which greatly facilitate the theoretical treatment. In *disordered materials*, this regularity is lacking – either so that the components of an alloy are placed at random in the regular lattice positions, or so that there is no lattice whatsoever as for instance in glass. It is exceedingly difficult to treat such materials theoretically. In 1958, Anderson published a paper in which he showed under what conditions an electron in a disordered system can either move through the system as a whole, or be more or less tied to a specific position as a localized electron. It was Mott who several years later called the attention of particularly the expert metal physicists to this paper, which has become one of the cornerstones in our understanding of, among other things, the electric conductivity in disordered systems. Mott and Anderson have in a series of papers created a multitude of new concepts which have turned out to be central for the understanding of disordered materials. Their ideas have to a large extent been experimentally verified and they have in this way laid the foundation for important technical developments.

The electric properties of crystals are described by the so-called band theory which gives a classification with respect to the conductivity in metals, semiconductors, and insulators. This theory is not universally valid, however, and a famous exception is provided by nickel oxide, which according to band theory ought to be a metallic conductor but in reality is an insulator. Mott has shown how this can be explained by means of a refined theory which takes the *electron-electron interaction* into account. This led also to the study of the so-called Mott transitions, by which certain metals can become insulators when the electronic density decreases by separating the atoms from each other in some convenient way.

... This year's prize puts the emphasis – on their work concerning *electron-electron interaction* and the *coupling between the motions of the electrons and the atomic nuclei in magnetic and disordered materials*, where they – particularly in the treatment of and the emphasis on localized electronic states – have gone far beyond the conventional theories, with direct importance for experiments and technology.

John H. Van Vleck – 1977 Nobel Lecture, December 8, 1977. *Quantum Mechanics the key to understanding Magnetism.*

[https://www.nobelprize.org/uploads/2018/06/vleck-lecture.pdf]

> [The Nobel Prize in Physics 1977 was awarded jointly to Philip Warren Anderson, Sir Nevill Francis Mott and John Hasbrouck Van Vleck "for their fundamental theoretical investigations of the electronic structure of magnetic and disordered systems".
>
> The electrical and magnetic properties of different materials are determined by how the electrons move about in relation to the atomic nucleus. When an atom from a foreign substance is inserted into a crystalline structure, the crystal's properties can be altered. During the 1930s John Van Vleck developed theories about how electrical fields in a crystal affect a foreign atom and how such an atom can be bound to nearby atoms through its electrons. He also showed how the interaction between the electron's movements can create local magnetic moments in crystals. [John H. Van Vleck – Facts. NobelPrize.org. https://www.nobelprize.org/prizes/physics/1977/vleck/facts/.]

In his Nobel Prize lecture Van Vleck provided a brief history of the evolution of the quantum theory of magnetic materials.

> The existence of *magnetic materials* has been known almost since prehistoric times, but only in the 20th century has it been understood how and why the magnetic susceptibility is influenced by chemical composition or crystallographic structure. In the 19th century the pioneer work of Oersted, Ampere, Faraday and Joseph Henry revealed the intimate connection between electricity and magnetism. Maxwell's classical field equations paved the way for the wireless telegraph and the radio. At the turn of the present century Zeeman and Lorentz received the second Nobel Prize in physics for respectively observing and explaining in terms of classical theory the so-called normal Zeeman effect. The other outstanding early attempt to understand magnetism at the atomic level was provided by the semi-empirical theories of Langevin and Weiss. To account for *paramagnetism*, Langevin (1) in 1905 assumed in a purely ad hoc fashion that an atomic or molecular magnet carried a permanent moment μ, whose spatial distribution was determined by the Boltzmann factor. It seems today almost incredible that this elegantly simple idea had not occurred earlier to some other physicist inasmuch as Boltzmann had developed his celebrated statistics over a quarter of a century earlier. With the Langevin model, the average

magnetization resulting from N elementary *magnetic dipoles* of *strength* μ in a *field* H is given by the expression

$$M = \ldots = NL\ (\mu H/kT), \text{ where } L(x) = \coth x - 1/x. \qquad (1)$$

At ordinary temperatures and field strengths, the argument x of the *Langevin function* can be treated as small compared with unity. Then $L(x) = 1/3\ x$, and Eq. (1) becomes

$$M = N\ \mu^2/3kT\ H \qquad (2)$$

so that the *magnetic susceptibility* $x = M/H$ is inversely proportional to the temperature, a relation observed experimentally for oxygen ten years earlier by Pierre Curie and hence termed Curie's law.

To explain *diamagnetism*, Langevin took into account the Larmor precession of the electrons about the magnetic field, and the resulting formula for the *diamagnetic susceptibility* is

$$x = -\ Ne^2/6mc^2\ \Sigma\ (r_i)^2 \qquad (3)$$

where $(r_i)^2$ is the *mean square radius* of an *electron orbit*, and the summation extends over all the electrons in the atom. The important thing about (3) is that, in substantial agreement with experiment, it gives a *diamagnetic susceptibility* independent of temperature, provided the size of the orbits does not change.

Two years later, in 1907, Pierre Weiss, another French physicist, took the effective field acting on the atom or molecule to be the applied field augmented by a mysterious internal or molecular field proportional to the intensity of magnetization. The argument of the Langevin function then becomes $\mu(H + qM)/kT$ rather than $\mu H/kT$ and in place of (2) one has

$$x = M/H = N\mu^2/3k(T - T_c) \text{ where } T_c = Nq\mu^2/3k \qquad (4)$$

Since the right side of (4) becomes infinite for $T = T_c$, the Weiss model predicts the existence of a Curie point below which *ferromagnetism* sets in. This model also describes qualitatively quite well many *ferromagnetic* phenomena. Despite its many successes there was *one insuperable difficulty* from the standpoint of classical electrodynamics. Namely the coefficient q of the molecular field qM should be of the order $4\pi/3$ whereas it had to be of the order 10^3 to describe the observed values of T_c.

There was, moreover, an *even worse difficulty*. If one applies classical dynamics and statistical mechanics consistently, a very simple calculation, which can be made in only a few lines but I shall not reproduce it here, shows that *the diamagnetic and paramagnetic contributions to the susceptibility exactly cancel.* Thus, *there should be no magnetism at all.* This appears to have been first pointed out by Niels Bohr in his doctor's dissertation in 1911, perhaps the most deflationary publication of all time in physics. This may be one reason why Bohr broke with tradition and came forth with his remarkable theory of the hydrogen spectrum in 1913. That year can be regarded as the debut of what is called the *old quantum theory* of atomic structure, which utilized classical mechanics supplemented by quantum conditions. In particular *it quantized angular momentum* and hence the *magnetic moment* of the *atom*, as was verified experimentally in the molecular beam experiments of Stern and Gerlach. Hence there was no longer the statistical continuous distribution of values of the dipole moment which was essential to the proof of zero magnetism in classical theory. When Langevin assumed that the *magnetic moment* of the atom or molecule had a fixed value μ, he was quantizing the system without realizing it, just as in Moliere's *Bourgeois Gentilhomme*, Monsieur Jourdain had been writing prose all his life, without appreciating it, and was overjoyed to discover he had been doing anything so elevated. *Magnetism could be understood qualitatively in terms of incomplete shells of electron orbits*, and a sentence of Bohr which I like to quote reads "In short, an examination of the magnetic properties and colors of the long periods gives us a striking illustration of how a wound in the otherwise symmetrical inner structure of the atom is first created and then healed." However, with the passage of time it became increasingly clear that *the old quantum theory could give quantitatively correct results for energy levels or spectral frequencies only in hydrogen.* One historian of science has referred to the early 1920's as the crisis in quantum theory, but I would characterize this era as one of increasing disillusion and disappointment in contrast to the hopes which were so high in the years immediately following 1913.

The advent of quantum mechanics in 1926 furnished at last the real key to the quantitative understanding of magnetism, I need not elaborate on the miraculous coincidence of three developments, the discovery of the matrix form of quantum mechanics by Heisenberg and Born, the alternative but equivalent wave mechanical form by de Broglie and Schrödinger, and the introduction of *electron spin* by Uhlenbeck and Goudsmit. *A quantum mechanics without spin and the Pauli exclusion principle would not have been enough - one wouldn't have been able to understand even the structure of the periodic table or most magnetic phenomena.* Originally *spin* was a sort of appendage to the mathematical framework, but in Dirac 1928 synthesized everything in his remarkable four first order simultaneous equations. To stress the importance of the quantum mechanical revolution, I cannot do better than to quote an often-mentioned sentence from one of Dirac's early papers, which reads "*The general theory of quantum mechanics is now almost complete. The underlying physical laws necessary for the mathematical theory of a large part of physics and all of chemistry are thus completely known*".

With at last the key available for the proper analysis of what was going on inside the atom, it was natural that more than one physicist would try applying it to a particular problem. So it is not surprising that four different researchers independently calculated and reported in practically simultaneous publications the *susceptibility* of a rotating *diatomic molecule* carrying a *permanent dipole moment*, which could be either electric or magnetic depending on whether one was interested in an electric or magnetic susceptibility. (I was one of the four. The others were Kronig, Manneback, and Miss Mensing working in collaboration with Pauli. The new mechanics happily restored the factor in the Langevin formula) (or the corresponding Debye expression in the electric case), as shown in Table I. *Thus was ended the confusion of the old quantum theory*, where half quanta worked better in band spectra even though whole integers were required with rational application of Bohr's 1913 ideas.

There are three common *paramagnetic* gases, viz. O_2, NO_2 , and NO. I shall discuss NO first as its behavior is the most interesting of the three. In 1926 Robert Mulliken, who has a sixth sense for deducing molecular energy levels from band spectra, had decided that the ground state of the NO molecule was a $^2\Pi$ state, whose two components were separated by about 122 cm^{-1} but he wasn't sure whether the doublet was regular rather than inverted. I tried calculating the *susceptibility* of NO on the basis of Mulliken's energy levels and found that the observed *susceptibility* at room temperatures could be explained on the basis that the doublet was regular, i.e. the $^2\Pi_{1/2}$ component lower than the $^2\Pi_{3/2}$. I wasn't entirely convinced that the agreement was real rather than spurious, as molecular quantum mechanics was then in its infancy. If the theory was correct there should be deviations from Curie's law, and so measurements on the *susceptibility* as a function of temperature should be decisive. To my surprise, experiments to test this prediction were performed in 1929 at three different laboratories in different parts of the world, with each going to a lower temperature than the preceding. As shown in Fig. 1, the agreement with theory was gratifying. …

…

In 1930 and 1931 a great deal of my time went into writing my book on the *Theory of Electric and Magnetic Susceptibilities*, which appeared in 1932 [Oxford University Press]. In this volume I aimed to include the major theoretical developments which had taken place up to the time of writing. Besides the things which I have already mentioned, there were other major developments in the theory of magnetism in the early days of quantum mechanics. Heisenberg took the mystery out of the then twenty-year-old Weiss molecular field. [Heisenberg, W. (September, 1928). *Zur Theory of Ferromagnetismus*. (On the theory of ferromagnetism.) See above.] He showed that it arose from *exchange effects* connecting the different magnetic atoms, which had the effect of introducing the needed *strong coupling between the spins*. Other notable theoretical developments prior to 1932 included Landau's paper on the *diamagnetism* of *free electrons*, in which he showed that *spinless* free electrons had a small susceptibility of *diamagnetic* sign, in contrast to the zero result of classical mechanics. [Landau, L. (1930). *Zeit. Phys*., 64, 629.] Pauli showed that the *spin moment* of conduction electrons gives rise to only a small *paramagnetic* susceptibility practically independent of temperature. [Pauli, W. (1927). Über Gasentartung und Paramagnetismus. (On Gas Degeneration and Paramagnetism.) *Zeit. Phys*., 41, 81-102; https://doi.org/10.1007/BF01391920.] This paper was notable because it was the first application of Fermi-Dirac statistics to the solid state. If one used the Boltzmann statistics one would have a large susceptibility obeying Curie's law.

On the other hand, there were some important developments which arrived just a little too late for me to include them in my volume. Néel's first paper on *antiferromagnetism* appeared in 1932, and in later years he introduced an important variant called *ferrimagnetism*, in which the anti-parallel dipoles are of unequal strength, so that they do not compensate and the resulting behavior can be *ferromagnetic*. There was also Peirls' theoretical explanation of the de Haas-van-Alphen effect, and Bloch's 1932 paper on the width of the boundaries (now called Bloch walls) separating the elementary domains in *ferromagnetic* materials. The corresponding *domain structure* was explained and elaborated by Landau and Lifschitz two years later.

In 1930 I held a Guggenheim fellowship for study and travel in Europe. I spent most of the time in Germany, but by far the most rewarding part of the trip scientifically was a walk which I took with Kramers along one of the canals near Utrecht. He told me about his own theorem on *degeneracy* in molecules with an odd number of electrons and also of Bethe's long paper concerned with the application of group theory to the determination of the *quantum mechanical energy levels* of atoms or ions exposed to a crystalline electric field, and in my book I referred to the role of the crystalline field only in a qualitative way, *stressing the fact that it could largely suppress the orbital part of the magnetic moment in salts of the iron group*. In the process of writing, I did not have the time or energy to attempt quantitative numerical computations. I was most fortunate when, beginning in the fall of 1931 I had two post-doctoral students from England, namely William (now Lord) Penney, and Robert Schlapp. I suggested to these two men that they make calculations respectively on *salts of the rare earth* and of the *iron group*. The basic idea of the crystalline field potential is an extremely simple one, namely that the magnetic ion is exposed not just to the applied magnetic field but experiences in addition a static field which is regarded as an approximate representation of the forces exerted upon it by other atoms in the crystal. The form of the crystalline *potential* depends on the type of crystalline symmetry. …

…

When applied to the *iron group* the results of *crystal field theory* are particularly striking and form the basis of much of what may be called modern *magnetochemistry.* The crystalline *potential* is much larger than for the rare earths and *is so powerful that it quenches a large part of the orbital part of the magnetic moment even at room temperatures.* Schlapp found that the magnetic behavior in the iron group required a large *crystalline field* of nearly (but usually not entirely) *cubic symmetry.*

…

So far, I have not said much about *ferromagnetism*, partly because more of my own work has been in *paramagnetism*, but mainly because most ferromagnetic metals are very complicated since they are conductors. Over the years there have been arguments ad infinitum as to which is the best model to use, each researcher often pushing his own views with the ardor of a religious zealot. Heisenberg's original model was one in which *the spins responsible for the ferromagnetism did not wander from atom to atom*, whereas in the band picture developed by Stoner *the electrons carrying a free spin can wander freely through the metal* without any correlation in their relative positions, as the *exchange effects* are approximated by an uncorrelated molecular field.

> [The *Stoner model* is a simplified model of a solid which is formulated in terms the dimensionless *density of spin up (down) electrons* and the *dispersion relation of spinless electrons* where the electron-electron interaction is disregarded. It can be used to calculate the total energy of the system as a function of its polarization.
>
> The *Heisenberg model* describes magnetic interactions in terms of *spin exchange* between neighboring magnetic moments. It focuses on the alignment of spins in a crystal lattice, considering the *exchange interaction energy*.]

Undoubtedly the truth is between the two extremes, and I have always favored as a first approximation a sort of compromise model, which may be called that of minimum polarity. In nickel for instance, this model there is continual interchange of electrons between the configurations d^{10} and d^9 but no admixture of d^8, d^7, etc. as then the correlation energy is increased.

…

Jaščur, M. (October, 2013). Quantum Theory of Magnetism.

Pavol Jozef Šafárik University. Košice, Slovakia.

http://www.upjs.sk/pracoviska/univerzitna-kniznica/e-publikacia/#pf (Slovak); English translation; https://www.upjs.sk/public/media/5596/Qauntum-Theory-of-Magnetism.pdf.

Preface

...

The text consists of four chapters that form closed themes of the introductory *quantum theory of magnetism.* In the first part, we explain the *quantum origin of magnetism*, illustrating the appearance of the *exchange interaction* for the case of hydrogen molecule. The second chapter is devoted to the application of the *Bogolyubov inequality* to the *Heisenberg model* and it represents the part which is usually missing in the textbooks of magnetism. The third chapter deals with the standard *spin wave-theory* including the *Bloch and Holstein-Primakoff* approach. Finally, we discuss the application of the *Jordan-Wigner* transformation to one-dimensional spin-1/2 XY model.

...

Chapter 1 Many-Body Systems

1.1 *Systems of Identical Particles*

Magnetism is a many-body phenomenon and its origin can only be explained within quantum physics. In this part, we will investigate some important properties of the wave functions of quantum systems, consisting of many identical particles that are important to understand magnetic properties in the solid state. In general, *the Hamiltonian of the many-body system depends on the time, coordinates and spin variables of all particles*. However, in this part we will mainly investigate the *symmetry properties of the wave-function* that are independent on the time and therefore we can exclude the time from our discussion. For simplicity we neglect also spin-degrees of freedom.

Let us consider a system of N identical particles described by the many-body Hamiltonian $\hat{H}(\mathbf{r}_1, \mathbf{r}_2, \ldots, \mathbf{r}_N,)$ which remains unchanged with respect to the transposition of arbitrary two particles in the system. This fact can be mathematically expressed as

$$\hat{H}(\mathbf{r}_1, \mathbf{r}_2, \ldots, \mathbf{r}_j, \ldots, \mathbf{r}_k, \ldots, \mathbf{r}_N) = \hat{H}(\mathbf{r}_1, \mathbf{r}_2, \ldots, \mathbf{r}_k, \ldots, \mathbf{r}_j, \ldots, \mathbf{r}_N) \quad (1.1)$$

or

$$\hat{P}_{jk}\, \hat{H}(\mathbf{r}_1, \mathbf{r}_2, \ldots, \mathbf{r}_j, \ldots, \mathbf{r}_k, \ldots, \mathbf{r}_N) = \hat{H}(\mathbf{r}_1, \mathbf{r}_2, \ldots, \mathbf{r}_k, \ldots, \mathbf{r}_j, \ldots, \mathbf{r}_N)\, \hat{P}_{jk}. \quad (1.2)$$

The previous equation defines the *transposition operator* $\hat{P}_{jk}$ of the jth and kth particle. Since a mutual interchange of arbitrary particles in the system does not change its state, then the total wavefunction of the system obeys the following relation

$$\hat{P}_{jk}\, \Psi(\mathbf{r}_1, \mathbf{r}_2, \ldots, \mathbf{r}_j, \ldots, \mathbf{r}_k, \ldots, \mathbf{r}_N) = e^{i\alpha}\, \Psi(\mathbf{r}_1, \mathbf{r}_2, \ldots, \mathbf{r}_k, \ldots, \mathbf{r}_j, \ldots, \mathbf{r}_N). \quad (1.3)$$

and

$$\hat{P}^2_{jk}\, \Psi(\mathbf{r}_1, \mathbf{r}_2, \ldots, \mathbf{r}_j, \ldots, \mathbf{r}_k, \ldots, \mathbf{r}_N, t) = e^{2i\alpha}\, \Psi(\mathbf{r}_1, \mathbf{r}_2, \ldots, \mathbf{r}_j, \ldots, \mathbf{r}_k, \ldots, \mathbf{r}_N), \quad (1.4)$$

where the real parameter α apparently satisfies equation

$$e^{2i\alpha} = 1 \text{ or } e^{i\alpha} = \pm 1. \quad (1.5)$$

The last equation implies that the total wavefunction either remains the same, or changes its sign when two arbitrary particles are interchanged in the system. In the first case, the total wave function is called as a *symmetric* function and in the later one as an *anti-symmetric* function. The symmetry of the wavefunction of an ensemble of identical particles is exclusively given by the type of particles and does not depend on external conditions. In fact, *it has been found that all particles obeying the Pauli exclusion principle have the antisymmetric wavefunction and all other particles have the symmetric wavefunction.* Unfortunately, the direct theoretical proof of this statement is impossible for strongly interacting particles, since in real many-body systems is not possible to calculate exactly the total wavefunction. On the other hand, the situation becomes tractable for the case of non-interacting or weakly interacting particle. Therefore, *we at first will investigate the system of spinless non-interacting identical particles and then we clarify the role of the interaction and spin.*

In the case of *noninteracting particles* the total Hamiltonian of the system can be expressed as a sum of single-particle Hamiltonians, i.e.

$$\hat{H} = \sum_j^N \hat{H}_j, \qquad (1.6)$$

where N denotes the number of particles. It is also well known that *the total wavefunction of non-interacting particles can be written as a product of the single-particle wavefunctions*, namely,

$$\Psi(\mathbf{r}_1, \mathbf{r}_2, \ldots, \mathbf{r}_j, \ldots, \mathbf{r}_k, \ldots, \mathbf{r}_N) = \prod_{j=1}^N \phi_{nj}(\mathbf{r}_j), \qquad (1.7)$$

where n_j represents the set of all *quantum numbers* characterizing the relevant *quantum state* of the jth particle. *The single-particle functions $\phi_{nj}(\mathbf{r}_j)$ are of course the solutions of the stationary* ***Schrodinger equation***

$$\hat{H}_j\phi_{nj}(\mathbf{r}_j) = \varepsilon_{nj}\, \phi_{nj}(\mathbf{r}_j). \qquad (1.8)$$

Here ε_{nj} denote the *eigenvalues* of the single-particle Hamiltonian $\hat{H}_j$, so that the eigenvalue of the total Hamiltonian is given by $E_{n1,n2\ldots,nN} = \sum_j \varepsilon_{nj}$. *The symmetric wavefunction describing whole system* can be written in the form

$$\Psi(\mathbf{r}_1, \mathbf{r}_2, \ldots, \mathbf{r}_j, \ldots, \mathbf{r}_k, \ldots, \mathbf{r}_N) = A\sum_P \phi_{n1}(\mathbf{r}_1)\phi_{n2}(\mathbf{r}_2)\ldots\, \phi_{nN}(\mathbf{r}_N), \qquad (1.9)$$

where A denotes the normalization constant and summation is performed over all possible permutation of the particles. Similarly for the *antisymmetric* case we obtain the form

$$\Psi(\mathbf{r}_1, \mathbf{r}_2, \ldots, \mathbf{r}_j, \ldots, \mathbf{r}_k, \ldots, \mathbf{r}_N) = 1\sqrt{N!}\, \begin{vmatrix} \phi_{n1}(\mathbf{r}_1) & \phi_{n1}(\mathbf{r}_2) & \ldots & \phi_{n1}(\mathbf{r}_N) \\ \phi_{n2}(\mathbf{r}_1) & \phi_{n2}(\mathbf{r}_2) & \ldots & \phi_{n2}(\mathbf{r}_N) \\ \cdot & \cdot & & \cdot \\ \cdot & \cdot & & \cdot \\ \phi_{nN}(\mathbf{r}_1) & \phi_{nN}(\mathbf{r}_2) & \ldots & \phi_{nN}(\mathbf{r}_N) \end{vmatrix} \qquad (1.10)$$

which is usually referred to as *the Slater determinant*. It is clear, that *interchanging of two particles in the system corresponds to the interchange of relevant rows in the determinant* (1.10), which naturally leads to the sign change of the wavefunction. Moreover, *if two or more particles occupy the same state, then two or more rows in the determinant (1.10) are equal and consequently, the resulting wavefunction is equal to zero in agreement with the Pauli exclusion principle*.

Until now, we have considered a simplified system of *non-interacting particles* and we have completely neglected *spin degrees of freedom*. However, the situation in realistic experimental systems is much more complicated, since *each particle has a spin* and moreover, the interactions between particles often also substantially influence the behavior of the system. However, if we assume the case of weak interactions and if we neglect the spin-orbit interactions then the main findings discussed above remain valid and moreover, the existence of the *exchange interaction* can be clearly demonstrated.

Instead of developing an abstract and general theory for such a case, it is much more useful to study a typical realistic example of the hydrogen molecule, which illustrates principal physical mechanisms leading to the appearance of magnetism.

1.2 *Heitler-London Theory of Direct Exchange*

In the previous part, we have found that *the total wavefunction of the many-electron system must be anti-symmetric*. In this subsection, *we will demonstrate that the use of anti-symmetric wave functions leads to the purely quantum contribution to the energy of the system, which is called the exchange energy*. In fact, this energy initiates a certain ordering of the spins, i.e. *it may lead to the magnetic order in the system*. A similar effect we also obtain using the single-product wavefunctions, if we explicitly include the *exchange-interaction term* into Hamiltonian. *This effect was independently found by Heisenberg and Dirac in 1926 and it represent the modern quantum-mechanical basis for understanding magnetic properties in many real systems.*

We start our analysis with one of the simplest possible system, namely, the H_2 molecule. In the *hydrogen molecule* two electrons interact with each other and with the nuclei of the atoms. ... considering this geometry, we can write the Hamiltonian of the H_2 molecule in the form

$$\hat{H} = \hat{H}_1 + \hat{H}_2 + \hat{W} + \hat{H}_{LS} \quad (1.11)$$

where

$$\hat{H}_1 = -\hbar^2/2m\, \Delta_1 - e_0^2/r_{a1} \quad (1.12)$$

$$\hat{H}_2 = -\hbar^2/2m\, \Delta_2 - e_0^2/r_{b2} \quad (1.13)$$

$$\hat{W} = e_0 2/r_{ab} + e_0^2/r_{12} - e_0^2/r_{b1} - e_0^2/r_{a2} \quad (1.14)$$

and we have used the abbreviation $e_0 = e/(4\pi\varepsilon_0)$. The terms $\hat{H}_1$ and $\hat{H}_2$ in (1.11) describe the situation when the two hydrogen atoms are isolated. *The operator $\hat{W}$ describes the interaction between two cores*, electrons, as well as between electrons and relevant nuclei. Finally, the last term represents the *spin-orbit interaction*, which is assumed to be small so that we can separate the orbital and spin degrees of freedom. To solve the problem of the *hydrogen molecule*, we apply the first-order perturbation theory neglecting the *spin-orbit coupling* and taking the term $\hat{W}$ as a small perturbation.

As it is usual in the perturbation theory, we at first solve the unperturbed problem, i.e. the system of two non-interacting hydrogen atoms. Thus, we have to solve the ***Schrodinger equation***

$$(\hat{H}_1 + \hat{H}_2)\Psi = U_0\Psi \qquad (1.15)$$

As we already have noted above, ***we will assume that the total wave function can be written as a product of the orbital and spin functions***, i.e. $\Psi = \varphi(\mathbf{r}_1, \mathbf{r}_2)\chi(s^z_1, s^z_2)$. Another very important point to be emphasized here is that our simplified Hamiltonian does not explicitly depend on the *spin* variables. Consequently, *we can at first evaluate only the problem with orbital functions and at the end of the calculation we can multiply the result by an appropriate spin function to ensure the anti-symmetry of the total wavefunction.* In order to express this situation mathematically, we introduce the following notations

- $\phi_\alpha(\mathbf{r}_i)$, where $\alpha = a$, b; i = 1, 2 represents the *orbital wave function* of the isolated hydrogen atom when the ith electron is localized closely to the αth nucleus,
- $\xi_\gamma(i)$, where $\gamma = \uparrow, \downarrow$; i = 1, 2 is the *spin function* describing the spin up or spin down of the ith electron.

Now, let us proceed with the discussion of the *orbital functions* of the system described by Eq. (1.15). Since the *orbital functions* $\phi_\alpha(\mathbf{r}_i)$ are *eigenfunctions* of the relevant one-atom ***Schrodinger equation*** with the same *eigenvalue* $E_0 = -13.55$eV, it is clear that the ground state of the *two noninteracting hydrogen atoms* has the energy $U_0 = 2E_0$ and it is doubly degenerated. This is so-called *exchange degeneracy* and two *wavefunctions* corresponding to this U_0 can be expressed as $\varphi_1(\mathbf{r}_1, \mathbf{r}_2) = \phi_a(\mathbf{r}_1)\phi_b(\mathbf{r}_2)$ and $\varphi_2(\mathbf{r}_1, \mathbf{r}_2) = \phi_a(\mathbf{r}_2)\phi_b(\mathbf{r}_1)$.
If we include into the Hamiltonian also the *interaction term* $\hat{W}$ then the situation becomes much more complex and the *eigenfunctions* and *eigenvalues* cannot be found exactly. In the spirit of the perturbation theory, we will assume that the *wavefunction* of the interacting system can be expressed in the form

$$\varphi(\mathbf{r}_1, \mathbf{r}_2) = c_1\varphi_1(\mathbf{r}_1, \mathbf{r}_2) + c_2\varphi_2(\mathbf{r}_1, \mathbf{r}_2) = c_1\phi_a(\mathbf{r}_1)\phi_b(\mathbf{r}_2) + c_2\phi_a(\mathbf{r}_2)\phi_b(\mathbf{r}_1) \qquad (1.16)$$

where c_1 and c_2 are the constant that will be determined later. The system is now described by the following ***Schrodinger equation***

$$(\hat{H}_1 + \hat{H}_2 + \hat{W})[c_1\varphi_1(\mathbf{r}_1, \mathbf{r}_2) + c_2\varphi_2(\mathbf{r}_1, \mathbf{r}_2] = E[c_1\varphi_1(\mathbf{r}_1, \mathbf{r}_2) + c_2\varphi_2(\mathbf{r}_1, \mathbf{r}_2)]. \quad (1.17)$$

Multiplying the previous equation by $\varphi_1{}^*(\mathbf{r}_1, \mathbf{r}_2)$ and integrating over the space one obtains

$$\begin{aligned} &2c_1E_0 + 2c_2E_0 \iint \varphi^*{}_1(\mathbf{r}_1, \mathbf{r}_2)\varphi_2(\mathbf{r}_1, \mathbf{r}_2)dV_1dV_2 \\ &\qquad + 2c_1 \iint \varphi^*{}_1(\mathbf{r}_1, \mathbf{r}_2)\hat{W}\varphi_1(\mathbf{r}_1, \mathbf{r}_2)dV_1dV_2 \\ &\qquad + c_2 \iint \varphi^*{}_1(\mathbf{r}_1, \mathbf{r}_2)\hat{W}\varphi_2(\mathbf{r}_1, \mathbf{r}_2)dV_1dV_2 \\ &= c_1E + c_2E \iint \varphi^*{}_1(\mathbf{r}_1, \mathbf{r}_2)\varphi_2(\mathbf{r}_1, \mathbf{r}_2)dV_1dV_2. \end{aligned} \quad (1.18)$$

Similarly, by multiplying Eq. (1.17) by $\varphi{*}_2(\mathbf{r}_1, \mathbf{r}_2)$ and integrating over the space one finds

$$\begin{aligned} &c_1E_0 \iint \varphi^*{}_2(\mathbf{r}_1, \mathbf{r}_2)\varphi_1(\mathbf{r}_1, \mathbf{r}_2)dV_1dV_2 \\ &\qquad + c_1 \iint \varphi^*{}_2(\mathbf{r}_1, \mathbf{r}_2)\hat{W}\varphi_1(\mathbf{r}_1, \mathbf{r}_2)dV_1dV_2 + 2c_2E_0 \\ &\qquad + c_2 \iint \varphi^*{}_2(\mathbf{r}_1, \mathbf{r}_2)\hat{W}\varphi_2(\mathbf{r}_1, \mathbf{r}_2)dV_1dV_2 \\ &= c_1E \iint \varphi^*{}_2(\mathbf{r}_1, \mathbf{r}_2)\varphi_1(\mathbf{r}_1, \mathbf{r}_2)dV_1dV_2 + c_2E. \end{aligned} \quad (1.19)$$

Here one should notice that in deriving Eqs. (1.18) and (1.19) we have used *normalization conditions* for φ_1 and φ_2.

In order to express two previous equations in an abbreviated form one introduces the following quantities:

1. The *overlap integral*

$$S_0 = \int \phi^*{}_a(\mathbf{r}_i)\phi_b(\mathbf{r}_i)dV_i \qquad i = 1, 2. \quad (1.20)$$

2. The *Coulomb integral*

$$K = \iint \varphi^*{}_1(\mathbf{r}_1, \mathbf{r}_2)\hat{W}\varphi_1(\mathbf{r}_1, \mathbf{r}_2)dV_1dV_2 = \iint \varphi^*{}_2(\mathbf{r}_1, \mathbf{r}_2)\hat{W}\varphi_2(\mathbf{r}_1, \mathbf{r}_2)dV_1dV_2. \quad (1.21)$$

3. The *exchange integral*

$$A = \iint \varphi^*{}_1(\mathbf{r}_1, \mathbf{r}_2)\hat{W}\varphi_2(\mathbf{r}_1, \mathbf{r}_2)dV_1dV_2 = \iint \varphi^*{}_2(\mathbf{r}_1, \mathbf{r}_2)\hat{W}\varphi_1(\mathbf{r}_1, \mathbf{r}_2)dV_1dV_2. \quad (1.22)$$

Applying this notation we rewrite Eqs. (1.18) and (1.19) in the form

$$\begin{aligned} &c_1(2E_0 + K - E) + c_2(2E_0S_0{}^2 + A - ES_0{}^2) = 0 \\ &c_1(2E_0S_0{}^2 + A - ES_0{}^2) + c_2(2E_0 + K - E) = 0 \end{aligned} \quad (1.23)$$

This is a homogeneous set of equations which determines unknown coefficients c_1 and c_2. Of course, we search for non-trivial solutions that can be naturally determined from the equation

$$\begin{vmatrix} 2E_0 + K - E & 2E_0S_0^2 + A - ES_0^2 \\ 2E_0S_0^2 + A - ES_0^2 & 2E_0 + K - E \end{vmatrix} = 0 \quad (1.24)$$

From this equation we easily find two possible solutions for the *energy* E and the coefficients c_i, namely,

$$E_a = 2E_0 + (K - A)/(1 - S_0^2), \qquad c_1 = - c_2 \quad (1.25)$$

and

$$E_s = 2E_0 + (K + A)/(1 + S_0^2), \qquad c_1 = c_2. \quad (1.26)$$

Finally, from the *normalization condition* of the total wavefunction one obtains

$$1 = \iint \varphi^*(\mathbf{r}_1, \mathbf{r}_2)\varphi(\mathbf{r}_1, \mathbf{r}_2)dV_1dV_2 = 2|\, c_1\,|^2\,(1 \mp S^2) \quad (1.27)$$

or

$$c_1 = 1/\sqrt{\{2(1 \mp S_0^2)\}}, \quad (1.28)$$

where the − and + sign corresponds to E_a and E_s, respectively. Using (1.28) we can express the *orbital part of the total wavefunctions* as

$$\varphi(\mathbf{r}_1, \mathbf{r}_2) = 1/\sqrt{2(1 - S_0^2)}\; [\phi_a(\mathbf{r}_1)\phi_b(\mathbf{r}_2) - \phi_a(\mathbf{r}_2)\phi_b(\mathbf{r}_1)] \quad (1.29)$$

which is clearly the *anti-symmetric* expression and it corresponds to E_a. Similarly, the *symmetric orbital function* which corresponds to E_s is given by

$$\varphi(\mathbf{r}_1, \mathbf{r}_2) = 1/\sqrt{2(1 + S_0^2)}\; [\phi_a(\mathbf{r}_1)\phi_b(\mathbf{r}_2) + \phi_a(\mathbf{r}_2)\phi_b(\mathbf{r}_1)]. \quad (1.30)$$

Finally, we also can account for the spins of the electrons and construct the following *anti-symmetric wavefunctions*:

$$\Psi = 1/\sqrt{2(1 - S_0^2)} \begin{vmatrix} \phi_a(\mathbf{r}_1) & \phi_a(\mathbf{r}_2) \\ \phi_b(\mathbf{r}_1) & \phi_b(\mathbf{r}_2) \end{vmatrix} \xi_\uparrow(1)\xi_\uparrow(2), \qquad S = 1, S^z = 1 \quad (1.31)$$

$$\Psi = 1/\sqrt{2(1 - S_0^2)} \begin{vmatrix} \phi_a(\mathbf{r}_1) & \phi_a(\mathbf{r}_2) \\ \phi_b(\mathbf{r}_1) & \phi_b(\mathbf{r}_2) \end{vmatrix} \xi_\downarrow(1)\xi_\downarrow(2), \qquad S = 1, S^z = 1 \quad (1.32)$$

$$\Psi = 1/\sqrt{2(1 - S_0^2)} \begin{vmatrix} \phi_a(\mathbf{r}_1) & \phi_a(\mathbf{r}_2) \\ \phi_b(\mathbf{r}_1) & \phi_b(\mathbf{r}_2) \end{vmatrix} 1/\sqrt{2}[\xi_\uparrow(1)\,\xi_\downarrow(2) + \xi_\downarrow(1)\xi_\uparrow(2)], \qquad S = 1, S^z = -1 \quad (1.33)$$

and

$$\Psi = 1/\sqrt{2(1 + S_0^2)}\ [\phi_a(\mathbf{r}_1)\phi_b(\mathbf{r}_2) + \phi_b(\mathbf{r}_1)\phi_a(\mathbf{r}_2)]\ 1/\sqrt{2}\ |\ \xi_\uparrow(1)\ \xi_\uparrow(2)\ | \quad S = 0,\ S^z = 0$$
$$|\ \xi_\downarrow(1)\ \xi_\uparrow(2)\ |. \quad (1.34)$$

Here the *wavefunctions* (1.31) - (1.33) form a *triplet* (*threefold degenerate state*) with the *energy* E_a, the *total spin* S = 1 and the total zth component $S^z = \pm 1, 0$. Similarly, the *wavefunction* (1.34) describes a *singlet state* with *energy* E_s, the *total spin* S = 0 and $S^z = 0$. Of course, *both the triplet and singlet states can be the ground state of the system* depending on the sign of the exchange integral. In order to prove this statement, we calculate the *energy difference* between E_a and E_s, i.e.,

$$\Delta E = E_a - E_s = 2(KS_0^2 - A)/(1 - S_0^4) \approx -2A, \qquad (1.35)$$

where we have neglected the *overlap integral* which very small in comparison with other terms in the expression. It is clear from the previous equation that for A > 0 *the ground state of the system will be represented by the triplet state with parallel alignment of spins and for A < 0 the singlet state with anti-parallel orientation of the spins.* Of course, *the parallel (anti-parallel) spin orientation implies also parallel (anti-parallel) alignment of the magnetic moments, which is crucial for the observation of the ferromagnetic or antiferromagnetic ordering in the real systems.*

The situation in the *hydrogen molecule*, of course, differs from the real situation in *ionic crystals* in many respects. For example, the *electrons* in the *hydrogen molecule* occupy *1s orbitals*, while in the *ionic crystals* the *unpaired localized electrons* usually occupy *d or f orbitals*. In fact, *the calculation of the exchange integral using 3d orbital functions leads to the positive value of the A. Therefore, the direct exchange interactions leads to the ferromagnetic ordering in such systems.* Later it was discovered by Anderson, that *the antiferromagnetic ordering is caused by so-called indirect exchange interaction via a non-magnetic bridging atom*. The situation in real materials is usually much more complicated and other mechanism such as the Dzyaloshinkii-Moriya or Ruderman-Kittel-Kasuya-Yosida interactions can play an important role.

Although the calculation performed for the *hydrogen molecule* does not reflect the complex situation of the real magnetic systems, it is extremely important from the methodological point of view. In fact, it is clear from our calculation, that *the parallel or anti-parallel alignment of the magnetic moments appears due to a new contribution to the energy of the system which originates from the indistinguishability of quantum particles*. The other mechanisms leading to macroscopic magnetism have also quantum nature and cannot be included into the classical theory.

Thus, *we can conclude that the physical origin of magnetism can be correctly understood and described only within the quantum theory.*

Chapter 2 Bogolyubov Inequality and Its Applications

2.1 *General Remarks on Bogolyubov Inequality*

The *Bogolyubov inequality for the Gibbs free energy G of an interacting many-body system* described by the Hamiltonian $\hat{H}$ is usually written in the form

$$G \leq G_0 + \langle H - \hat{H}_0 \rangle_0 = \phi(\lambda_x, \lambda_y, \lambda_z, \ldots). \quad (2.1)$$

In this equation, the so-called trial Hamiltonian $\hat{H}_0 = \hat{H}_0(\lambda_x, \lambda_y, \lambda_z, \ldots)$ depends on some variational parameters λ_i that are naturally determined in the process of calculation and the symbol $\langle \ldots \rangle_0$ stands for the usual ensemble average calculated with the trial Hamiltonian. Finally, G_0 denotes the *Gibbs free energy* of the trial system defined by

$$G_0 = G_0(\lambda_x, \lambda_y, \lambda_z, \ldots) = -1/\beta \ln Z_0. \quad (2.2)$$

where $\beta = 1/k_B T$ and the partition function Z_0 is given by

$$Z_0 = \mathrm{Tr}\, e^{-\beta \hat{H}_0}. \quad (2.3)$$

In general, there is a remarkable freedom in defining the trial Hamiltonian. In fact, the only limitations to be taken into account follow from the two obvious requirements:

1. The trial Hamiltonian should naturally represent a simplified physical model of the real system.
2. The expression of G_0 must be calculated exactly, in order to obtain a closed-form formula for the r.h.s of Eqs. (2.1).

2.2 *Mean-Field Theory of the Spin-1/2 Anisotropic Heisenberg Model*

2.2.1 *General Formulation*

The accurate theoretical analysis of the Heisenberg model is extremely hard due to interaction terms in the Hamiltonian including *non-commutative spin operators*. Probably the simplest analytic theory applicable to the model is the *standard mean-field* approach. Although, the *mean-field theory* can mathematically be formulated in many different ways, *we will develop in this text an approach based on the Bogolyubov inequality for the free energy of the system.* The main advantage of this formulation is its completeness (we can derive analytic formulas for all thermodynamic quantities of interest) and a possibility of further generalizations, for example, an extension to the Oguchi approximation.

In this part we will study the spin-1/2 anisotropic Heisenberg model on a crystal lattice described by the Hamiltonian

$$\hat{H} = -\sum_{i,j} (J_x \hat{S}^x_i \hat{S}^x_j + J_y \hat{S}^y_i \hat{S}^y_j + J_z \hat{S}^z_i \hat{S}^z_j)$$
$$- g\mu_B \sum_i (H_x \hat{S}^x_i + H_y \hat{S}^y_i + H_z \hat{S}^z_i) \qquad (2.4)$$

where g is the *Lande factor*, μ_B is the *Bohr magneton*, J_α and H_α, $\alpha = x, y, z$ denote the spatial components of the *exchange interaction* and external *magnetic field*, respectively. One should note here that *J_α is assumed to be positive for the ferromagnetic systems and negative for the antiferromagnetic ones*. Moreover, it is well-known that the absolute value of *exchange integrals* very rapidly decreases with the distance, thus we can restrict the summation in the first term of (2.4) to the *nearest-neighboring pairs of atoms on the lattice*. Finally, the spatial components of the *spin-1/2 operators* are given by

$$\hat{S}^x_k = \tfrac{1}{2} \begin{pmatrix} 0 & 1 \\ 1 & 0 \end{pmatrix}_k, \qquad \hat{S}^y_k = \tfrac{1}{2} \begin{pmatrix} 0 & -i \\ i & 0 \end{pmatrix}_k, \qquad \hat{S}^x_k = \tfrac{1}{2} \begin{pmatrix} 1 & 0 \\ 0 & -1 \end{pmatrix}_k, \qquad (2.5)$$

To proceed further we choose the trial Hamiltonian in the form

$$\hat{H}_0 = -\sum_k^N (\lambda_x \hat{S}^x_k + \lambda_y \hat{S}^y_k + \lambda_z \hat{S}^z_k), \qquad (2.6)$$

where N denotes the total number of the lattice sites and λ_i are variational parameters to be determined be minimizing of the r.h.s. of the *Bogolyubov inequality.* After substituting (2.5) into (2.6), we can apparently rewrite the previous equation as follows

$$\hat{H}_0 = \sum_{k=1}^N \hat{H}_{0k}, \qquad (2.7)$$

where the site Hamiltonian $\hat{H}_{0k}$ is given by

$$\hat{H}_{0k} = -\tfrac{1}{2} \begin{pmatrix} \lambda_z & \lambda_x - i\lambda_y \\ \lambda_x + i\lambda_y & -\lambda_z \end{pmatrix}_k. \qquad (2.8)$$

In order to evaluate the terms entering the *Bogolyubov inequality*, we first have to calculate the *partition function* Z_0. Taking into account the *commutation relation* $[\hat{H}_{0k}, \hat{H}_{0\ell}] = 0$ for $k \neq \ell$, we obtain the following expression

$$Z_0 = \mathrm{Tr} \exp(-\beta \sum_{k=1}^N \hat{H}_{0k}) = \prod_{k=1}^N \mathrm{Tr}_k \exp(-\beta \hat{H}_{0k}), \qquad (2.9)$$

where Tr_k denotes the *trace* of the relevant operator related to kth lattice site. Now, after setting Eq. (2.8) into (2.9) one obtains

$$Z_0 = \prod_{k=1}^N \mathrm{Tr}_k \exp \begin{Bmatrix} \beta/2\, \lambda_z & \beta/2\, (\lambda_x - i\lambda_y) \\ \beta/2\, (\lambda_x + i\lambda_y) & -\beta/2\, \lambda_z \end{Bmatrix}_k. \qquad (2.10)$$

$$= [\mathrm{Tr}_k \exp \{ \beta/2\, \lambda_z \quad \beta/2\, (\lambda_x - i\lambda_y)\}]^N$$
$$\{\beta/2\, (\lambda_x + i\lambda_y) \quad -\beta/2\, \lambda_z \quad)\}_k].$$

[The *trace* of a square matrix A, denoted tr(A), is defined to be the sum of elements on the main diagonal (from the upper left to the lower right) of A. The trace is only defined for a square matrix ($n \times n$). It can be proven that *the trace of a matrix is the sum of its eigenvalues* (counted with multiplicities). It can also be proven that tr(AB) = tr(BA) for any two matrices A and B of appropriate sizes. This implies that similar matrices have the same *trace*.]

The crucial point for further calculation is the evaluation of exponential function in previous equation. Let us note that in performing this task, we will not follow the usual approach based on the diagonalization of modified matrix (2.8) entering the argument of exponential. Instead, we will perform all calculations in the real space by applying the following matrix form of the *Cauchy integral formula*

...

After introducing notation $\langle S^{\wedge\alpha}{}_j\rangle_0 = m_\alpha$, $\alpha = x, y, z$, we can rewrite the r.h.s of the *Bogolyubov inequality* in the form:

$$\phi(\lambda_x, \lambda_y, \lambda_z) = \ldots \qquad (27)$$

...

Having obtained the last equation, we are able to identify the physical meaning of the parameters m_x, m_y, m_z and consequently also the meaning of variational parameters λ_x, λ_y, λ_z.

For this purpose, we at first recall that the spatial component of the *total magnetization* M_α is defined as

$$M_\alpha = -\,(\partial G/\partial H_\alpha)_\beta = -\,g\mu_B\, (\partial G/\partial h_\alpha)_\beta. \qquad (2.33)$$

After a straightforward calculation we obtain the following very simple expressions

$$M_x/Ng\mu_B = m_x = \langle S^{\wedge x}{}_j\rangle 0, \qquad M_y/Ng\mu_B = m_y = \langle S^{\wedge y}{}_j\rangle 0, \qquad M_z/Ng\mu_B = m_z = \langle S^{\wedge z}{}_j\rangle 0, \qquad (2.34)$$

which clearly indicate that *m_α represents the spatial components of the magnetization*. Consequently, the parameters λ_α have the meaning of the molecular-field components acting on one atom in the lattice. Thus, *our formulation is nothing but the standard mean-field theo*ry. The main advantage of our approach is that this formulation provides a close-form analytical expression for the *Gibbs free energy* which is of principal importance for finding stability conditions of various physical quantities.

2.2.2 *Thermodynamic Properties of the Heisenberg Model*

…

Chapter 3 Spin-Wave Theory

3.1 *Bloch Spin-Wave theory of ferromagnets*

In this part we will discuss the concept of *spin waves* which enables us to study some physical properties of magnetic systems at low temperatures. This theory was originally introduced by F. Bloch [Bloch, F. (1930). *Zeit. Phys.*, 61 206] and its main assumption is that the *ground state* of the system is ordered and the excited states are described as a collection of *spin waves*. At low temperatures it is reasonable to expect the small amplitudes of the *spin waves*, thus the interaction among *spin waves* can be neglected. We will consider in this part an isotropic Heisenberg model defined by the Hamiltonian

$$\hat{H} = -J \sum_{i,j} \hat{\mathbf{S}}_i \hat{\mathbf{S}}_j - g\mu_B H \sum_i \hat{S}^z_i. \qquad (3.1)$$

Here $\hat{\mathbf{S}}_i$ represents a vector *spin-operator* at the ith site of the lattice. This operator is naturally defined as $\hat{\mathbf{S}}_i = (\hat{S}^x_i, \hat{S}^y_i, \hat{S}^z_i)$ where $\hat{S}^\alpha_i$ (α = x, y, z) represent spatial components of the standard *spin operators*, g is the *Lande factor*, μB is the *Bohr magneton* and H denotes the *external magnetic field* which is applied along z axis. J is the *exchange integral* which is assumed to be positive since we are going to investigate the *ferromagnetic* systems. …

…

3.2 *Holstein-Primakoff theory of ferromagnets*

The *Bloch spin-wave theory* discussed in previous chapter represents a very simple approach in which we completely neglect the interaction among *spin waves*. This deficiency can be eliminated using an alternative formulation based on *magnon* variables (or *creation-annihilation operators*) developed by Holstein and Primakoff [Holstein, T. & Primakoff, H. (1940). *Phys. Rev.*, 58, 1908]. *In this part we apply The Holstein-Primakoff theory to case of an anisotropic quantum Heisenberg model* described by the Hamiltonian

$$\hat{H} = -J \sum_{i,j} (\hat{S}^x_i \hat{S}^x_j + \hat{S}^y_i \hat{S}^y_j + \hat{S}^z_i \hat{S}^z_j) - g\mu_B H \sum_i \hat{S}^z_i. \qquad (3.27)$$

The meaning of all symbols in previous Eq. (3.27) is the same as in previous text.

In order to apply a *spin-wave* picture to analyze magnetic properties of the model under investigation, one has to perform three subsequent mathematical transformation. *At first, we express the components of the spin operators through spin-lowering and spin raising operators, then we introduce so-called creation and annihilation operators and finally, we perform a Fourier transform of the relevant all creation and annihilation operators entering the Hamiltonian.* The *eigenvalues* of the final Hamiltonian then enable us to determine relevant physical quantities applying standard relations of statistical mechanics.

…

3.3 *Holstein-Primakoff theory of antiferromagnets*

The *Holstein-Primakoff theory* can be extended also to the case of the *antiferromagnetic Heisenberg model* which can describe many real materials [de Jongh, L. J. & Miedema, A. R. (1974). *Adv. Phys*., 23, 2.]. On the other hand, the generalization of the Holstein-Primakoff for *quantum antiferromagnets* is very interesting because such systems exhibit very interesting magnetic properties that differ from quantum ferromagnets in many respects. *Our aim in this subsection is to investigate the antiferromagnetic Heisenberg model spin system consisting of two interpenetrating sublattices a and b*, which is described by the Hamiltonian

$$\hat{H} = -J \sum_{i,j} \hat{S}_{ai}\hat{S}_{bj} - g\mu_B h_a \sum_i \hat{S}^{z}_{ai} + g\mu_B h_a \sum_j \hat{S}^{z}_{bj}, \tag{3.73}$$

where $J < 0$ is the *exchange interaction* which *couples the nearest neighbors on the lattice* and the quantity $H_a > 0$ represents an *external magnetic field* which is parallel to the z axis. In what follows, we will treat the problem by applying the Holstein-Primakoff transformation for each sublattice separately. At first we express the spin rising and lowering sublattice operators with the help of creation-annihilation operators as follows …

…

Chapter 4 Jordan-Wigner Transformation for the XY Model

In previous parts of this text we have discussed how to apply the bosonization technique within the Holstein-Primakoff approach to the ferromagnetic and antiferromagnetic spin systems. In the present chapter we will investigate a similar approach, which is known as a method of *fermioniozation*. In particular, we will discuss application of the *Wigner-Jordan transformation* to the isotropic quantum XY model in order to clarify important points of this method.

The main idea of this approach is to transform the Hamiltonian describing a spin system by the use of new operators obeying the fermion anticommutation rules.

Before we start our analysis, it is useful to note that the spectrum of the XY model was exactly found by H. Bethe in 1931. [Bethe, H. (1931). *Zeit. Phys.*, 71, 205.] Bethe's approach is of great importance, however, it is quite involved and rather abstract, thus it is difficult to understand even such basic properties as long-range order. *A much more natural approach to the problem of interacting spin-1/2 systems was originally introduced in 1928 by Jordan and Wigner* [Jordan, P. & Wigner, Z. (1928). *Zeit. Phys.*, 47, 631.], who invented *simple mathematical transformations converting spin-1/2 systems into problems of interacting (and in some cases even non-interacting) spinless fermions*. In fact, the XY model which is a special case of the Heisenberg Hamiltonian, reduces to a free theory of *spinless fermions* under the *Jordan-Wigner transformations*. Another reason for choosing the XY model, is that the low-energy properties of the full *anti-ferromagnetic* Heisenberg chain, such as the presence of gapless excitations and absence of a long-range order are very similar to those of the XY model (see Tsvelik, A. (1995). *Quantum Field Theory in Condensed Matter Physics*, Cambridge, Cambridge University Press and references therein).

We will study a linear chain of N spin-1/2 atoms interacting antiferromagnetically with their nearest neighbors. This system is described by the Hamiltonian

$$\hat{H} = J \sum_{j=1}^{N} (\hat{S}^{x}_{j} \hat{S}^{x}_{j+1} + \hat{S}^{y}_{j} \hat{S}^{y}_{j+1}), \qquad (4.1)$$

where $J > 0$ represents the *exchange interaction* and $\hat{S}^{\alpha}_{i}$ are *spin-1/2 operators* obeying the usual commutation relations (3.3). As usually, we assume cyclic boundary conditions, with $\hat{S}^{\alpha}_{N+1} = \hat{S}^{\alpha}_{1}$, $\alpha = x, y$. Of course, for $J < 0$ the *ferromagnetic* case of (4.1) is obtained. Thus, if we solve the *antiferromagnetic* problem exactly, then we can immediately obtain also the solution of the *ferromagnetic* XY chain. Similarly, as in the case of *bozonization*, we at first introduce *spin-raising* and *spin-lowering* operators (3.13) and rewrite the Hamiltonian as a bilinear form in these operators …

…

Conclusion

In this work we have discussed some theoretical tools that are frequently used to investigate the localized *quantum spin models*.

The first approach is the *standard mean-field theory*, which represents the simplest possible technique usually applied to understand basic quantitative features of various theoretical models of *quantum magnetism*. Although this method is included in many textbooks, the formulation is usually very simplified and frequently the authors do not derive the Gibbs free energy within this method. To avoid this problem, *we have used an elegant formulation based on the Bogolyubov inequality* which enables us to extend our approach to formulate more accurate theories (for example the Oguchi approximation or Constant Coupling Method). One should also emphasize that in deriving the partition function *we have applied the Cauchy integral formula, which enables to calculate exponential function with a matrix argument*. This is an unknown trick which can be used in calculations in many branches of theoretical physics.

In Chapter 3 *we have discussed the standard spin-wave theory separately for ferromagnetic and antiferromagnetic materials*. We have shown in detail how to introduce the *boson operators* that enable to calculate several physical quantities such as magnetization, internal energy and magnetic contribution to the specific heat.

The last part is dedicated to the method of *fermionization*. Here we have discussed the application of the Jordan-Wigner transformation to the quantum XY model with spin 1/2. *This approach represents another standard theoretical approach which is used to investigate quantum magnetism.*

It is clear from the content of this supporting text that we have covered just one special part of *quantum theory of magnetism*, which is, however, discussed elementary enough. In order to understand further interesting features of *quantum magnetism* it is necessary to study further textbooks, for example Kittel, C. (1987). *Quantum Theory of Solids*, New York, John Wiley and Sons; Yosida, K. (1998). *Theory of Magnetism*, Berlin, Springer; Mohn, P. (2003). *Magnetism in the Solid State*, Berlin, Springer; Majlis, N. (2007). *The Quantum Theory of Magnetism*, Singapore, World Scientific.

PART IV Quantum Entanglement.

Quantum entanglement is the phenomenon of a *group of particles* being generated, interacting, or sharing spatial proximity *in such a way that the quantum state of each particle of the group cannot be described independently of the state of the others*, including when the particles are separated by a large distance. The topic of *quantum entanglement* is at the heart of the disparity between classical and quantum physics: *entanglement* is a primary feature of quantum mechanics not present in classical mechanics.

Measurements of physical properties such as position, momentum, spin, and polarization performed on *entangled* particles can, in some cases, be found to be perfectly correlated. For example, if a pair of *entangled* particles is generated such that their *total spin* is known to be zero, and *one particle is found to have clockwise spin on a first axis, then the spin of the other particle, measured on the same axis, is found to be anticlockwise*. However, this behavior gives rise to seemingly paradoxical effects: *any measurement of a particle's properties results in an apparent and irreversible wave function collapse of that particle and changes the original quantum state*. With *entangled* particles, such measurements affect the *entangled* system as a whole.

Such phenomena were the subject of a 1935 paper by Albert Einstein, Boris Podolsky, and Nathan Rosen, [Einstein, A., Podolsky, B. & Rosen, N. (May, 1935). Can Quantum-Mechanical Description of Physical Reality Be Considered Complete? *Phys. Rev*., 47, 10, 777-80], describing what came to be known as the *Einstein–Podolsky–Rosen paradox* (*EPR paradox*). Einstein considered such behavior impossible, as it violated the local realism view of causality and argued that the accepted formulation of quantum mechanics must therefore be incomplete.

However, the three scientists did not coin the word *entanglement*, nor did they generalize the special properties of the *quantum state* they considered. Einstein later famously derided *entanglement* as "spukhafte Fernwirkung" or "spooky action at a distance." [Letter from Einstein to Max Born, 3 March 1947; The Born-Einstein Letters; Correspondence between Albert Einstein and Max and Hedwig Born from 1916 to 1955, Walker, New York, 1971]

Following the EPR paper, Erwin Schrödinger, who had discussed the phenomenon as early as 1932, wrote a letter to Einstein in German in which he used the word *Verschränkung* (translated by himself as *entanglement*) "to describe the correlations between two particles that interact and then separate, as in the EPR experiment." Schrödinger shortly thereafter published a seminal paper defining and discussing the notion of "*entanglement*". [Schrödinger, E. (1935). Discussion of probability relations between separated systems. *Mathematical Proceedings of the Cambridge Philosophical Society*, 31, 4, 555–63. See below.] In the paper, he recognized the importance of the concept, and stated: "I would not call [*entanglement*] one but rather *the characteristic trait of quantum mechanics*, the one that enforces its entire departure from classical lines of thought."

Meaning of entanglement

An entangled system is defined to be one whose quantum state cannot be factored as a product of states of its local constituents; that is to say, they are not individual particles but are an inseparable whole. In entanglement, one constituent cannot be fully described without considering the other(s).

The state of a ***composite system*** *is always expressible as a sum, or* ***superposition****, of products of states of local constituents; it is* ***entangled*** *if this sum cannot be written as a single product term.*

Quantum superposition is a fundamental principle of quantum mechanics that states that linear combinations of solutions to the Schrödinger equation are also solutions of the Schrödinger equation. This follows from the fact that the Schrödinger equation is a linear differential equation in time and position. More precisely, the state of a system is given by a linear combination of all the eigenfunctions of the Schrödinger equation governing that system.

As an example of ***entanglement***: *a subatomic particle decays into an entangled pair of other particles*. The decay events obey the various conservation laws, and as a result, the measurement outcomes of one daughter particle must be highly correlated with the measurement outcomes of the other daughter particle (*so that the total momenta, angular momenta, energy, and so forth remains roughly the same before and after this process*). For instance, a *spin-zero* particle could decay into a pair of *spin-1/2* particles. Since the *total spin* before and after this decay must be zero (*conservation of angular momentum*), whenever the first particle is measured to be *spin up* on some axis, the other, when measured on the same axis, is always found to be *spin down*. (This is called the *spin anti-correlated case*; and if the prior probabilities for measuring each *spin* are equal, the pair is said to be in the *singlet state*.)

The above result may or may not be perceived as surprising. A classical system would display the same property, and a *hidden variable* theory would certainly be required to do so, based on *conservation of angular momentum* in classical and quantum mechanics alike. *The difference is that a classical system has definite values for all the observables all along, while the quantum system does not*. In a sense to be discussed below, ***the quantum system considered here seems to acquire a probability distribution for the outcome of a measurement of the spin along any axis of the other particle upon measurement of the***

first particle. This probability distribution is in general different from what it would be without measurement of the first particle. This may certainly be perceived as surprising in the case of *spatially separated entangled particles*.

According to some interpretations of quantum mechanics, the effect of one measurement occurs instantly. Other interpretations which do not recognize *wavefunction* collapse dispute that there is any "effect" at all. However, all interpretations agree that *entanglement* produces correlation between the measurements, and that the mutual information between the *entangled* particles can be exploited. The use of *entanglement* in communication, computation and quantum radar is an active area of research and development.

Paradox

The paradox is that a measurement made on either of the particles apparently collapses the state of the entire entangled system—and does so instantaneously, before any information about the measurement result could have been communicated to the other particle (assuming that information cannot travel faster than light) and hence assured the "proper" outcome of the measurement of the other part of the *entangled* pair. In the Copenhagen interpretation, *the result of a spin measurement on one of the particles is a collapse (of wave function) into a state in which each particle has a definite spin (either up or down)* along the axis of measurement. The outcome is taken to be random, with each possibility having a probability of 50%. However, if both *spins* are measured along the same axis, they are found to be *anti-correlated.* This means that the random outcome of the measurement made on one particle seems to have been transmitted to the other, so that it can make the "right choice" when it too is measured.

The distance and timing of the measurements can be chosen so as to make the interval between the two measurements spacelike, hence, *any causal effect connecting the events would have to travel faster than light*.

> [According to the principles of *special relativity*, it is not possible for any information to travel between two such measuring events. It is not even possible to say which of the measurements came first. For two spacelike separated events x_1 and x_2 there are inertial frames in which x_1 is first and others in which x_2 is first. Therefore, the correlation between the two measurements cannot be explained as one measurement determining the other: different observers would disagree about the role of cause and effect.]

Hidden variables theory

A possible resolution to the paradox is to assume that quantum theory is incomplete, and the result of measurements depends on predetermined "*hidden variables*". The state of the particles being measured contains some hidden variables, whose values effectively determine, right from the moment of separation, what the outcomes of the *spin* measurements are going to be. This would mean that each particle carries all the required information with it, and nothing needs to be transmitted from one particle to the other at the time of measurement. Einstein and others believed this was the only way out of the paradox, and the accepted quantum mechanical description (with a random measurement outcome) must be incomplete.

The first experiment that verified Einstein's spooky action at a distance (*entanglement)* was successfully corroborated in a lab by Chien-Shiung Wu and colleague I. Shaknov in 1949, and was published on New Year's Day in 1950. The result specifically proved the quantum correlations of a pair of *photons*. In experiments in 2012 and 2013, polarization correlation was created between *photons* that never coexisted in time. The authors claimed that this result was achieved by *entanglement swapping* between two pairs of entangled *photons* after measuring the polarization of one *photon* of the early pair, and that it proves that quantum non-locality applies not only to space but also to time.

Violations of Bell's inequality

Despite the interest, the weak point in EPR's argument was not discovered until 1964, when John Stewart Bell proved that one of their key assumptions, *the principle of locality*, as applied to the kind of hidden variables interpretation hoped for by EPR, *was mathematically inconsistent with the predictions of quantum theory*. Specifically, Bell demonstrated an upper limit, seen in *Bell's inequality*, regarding the strength of correlations that can be produced in any theory obeying *local realism*, and showed that quantum theory predicts violations of this limit for certain *entangled* systems. [Bell, J. S. (1964). On the Einstein-Poldolsky-Rosen paradox. *Physics Physique Физика*, 1, 3, 195–200; doi:10.1103/PhysicsPhysiqueFizika.1.195. See below.]

Local hidden variable theories fail, however, when measurements of the spin of entangled particles along different axes are considered. If a large number of pairs of such measurements are made (on a large number of pairs of *entangled* particles), then statistically, if the local realist or hidden variables view were correct, the results would always satisfy *Bell's inequality*.

The fundamental issue about measuring *spin* along different axes is that these measurements cannot have definite values at the same time—they are incompatible in the sense that these measurements' maximum simultaneous precision is constrained by the *uncertainty principle*. This is contrary to what is found in classical physics, where any number of properties can be measured simultaneously with arbitrary accuracy. It has been proven mathematically that compatible measurements cannot show *Bell-inequality-violating* correlations, and thus *entanglement is a fundamentally non-classical phenomenon. Bell's inequality* was experimentally testable, and there were numerous efforts to test it. A number of experiments have shown in practice that *Bell's inequality is not satisfied.*

An early experimental breakthrough was due to Carl Kocher, who in 1967 presented an apparatus in which two *photons* successively emitted from a calcium atom were shown to be *entangled* – the first case of *entangled* visible light. The two *photons* passed diametrically positioned parallel polarizers with higher probability than classically predicted but with correlations in quantitative agreement with quantum mechanical calculations. He also showed that the correlation varied as the squared cosine of the angle between the polarizer settings and decreased exponentially with time lag between emitted photons. Kocher's apparatus, equipped with better polarizers, was later used by Stuart Freedman and John Clauser in 1972, who confirmed the cosine-squared dependence and use it to demonstrate a violation of *Bell's inequality* for a set of fixed angles. All these experiments have shown agreement with quantum mechanics rather than the principle of local realism. *Schrödinger was proved correct and Einstein was [again] wrong*, when the counterintuitive predictions of quantum mechanics were verified in tests where polarization or *spin* of *entangled particles* were measured at separate locations, statistically violating *Bell's inequality*.

In three independent experiments in 2013, it was shown that classically communicated separable *quantum states* can be used to carry *entangled* states. For decades, each had left open at least one loophole by which it was possible to question the validity of the results. In earlier tests, it could not be ruled out that the result at one point could have been subtly transmitted to the remote point, affecting the outcome at the second location. However, in 2015, the first loophole-free Bell test was held by Ronald Hanson of the Delft University of Technology, that simultaneously closed both the detection and locality loopholes, confirming the violation of the *Bell inequality*. This experiment ruled out a large class of local realism theories with certainty.

So-called "loophole-free" Bell tests have since been performed where the locations were sufficiently separated that communications at the speed of light would have taken longer—in one case, 10,000 times longer—than the interval between the measurements.

Quantum entanglement has been demonstrated experimentally with photons, electrons, top quarks, molecules and even small diamonds.

> [The *top quark*, sometimes also referred to as the truth quark, (symbol: t) is the most massive of all observed elementary particles. It derives its mass from its coupling to the Higgs Boson. This coupling is very close to unity; in the Standard Model of particle physics, it is the largest (strongest) coupling at the scale of the weak interactions and above. Like all other quarks, the top quark is a fermion with spin ½ and participates in all four fundamental interactions: gravitation, electromagnetism, weak interactions, and strong interactions.]

In 2022, the Nobel Prize in Physics was awarded to Alain Aspect, John Clauser, and Anton Zeilinger "for experiments with *entangled photons*, establishing the violation of *Bell inequalities* and pioneering quantum information science".

Naturally entangled systems

The electron shells of multi-electron atoms always consist of entangled electrons. The correct ionization energy can be calculated only by consideration of *electron entanglement.*

Methods of creating entanglement

Quantum systems can become entangled through various types of interactions. Entanglement is broken when the entangled particles decohere through interaction with the environment; for example, when a measurement is made.

Entanglement is usually created by direct interactions between subatomic particles. These interactions can take numerous forms. One of the most commonly used methods is spontaneous parametric down-conversion to generate a pair of *photons entangled in polarization.* Other methods include the use of a fiber coupler to confine and mix *photons*, *photons* emitted from decay cascade of the bi-exciton in a quantum dot, the use of the Hong–Ou–Mandel effect, etc.

Quantum entanglement of a *particle* and its *antiparticle*, such as an *electron* and a *positron*, can be created by partial overlap of the corresponding quantum wave functions in *Hardy's interferometer.* In the earliest tests of *Bell's theorem*, the *entangled particles* were generated using *atomic cascades.*

It is also possible to create *entanglement* between quantum systems that never directly interacted, through the use of *entanglement swapping.* Two independently prepared, identical particles may also be *entangled* if their *wave functions* merely spatially overlap, at least partially.

Since 2016, various companies, for example IBM and Microsoft, have created *quantum computers* that allowed developers and tech enthusiasts to freely experiment with concepts of quantum mechanics including *quantum entanglement.*

Emergence of time from quantum entanglement

There is a fundamental conflict, referred to as the problem of time, between the way the concept of time is used in quantum mechanics, and the role it plays in general relativity. In standard quantum theories time acts as an independent background through which states evolve, with the Hamiltonian operator acting as the generator of infinitesimal translations of quantum states through time.

In contrast, *general relativity* treats time as a dynamical variable which relates directly with matter and moreover requires the Hamiltonian constraint to vanish. [In quantized *general relativity*, the quantum version of the Hamiltonian constraint using metric variables, leads to the Wheeler–DeWitt equation:

$$H^\wedge(x) \mid \psi\rangle = 0$$

where $H^\wedge(x)$ is the Hamiltonian constraint and $\mid \psi \rangle$ stands for the *wave function* of the universe. The operator $H^\wedge$ acts on the Hilbert space of *wave functions*, but it is not the same Hilbert space as in the *nonrelativistic* case. *This Hamiltonian no longer determines the evolution of the system because the Schrödinger equation:* $H^\wedge\mid \psi \rangle = i\hbar\, \partial/\partial t \mid \psi \rangle$*, ceases to be valid.* This property is known as timelessness. *Various attempts to incorporate time in a fully quantum framework have been made*, starting with the Page and Wootters mechanism and other subsequent proposals.

The emergence of time was also proposed as arising from quantum correlations between an evolving system and a reference quantum clock system, the concept of system-time entanglement is introduced as a quantifier of the actual distinguishable evolution undergone by the system.

Emergent gravity

Based on AdS/CFT correspondence, Mark Van Raamsdonk suggested that spacetime arises as an emergent phenomenon of the quantum degrees of freedom that are entangled and live in the boundary of the space-time. Induced gravity can emerge from the entanglement first law.

Entanglement as a resource

In quantum information theory, entangled states are considered a 'resource', i.e., something costly to produce and that allows implementing valuable transformations. The setting in which this perspective is most evident is that of "distant labs", i.e., two quantum systems labeled "A" and "B" on each of which arbitrary quantum operations can be performed, but which do not interact with each other quantum mechanically. The only interaction allowed is the exchange of classical information, which combined with the most general local quantum operations gives rise to the class of operations called LOCC (*local operations and classical communication*). These operations do not allow the production of *entangled states* between systems A and B. But if A and B are provided with a supply of *entangled states*, then these, together with LOCC operations can enable a larger class of transformations. For example, an interaction between a qubit of A and a qubit of B can be realized by first teleporting A's qubit to B, then letting it interact with B's qubit (which is now a LOCC operation, since both qubits are in B's lab) and then teleporting the qubit back to A. Two maximally entangled states of two qubits are used up in this process. Thus, *entangled states are a resource that enables the realization of quantum interactions* (or of quantum channels) in a setting where only LOCC are available, but they are consumed in the process. There are other applications where *entanglement* can be seen as a resource, e.g., private communication or distinguishing quantum states.

Applications

Entanglement has many applications in *quantum information theory*. With the aid of *entanglement*, otherwise impossible tasks may be achieved.

Among the best-known applications of *entanglement* are *superdense coding* and *quantum teleportation*.

Most researchers believe that *entanglement* is necessary to realize *quantum computing* (although this is disputed by some).

Entanglement is used in some protocols of *quantum cryptography*, but to prove the security of quantum key distribution (QKD) under standard assumptions does not require *entanglement*. However, the device independent security of QKD is shown exploiting *entanglement* between the communication partners.

Quantum communications

In 2016, China launched the world's first *quantum communications* satellite. The $100m Quantum Experiments at Space Scale (QUESS) mission was launched on 16 Aug 2016, from the Jiuquan Satellite Launch Center in northern China at 01:40 local time. For the next two years, the satellite – nicknamed "Micius" after the ancient Chinese philosopher – demonstrated the feasibility of *quantum communication* between Earth and space, and test *quantum entanglement* over unprecedented distances.

In the June 16, 2017, issue of *Science*, Yin et al. report setting a new *quantum entanglement* distance record of 1,203 km, demonstrating the survival of a *two-photon pair* and a violation of a *Bell inequality*, reaching a CHSH valuation of 2.37 ± 0.09, under strict Einstein locality conditions, from the Micius satellite to bases in Lijian, Yunnan and Delingha, Quinhai, increasing the efficiency of transmission over prior fiberoptic experiments by an order of magnitude.

Einstein, A., Podolsky, B. & Rosen, N. (May, 1935). Can Quantum-Mechanical Description of Physical Reality Be Considered Complete?

Phys. Rev., 47, 10, 777-80; https://doi.org/10.1103/PhysRev.47.777;
also at https://journals.aps.org/pr/pdf/10.1103/PhysRev.47.777.

Received 25 March 1935.

Institute for Advanced Study, Princeton, New Jersey.

This is known as the EPR paper. "In a complete theory there is an element corresponding to each element of reality. *A sufficient condition for the reality of a physical quantity is the possibility of predicting it with certainty, without disturbing the system*. In quantum mechanics in the case of two physical quantities described by non-commuting operators, *the knowledge of one precludes the knowledge of the other*. Then either (1) the description of reality given by the *wave function* in quantum mechanics is not complete or (2) these two quantities cannot have simultaneous reality. Consideration of the problem of making predictions concerning a system on the basis of measurements made on another system that had previously interacted with it leads to the result that if (1) is false then (2) is also false. One is thus led to conclude that the description of reality as given by a *wave function* is not complete." [Einstein was wrong, again.]

Abstract

In a complete theory there is an element corresponding to each element of reality. A sufficient condition for the reality of a physical quantity is the possibility of predicting it with certainty, without disturbing the system. In quantum mechanics in the case of two physical quantities described by *non-commuting operators*, the knowledge of one precludes the knowledge of the other. Then either (1) the description of reality given by the wave function in quantum mechanics is not complete or (2) these two quantities cannot have simultaneous reality. Consideration of the problem of making predictions concerning a system on the basis of measurements made on another system that had previously interacted with it leads to the result that if (1) is false then (2) is also false. One is thus led to conclude that the description of reality as given by a *wave function* is not complete.

1.

Any serious consideration of a physical theory must take into account the distinction between the objective reality, which is independent of any theory, and the physical concepts with which the theory operates. These concepts are intended to correspond with the objective reality, and by means of these concepts we picture this reality to ourselves.

In attempting to judge the success of a physical theory, we may ask ourselves two questions: (1) "Is the theory correct?" and (2) "Is the description given by the theory complete?" It is only in the case in which positive answers may be given to both of these questions, that the concepts of the theory may be said to be satisfactory. The correctness of the theory is judged by the degree of agreement between the conclusions of the theory and human experience. This experience, which alone enables us to make inferences about reality, in physics takes the form of experiment and measurement. It is the second question that we wish to consider here, as applied to quantum mechanics.

Whatever the meaning assigned to the term *complete*, the following requirement for a complete theory seems to be a necessary one: *every element of the physical reality must have a counterpart in the physical theory*. We shall call this the *condition of completeness*. The second question is thus easily answered, as soon as we are able to decide what are the elements of the physical reality.

The elements of the physical reality cannot be determined by *a priori* philosophical considerations, but must be found by an appeal to results of experiments and measurements. A comprehensive definition of reality is, however, unnecessary for our purpose. We shall be satisfied with the following criterion, which we regard as reasonable. *If, without in any way disturbing a system, we can predict with certainty (i.e., with probability equal to unity) the value of a physical quantity, then there exists an element of physical reality corresponding lo this physical quantity*. It seems to us that this criterion, while far from exhausting all possible ways of recognizing a physical reality, at least provides us with one such way, whenever the conditions set down in it occur. Regarded not as a necessary, but merely as a sufficient, condition of reality, this criterion is in agreement with classical as well as quantum-mechanical ideas of reality.

To illustrate the ideas involved let us consider the quantum-mechanical description of the behavior of a particle having a single degree of freedom. The fundamental concept of the theory is the concept of *state*, which is supposed to be completely characterized by the *wave function* ψ, which is a function of the variables chosen to describe the particle's behavior. Corresponding to each physically observable quantity A there is an operator, which may be designated by the same letter.

If ψ is an *eigenfunction* of the operator A, that is, if

$$\psi' \equiv A\psi = a\psi, \tag{1}$$

where a is a number, then the physical quantity A has with certainty the value a whenever the particle is in the *state* given by ψ. In accordance with our criterion of reality, for a particle in the *state* given by ψ for which Eq. (1) holds, there is an element of physical reality corresponding to the physical quantity A. Let, for example,

$$\psi = e^{(2\pi i/h)\, p0x}, \qquad (2)$$

where h is Planck's constant, p_0 is some constant number, and x the independent variable. Since the operator corresponding to the *momentum* of the particle is

$$p = (h/2\pi i)\, \partial/\partial x, \qquad (3)$$

we obtain

$$p' = p\psi = (h/2\pi i)\, \partial\psi/\partial x = p_0\psi \qquad (4)$$

Thus, in the *state* given by Eq. (2), the *momentum* has certainly the value p_0. It thus has meaning to say that the *momentum* of the particle in the *state* given by Eq. (2) is real.

On the other hand, if Eq. (1) does not hold, we can no longer speak of the physical quantity A having a particular value. This is the case, for example, with the coordinate of the particle. The operator corresponding to it, say q, is the operator of multiplication by the independent variable. Thus

$$q\psi = x\psi \neq a\psi. \qquad (5)$$

In accordance with quantum mechanics, we can only say that the relative probability that a measurement of the coordinate will give a result lying between a and b is

$$P(a, b) = \int_a^b \psi^{\cdot}\psi \, dx = \int_a^b dx = b - a. \qquad (6)$$

Since this probability is independent of a, but depends only upon the difference $b - a$, we see that all values of the coordinate are equally probable.

A definite value of the coordinate, for a particle in the *state* given by Eq. (2), is thus not predictable, but may be obtained only by a direct measurement. Such a measurement however disturbs the particle and thus alters its *state*. After the coordinate is determined, the particle will no longer be in the *state* given by Eq. (2). The usual conclusion from this in quantum mechanics is that *when the momentum of a particle is known, its coordinate has no physical reality.*

More generally, it is shown in quantum mechanics that, *if the operators corresponding to two physical quantities, say A and B, do not commute, that is, if AB ≠ BA, then the precise knowledge of one of them precludes such a knowledge of the other.* Furthermore, any attempt to determine the latter experimentally will alter the *state* of the system in such a way as to destroy the knowledge of the first.

From this follows that either (1) *the quantum-mechanical description of reality given by the wave function is not complete* or (2) *when the operators corresponding to two physical quantities do not commute the two quantifies cannot have simultaneous reality*. For if both of them had simultaneous reality —and thus definite values —these values would enter into the complete description, according to the condition of completeness. If then the *wave function* provided such a complete description of reality, it would contain these values; these would then be predictable. This not being the case, we are left with the alternatives stated.

In quantum mechanics it is usually assumed that the *wave function* does contain a complete description of the physical reality of the system in the *state* to which it corresponds. At first sight this assumption is entirely reasonable, for the information obtainable from a *wave function* seems to correspond exactly to what can be measured without altering the *state* of the system. We shall show, however, that this assumption, together with the criterion of reality given above, leads to a contradiction.

2.

For this purpose, let us suppose that we have two systems, I and II, which we permit to interact from the time t = 0 to t = T, after which time we suppose that there is no longer any interaction between the two parts. We suppose further that the *states* of the two systems before t = 0 were known. We can then calculate with the help of ***Schrodinger's equation*** the state of the combined system I + II at any subsequent time; in particular, for any t > T.

> [The ***Schrödinger equation*** is a partial differential equation that governs the wave function of a *non-relativistic* quantum-mechanical system. Its discovery was a significant landmark in the development of quantum mechanics. It is named after Erwin Schrödinger, an Austrian physicist, who postulated the equation in 1925 and published it in 1926, forming the basis for the work that resulted in his Nobel Prize in Physics in 1933.
>
> Conceptually, the Schrödinger equation is the quantum counterpart of Newton's second law in classical mechanics. Given a set of known initial conditions, Newton's second law makes a mathematical prediction as to what path a given physical system will take over time. The Schrödinger equation gives the evolution over time of the wave function, the quantum-mechanical characterization of an isolated physical system. The equation was postulated by Schrödinger based on a postulate of Louis de Broglie that all matter has an associated matter wave. The

equation predicted bound states of the atom in agreement with experimental observations.

The Schrödinger equation is not the only way to study quantum mechanical systems and make predictions. Other formulations of quantum mechanics include *matrix mechanics*, introduced by Werner Heisenberg, and the *path integral formulation*, developed chiefly by Richard Feynman. When these approaches are compared, the use of the Schrödinger equation is sometimes called "wave mechanics".

The equation given by Schrödinger is nonrelativistic because it contains a first derivative in time and a second derivative in space, and therefore space and time are not on equal footing.

> [*Paul Dirac incorporated* ***special relativity*** *and quantum mechanics into a single formulation that simplifies to the Schrödinger equation in the non-relativistic limit.* This is the ***Dirac equation***, which contains a single derivative in both space and time. Another partial differential equation, the Klein–Gordon equation, led to a problem with probability density even though it was a relativistic wave equation. The probability density could be negative, which is physically unviable. This was fixed by Dirac by taking the so-called square root of the Klein–Gordon operator and in turn introducing Dirac matrices. In a modern context [???], the Klein–Gordon equation describes spin-less particles, while the Dirac equation describes spin-1/2 particles.]

The most general form is the *time-dependent Schrödinger equation*, which gives a description of a system evolving with time:

$$i\hbar d/dt\, |\Psi(t)\rangle = H\hat{}\,|\Psi(t)\rangle$$

where t is time, $|\Psi(t)\rangle$ is the state vector of the quantum system, and H^ is an observable, the Hamiltonian operator.

> [The term "*Schrödinger equation*" can refer to both the general equation, or the specific *nonrelativistic* version. The general equation is indeed quite general, used throughout quantum mechanics, for everything from the Dirac equation to quantum field theory, by plugging in diverse expressions for the Hamiltonian. The specific *nonrelativistic* version is an approximation that yields accurate results in many situations, but only to a certain extent (see relativistic quantum mechanics and relativistic quantum field theory).]

To apply the Schrödinger equation, write down the Hamiltonian for the system, accounting for the *kinetic* and *potential energies* of the particles constituting the system, then insert it into the *Schrödinger equation.* The resulting partial differential equation is solved for the wave function, which contains information about the system. In practice, the square of the absolute value of the wave function at each point is taken to define a *probability density function*. For example, given a wave function in position space $\Psi(x, t)$ as above, we have

$$\Pr(x, t) = |\Psi(x, t)|^2.$$

Given the Schrödinger equation

$$\hat{H}|n\rangle = E_n|n\rangle,$$

where $|n\rangle$ indexes the set of eigenstates of the Hamiltonian with energy eigenvalues E_n, we see immediately that

$$\hat{H}(|n\rangle + |n'\rangle) = E_n|n\rangle + E_{n'}|n'\rangle,$$

where

$$|\Psi\rangle = |n\rangle + |n'\rangle$$

is a solution of the Schrödinger equation but is not generally an eigenstate because E_n and $E_{n'}$ are not generally equal. *We say that $|\Psi\rangle$ is made up of a* ***superposition*** *of energy eigenstates.*

Now consider the more concrete case of *an electron that has either spin up or down.* We now index the eigenstates with the spinors in the $\hat{z}$ basis:

$$|\Psi\rangle = c_1|\uparrow\rangle + c_2|\downarrow\rangle,$$

where $|\uparrow\rangle$ and $|\downarrow\rangle$ denote *spin-up and spin-down states respectively*. As previously discussed, the magnitudes of the complex coefficients give the ***probability of finding the electron in either definite spin state:***

$$P(|\uparrow\rangle) = |c_1|^2,$$
$$P(|\downarrow\rangle) = |c_2|^2,$$
$$P_{total} = P(|\uparrow\rangle) + P(|\downarrow\rangle) = |c_1|^2 + |c_2|^2 = 1,$$

where the probability of finding the particle with either spin up or down is normalized to 1. Notice that c_1 and c_2 are complex numbers, so that

$$|\Psi\rangle = 3/5\ i|\uparrow\rangle + 4/5\ |\downarrow\rangle.$$

is an example of an ***allowed state***. We now get

$$P(|\uparrow\rangle) = |3i/5|^2 = 9/25,$$
$$P(|\downarrow\rangle) = |4/5|^2 = 16/25,$$
$$P_{total} = P(|\uparrow\rangle) + P(|\downarrow\rangle) = 9/25 + 16/25 = 1.$$

Quantum superposition is a fundamental principle of quantum mechanics that ***states that linear combinations of solutions to the Schrödinger equation are also solutions of the Schrödinger equation.*** This follows from the fact that the Schrödinger equation is a linear differential equation in time and position. More precisely, the state of a system is given by a linear combination of all the eigenfunctions of the Schrödinger equation governing that system.

An example is a qubit used in quantum information processing. The interference fringes in the double-slit experiment provide another example of the *superposition* principle.]

Let us designate the corresponding *wave function* by Ψ. We cannot, however, calculate the *state* in which either one of the two systems is left after the interaction. This, according to quantum mechanics, can be done only with the help of further measurements, by a process known as the *reduction of the wave packet*. Let us consider the essentials of this process.

Let a_1, a_2, a_3, … be the *eigenvalues* of some physical quantity A pertaining to system I and $u_1(x_1)$, $u_2(x_1)$, $u_3(x_1)$, … the corresponding *eigenfunctions*, where x_1 stands for the variables used to describe the first system. Then Ψ, considered as a function of x_1, can be expressed as

$$\Psi(x_1, x_2) = \Sigma_{n=1}^{\infty} \psi_n(x_2)u_n(x_1), \qquad (7)$$

where x_2 stands for the variables used to describe the second system. Here $\psi_n(x_2)$ are to be regarded merely as the coefficients of the expansion of Ψ into a series of orthogonal functions $u_n(x_1)$. Suppose now that the quantity A is measured and it is found that it has the value a_k. It is then concluded that after the measurement the first system is left in the *state* given by the *wave function* $u_k(x_1)$, and that the second system is left in the *state* given by the wave function $\psi_k(x_2)$. This is the process of *reduction of the wave packet*; the wave packet given by the infinite series (7) is reduced to a single term $\psi_k(x_2)u_k(x_1)$.

The set of functions $u_n(x_1)$ is determined by the choice of the physical quantity A. If, instead of this, we had chosen another quantity, say B, having the *eigenvalues* b_1, b_2, b_3, … and *eigenfunctions* $v_1(x_1)$, $v_2(x_1)$, $v_3(x_1)$, … we should have obtained, instead of Eq. (7), the expansion

$$\Psi(x_1, x_2) = \Sigma_{n=1}^{\infty} \varphi_n(x_2)v_n(x_1), \qquad (8)$$

where φ_n's are the new coefficients. If now the quantity B is measured and is found to have the value b_r, we conclude that after the measurement the first system is left in the *state* given by $v_r(x_1)$ and the second system is left in the *state* given by $\varphi_r(x_2)$.

We see therefore that, as a consequence of two different measurements performed upon the first system, the second system may be left in *states* with two different *wave functions*. On the other hand, since at the time of measurement the two systems no longer interact, no real change can take place in the second system in consequence of anything that may be done to the first system. This is, of course, merely a statement of what is meant by the absence of an interaction between the two systems. Thus, *it is possible to assign two different wave functions* (in our example ψ_k and φ_r) *to the same reality* (the second system after the interaction with the first).

Now, it may happen that the two *wave functions*, ψ_k and φ_r, are *eigenfunctions* of two *non-commuting operators* corresponding to some physical quantities P and Q, respectively. That this may actually be the case can best be shown by an example. Let us suppose that the two systems are two particles, and that

$$\Psi(x_1, x_2) = \int_{-\infty}^{\infty} e^{(2\pi i/h)(x1-x2+x0)p}\, dp, \qquad (9)$$

where x_0 is some constant. Let A be the momentum of the first particle; then, as we have seen in Eq. (4), its *eigenfunctions* will be

$$u_p(x_1) = e^{(2\pi i/h)\, p\, x1} \qquad (10$$

corresponding to the *eigenvalue* p. Since we have here the case of a continuous spectrum, Eq. (7)

$$[\Psi(x_1, x_2) = \Sigma_{n=1}^{\infty}\, \psi_n(x_2)u_n(x_1), \qquad (7)]$$

will now be written

$$\Psi(x_1, x_2) = \int_{-\infty}^{\infty} \psi_p(x_2)u_p(x_1)\, dp, \qquad (11)$$

where

$$\psi_p(x_2) = e^{-(2\pi i/h)(x2-x0)p}. \qquad (12)$$

This ψ_p however is the *eigenfunction* of the operator

$$P = (h/2\pi i)\, \partial/\partial x_2, \qquad (13)$$

corresponding to the *eigenvalue* – p of the *momentum* of the second particle. On the other hand, if B is the coordinate of the first particle, it has for *eigenfunctions*

$$v_x(x_1) = \delta(x_2 - x_1) \qquad (14)$$

corresponding to the *eigenvalue* x, where $\delta(x_2 - x_1)$ is the well-known *Dirac delta-function*. Eq. (8)

$$[\Psi(x_1, x_2) = \Sigma_{n=1}^{\infty}\, \varphi_n(x_2)v_n(x_1), \qquad (8)]$$

in this case becomes

$$\Psi(x_1, x_2) = \int_{-\infty}^{\infty} \varphi_x(x_2) v_x(x_1), \tag{15}$$

where

$$\varphi_x(x_2) = \int_{-\infty}^{\infty} e^{(2\pi i/h)(x - x2 + x0)p}\, dp = h\delta(x - x_2 + x_0). \tag{16}$$

This φ_x, however, is the *eigenfunction* of the operator

$$Q = x_2 \tag{17}$$

corresponding to the *eigenvalue* $x + x_0$ of the coordinate of the second particle. Since

$$PQ - QP = h/2\pi i, \tag{18}$$

we have shown that it is in general possible for ψ_k and φ_r to be *eigenfunctions* of two *non-commuting operators*, corresponding to physical quantities.

Returning now to the general case contemplated in Eqs. (7) and (8), we assume that ψ_k and φ_r are indeed *eigenfunctions* of some *non-commuting operators* P and Q, corresponding to the *eigenvalues* p_k and q_r, respectively. Thus, by measuring either A or B we are in a position to predict with certainty, and without in any way disturbing the second system, either the value of the quantity P (that is p_k) or the value of the quantity Q (that is q_r). In accordance with our criterion of reality, in the first case we must consider the quantity P as being an element of reality, in the second case the quantity Q is an element of reality. But, as we have seen, both *wave functions* ψ_k and φ_r belong to the same reality.

Previously we proved that either (1) the quantum-mechanical description of reality given by the *wave function* is not complete or (2) when the operators corresponding to two physical quantities do not commute the two quantities cannot have simultaneous reality. Starting then with the assumption that the *wave function* does give a complete description of the physical reality, we arrived at the conclusion that two physical quantities, with *non-commuting operators*, can have simultaneous reality. Thus, the negation of (1) leads to the negation of the only other alternative (2). *We are thus forced to conclude that the quantum-mechanical description of physical reality given by wave functions is not complete.*

One could object to this conclusion on the grounds that our criterion of reality is not sufficiently restrictive. Indeed, one would not arrive at our conclusion if one insisted that two or more physical quantities can be regarded as simultaneous elements of reality *only when they can be simultaneously measured or predicted.* On this point of view, since either one or the other, but not both simultaneously, of the quantities P and Q can be predicted, they are not simultaneously real. This makes the reality of P and Q depend upon the process of measurement carried out on the first system, which does, not disturb the second system in any way. No reasonable definition of reality could be expected to permit this.

While we have thus shown that the *wave function* does not provide a complete description of the physical reality, we left open the question of whether or not such a description exists. We believe, however, that such a theory is possible.

Erwin Rudolf Josef Alexander Schrödinger (August 12, 1887 – January 4, 1961).

Schrödinger was a Nobel Prize-winning Austrian-Irish physicist who developed a number of fundamental results in quantum theory: the ***Schrödinger equation*** provides a way to calculate the wave function of a system and how it changes dynamically in time.

In addition, he wrote many works on various aspects of physics: statistical mechanics and thermodynamics, physics of dielectrics, color theory, electrodynamics, general relativity, and cosmology, and he made several attempts to construct a unified field theory. In his book What Is Life? Schrödinger addressed the problems of genetics, looking at the phenomenon of life from the point of view of physics. He paid great attention to the philosophical aspects of science, ancient, and oriental philosophical concepts, ethics, and religion. He also wrote on philosophy and theoretical biology. In popular culture, he is most known for his "Schrödinger's cat" thought experiment.

Schrödinger was born in Erdberg, Vienna, Austria, on 12 August 1887, to Rudolf Schrödinger (cerecloth producer, botanist) and Georgine Emilia Brenda Schrödinger (née Bauer) (daughter of Alexander Bauer, professor of chemistry, TU Wien). He was their only child. His mother was of half Austrian and half English descent; his father was Catholic and his mother was Lutheran. He was also able to learn English outside school, as his maternal grandmother was British.

Between 1906 and 1910 (the year he earned his doctorate) Schrödinger studied at the University of Vienna under the physicists Franz S. Exner and Friedrich Hasenöhrl. He received his doctorate at Vienna under Hasenöhrl. He also conducted experimental work with Karl Wilhelm Friedrich "Fritz" Kohlrausch. In 1911, Schrödinger became an assistant to Exner. In 1914 Schrödinger achieved habilitation (venia legendi).

Between 1914 and 1918 he participated in war work as a commissioned officer in the Austrian fortress artillery (Gorizia, Duino, Sistiana, Prosecco, Vienna).

On 6 April 1920, Schrödinger married Annemarie (Anny) Bertel. Schrödinger suffered from tuberculosis and several times in the 1920s stayed at a sanatorium in Arosa. It was there that he formulated his wave equation.

In 1920 he became the assistant to Max Wien, in Jena, and in September 1920 he attained the position of ao. Prof. (ausserordentlicher Professor) in Stuttgart, roughly equivalent to reader (UK) or associate professor (US).

In 1921, he became o. Prof. (ordentlicher Professor, i.e. full professor), in Breslau (now Wrocław, Poland). In 1921, he moved to the University of Zürich. In the first years of his career Schrödinger became acquainted with the ideas of the old quantum theory, developed in the works of Max Planck, Albert Einstein, Niels Bohr, Arnold Sommerfeld, and others. This knowledge helped him work on some problems in theoretical physics, but the Austrian scientist at the time was not yet ready to part with the traditional methods of classical physics.

The first publications of Schrödinger about atomic theory and the theory of spectra began to emerge only from the beginning of the 1920s, after his personal acquaintance with Sommerfeld and Wolfgang Pauli and his move to Germany. In January 1921, Schrödinger finished his first article on this subject, about the framework of the Bohr-Sommerfeld effect of the interaction of electrons on some features of the spectra of the alkali metals. *Of particular interest to him was the introduction of relativistic considerations in quantum theory.*

In autumn 1922 he analyzed the electron orbits in an atom from a geometric point of view, using methods developed by the mathematician Hermann Weyl (1885–1955). This work, in which it was shown that quantum orbits are associated with certain geometric properties, was an important step in predicting some of the features of wave mechanics. Earlier in the same year he created the Schrödinger equation of the ***relativistic*** Doppler effect for spectral lines, based on the hypothesis of light quanta and considerations of energy and momentum. He liked the idea of his teacher Exner on the statistical nature of the conservation laws, so he enthusiastically embraced the articles of Bohr, Kramers, and Slater, which suggested the possibility of violation of these laws in individual atomic processes (for example, in the process of emission of radiation). Although the experiments of Hans Geiger and Walther Bothe soon cast doubt on this, the idea of *energy as a statistical concept* was a lifelong attraction for Schrödinger and he discussed it in some reports and publications.

In March 1926, Schrödinger published his first paper on wave mechanics and presented what is now known as the ***Schrödinger equation***. [Schrodinger, E. (March, 1926). Quantisierung als Eigenwertproblem. (Erste Mitteilung) (Quantization as an eigenvalue problem. (First communication).) *Ann. Physik*, 384, 4, 79, 261-376; https://doi.org/10.1002/andp.19263840404.]

> [The ***Schrödinger equation*** is a partial differential equation that governs the wave function of a *non-relativistic* quantum-mechanical system. Its discovery was a significant landmark in the development of quantum mechanics. It is named after Erwin Schrödinger, an Austrian physicist, who postulated the equation in 1925 and published it in 1926, forming the basis for the work that resulted in his Nobel Prize in Physics in 1933.
>
> Conceptually, the Schrödinger equation is the quantum counterpart of Newton's second law in classical mechanics. Given a set of known initial conditions,

Newton's second law makes a mathematical prediction as to what path a given physical system will take over time. The Schrödinger equation gives the evolution over time of the wave function, the quantum-mechanical characterization of an isolated physical system. The equation was postulated by Schrödinger based on a postulate of Louis de Broglie that all matter has an associated matter wave. The equation predicted bound states of the atom in agreement with experimental observations.

The Schrödinger equation is not the only way to study quantum mechanical systems and make predictions. Other formulations of quantum mechanics include *matrix mechanics*, introduced by Werner Heisenberg, and the *path integral formulation*, developed chiefly by Richard Feynman. When these approaches are compared, the use of the Schrödinger equation is sometimes called "wave mechanics".

The equation given by Schrödinger is nonrelativistic because it contains a first derivative in time and a second derivative in space, and therefore space and time are not on equal footing.

> [*Paul Dirac incorporated* ***special relativity*** *and quantum mechanics into a single formulation that simplifies to the Schrödinger equation in the non-relativistic limit.* This is the ***Dirac equation***, which contains a single derivative in both space and time. Another partial differential equation, the Klein–Gordon equation, led to a problem with probability density even though it was a relativistic wave equation. The probability density could be negative, which is physically unviable. This was fixed by Dirac by taking the so-called square root of the Klein–Gordon operator and in turn introducing Dirac matrices. In a modern context [???], the Klein–Gordon equation describes spin-less particles, while the Dirac equation describes spin-1/2 particles.]

The most general form is the *time-dependent Schrödinger equation*, which gives a description of a system evolving with time:

$$i\hbar d/dt\,|\Psi(t)\rangle = H\hat{}\,|\Psi(t)\rangle$$

where t is time, $|\Psi(t)\rangle$ is the state vector of the quantum system, and H^ is an observable, the Hamiltonian operator.

> [The term "Schrödinger equation" can refer to both the general equation, or the specific nonrelativistic version. The general equation is indeed quite general, used throughout quantum mechanics, for everything from the Dirac equation to quantum field theory, by plugging in diverse expressions for the Hamiltonian. The specific nonrelativistic version is an approximation that

yields accurate results in many situations, but only to a certain extent (see relativistic quantum mechanics and relativistic quantum field theory).]

To apply the Schrödinger equation, write down the Hamiltonian for the system, accounting for the *kinetic* and *potential energies* of the particles constituting the system, then insert it into the *Schrödinger equation.* The resulting partial differential equation is solved for the wave function, which contains information about the system. In practice, the square of the absolute value of the wave function at each point is taken to define a *probability density function.* For example, given a wave function in position space $\Psi(x, t)$ as above, we have

$$\Pr(x, t) = |\Psi(x, t)|^2.$$

Given the Schrödinger equation

$$\hat{H}|n\rangle = E_n|n\rangle,$$

where $|n\rangle$ indexes the set of eigenstates of the Hamiltonian with energy eigenvalues E_n, we see immediately that

$$\hat{H}(|n\rangle + |n'\rangle) = E_n|n\rangle + E_{n'}|n'\rangle,$$

where

$$|\Psi\rangle = |n\rangle + |n'\rangle$$

is a solution of the Schrödinger equation but is not generally an eigenstate because E_n and $E_{n'}$ are not generally equal. *We say that $|\Psi\rangle$ is made up of a* ***superposition*** *of energy eigenstates.*

Now consider the more concrete case of *an electron that has either spin up or down.* We now index the eigenstates with the spinors in the $\hat{z}$ basis:

$$|\Psi\rangle = c_1|\uparrow\rangle + c_2|\downarrow\rangle,$$

where $|\uparrow\rangle$ and $|\downarrow\rangle$ denote *spin-up and spin-down states respectively.* As previously discussed, the magnitudes of the complex coefficients give the ***probability of finding the electron in either definite spin state:***

$$P(|\uparrow\rangle) = |c_1|^2,$$
$$P(|\downarrow\rangle) = |c_2|^2,$$
$$P_{total} = P(|\uparrow\rangle) + P(|\downarrow\rangle) = |c_1|^2 + |c_2|^2 = 1,$$

where the probability of finding the particle with either spin up or down is normalized to 1. Notice that c_1 and c_2 are complex numbers, so that

$|\Psi\rangle = 3/5\, i|\uparrow\rangle + 4/5\, |\downarrow\rangle.$

is an example of an ***allowed state***. We now get

$P(|\uparrow\rangle) = |3i/5|^2 = 9/25,$
$P(|\downarrow\rangle) = |4/5|^2 = 16/25,$
$P_{total} = P(|\uparrow\rangle) + P(|\downarrow\rangle) = 9/25 + 16/25 = 1.$

Quantum superposition is a fundamental principle of quantum mechanics that ***states that linear combinations of solutions to the Schrödinger equation are also solutions of the Schrödinger equation.*** This follows from the fact that the Schrödinger equation is a linear differential equation in time and position. More precisely, the state of a system is given by a linear combination of all the eigenfunctions of the Schrödinger equation governing that system.

An example is a qubit used in quantum information processing. The interference fringes in the double-slit experiment provide another example of the *superposition* principle.]

In this paper, he gave a *"derivation" of the wave equation for time-independent systems and showed that it gave the correct energy eigenvalues for a hydrogen-like atom.* This paper has been universally celebrated as one of the most important achievements of the twentieth century and created a revolution in most areas of quantum mechanics and indeed of all physics and chemistry.

A second paper was submitted just four weeks later that solved the quantum harmonic oscillator, rigid rotor, and diatomic molecule problems and gave a new derivation of the Schrödinger equation. [Schrodinger, E. (1926). Quantisierung als Eigenwertproblem (Zweite Mitteilung). (Quantization as an eigenvalue problem. (Second communication).) *Ann. Physik*, 4, 79, 489-527.]

A third paper, published in May, showed the equivalence of his approach to that of Heisenberg and gave the treatment of the Stark effect. [Schrodinger, E. (1926). Quantisierung als Eigenwertproblem (Dritte Mitteilung: Störungstheorie, mit Anwendung auf den Starkeffekt der Balmerlinien). (Quantization as an eigenvalue problem. (Third communication: Perturbation theory, with application to the strong effect of Balmer lines).) *Ann. Physik,* 4, 80, 437-90.]

A fourth paper in this series showed how to treat problems in which the system changes with time, as in scattering problems. [Schrodinger, E. (1926). Quantisierung als Eigenwertproblem (Vierte Mitteilung). (Quantization as an eigenvalue problem. (Fourth communication).) *Ann. Physik,* 4, 81, 109-39.]

In this paper he introduced a complex solution to the wave equation in order to prevent the occurrence of fourth and sixth order differential equations. (*This was arguably the moment when quantum mechanics switched from real to complex numbers.*) When he introduced complex numbers in order to lower the order of the differential equations, something magical happened, and all of wave mechanics was at his feet. (He eventually reduced the order to one.)

These papers were his central achievement and were at once recognized as having great significance by the physics community. An account of the four papers in English was published in December of that year. [Schrodinger, E. (December, 1926). A Wave Theory of the Mechanics of Atoms and Molecules. *Phys. Rev.*, 28, 1049-70.]

Schrödinger was not entirely comfortable with the implications of quantum theory referring to his theory as "wave mechanics." He wrote about the probability interpretation of quantum mechanics, saying: "I don't like it, and I'm sorry I ever had anything to do with it."

In 1927, he succeeded Max Planck at the Friedrich Wilhelm University in Berlin. In 1933, Schrödinger decided to leave Germany because he disliked the Nazis' antisemitism. He became a Fellow of Magdalen College at the University of Oxford. Soon after he arrived, he received the Nobel Prize for the formulation of the Schrödinger equation, which he shared with Dirac.

His position at Oxford did not work out well; his unconventional domestic arrangements, sharing living quarters with two women, were not met with acceptance. In 1934, Schrödinger lectured at Princeton University; he was offered a permanent position there, but did not accept it. Again, his wish to set up house with his wife and his mistress may have created a problem. He had the prospect of a position at the University of Edinburgh but visa delays occurred, and in the end, he took up a position at the University of Graz in Austria in 1936. In the midst of these tenure issues in 1935, he published a brilliant rebuttal to the EPR paper [Schrödinger, E. (October, 1935). Discussion of probability relations between separated systems. See below.] and after extensive correspondence with Albert Einstein, proposed what is now called the Schrödinger's cat thought experiment.

In 1938, after the Anschluss, Schrödinger had problems in Graz because of his flight from Germany in 1933 and his known opposition to Nazism. He issued a statement recanting this opposition (he later regretted doing so and explained the reason to Einstein). However, this did not fully appease the new dispensation and the University of Graz dismissed him from his post for political unreliability. He suffered harassment and was instructed not to leave the country. He and his wife, however, fled to Italy. From there, he went to visiting positions in Oxford and Ghent University.

In the same year he received a personal invitation from Ireland's Taoiseach, Éamon de Valera – a mathematician himself – to reside in Ireland and agree to help establish an Institute for Advanced Studies in Dublin. When he migrated to Ireland in 1938, he obtained visas for himself, his wife and also another woman, Mrs. Hilde March. March was the wife of an Austrian colleague with whom Schrödinger had fathered a daughter in 1934. Schrödinger wrote personally to de Valera to obtain the visa for Mrs. March. In October 1939 the ménage à trois duly took up residence in Dublin. He moved to Kincora Road, Clontarf, Dublin and lived modestly. Schrödinger fathered two further daughters by two different women during his time in Ireland.

He became the Director of the School for Theoretical Physics in 1940 and remained there for 17 years. He became a naturalized Irish citizen in 1948, but also retained his Austrian citizenship. He wrote around 50 further publications on various topics, including his explorations of unified field theory.

In 1944, he wrote *What Is Life?*, which contains a discussion of negentropy and the concept of a complex molecule with the genetic code for living organisms. According to James D. Watson's memoir, *DNA, the Secret of Life*, Schrödinger's book gave Watson the inspiration to research the gene, which led to the discovery of the DNA double helix structure in 1953. Similarly, Francis Crick, in his autobiographical book *What Mad Pursuit*, described how he was influenced by Schrödinger's speculations about how genetic information might be stored in molecules.

Following his work on quantum mechanics, Schrödinger devoted considerable effort to working on a unified field theory that would unite gravity, electromagnetism, and nuclear forces within the basic framework of General Relativity, doing the work with an extended correspondence with Albert Einstein. In 1947, he announced a result, "Affine Field Theory," in a talk at the Royal Irish Academy, but the announcement was criticized by Einstein as "preliminary" and failed to lead to the desired unified theory. Following the failure of his attempt at unification, Schrödinger gave up his work on unification and turned to other topics

In 1956, he returned to Vienna to take up his appointment as Chair of Physics at the University of Vienna. At an important lecture during the World Energy Conference, he refused to speak on nuclear energy because of his skepticism about it and gave a philosophical lecture instead. During this period Schrödinger turned from mainstream quantum mechanics' definition of wave–particle duality and promoted the wave idea alone, causing much controversy. On 4 January 1961, Schrödinger died of tuberculosis, aged 73, in Vienna.

Schrödinger, E. (October, 1935). Discussion of probability relations between separated systems.

Mathematical Proceedings of the Cambridge Philosophical Society, 31, 4, 555–63; https://doi.org/10.1017/S0305004100013554; also at https://sci-hub.se/10.1017/S0305004100013554.

Communicated by Mr. M. Born.

Received August 14, 1935.

Read October 28, 1935.

"When two systems, of which we know the states by their respective representatives, enter into temporary physical interaction due to known forces between them, and when after a time of mutual influence, the systems separate again, then they can no longer be described in the same way as before, viz. by endowing each of them with a representative of its own. *I would not call that one but rather the characteristic trait of quantum mechanics*, the one that enforces its entire departure from classical lines of thought. *By the interaction the two representatives (or ψ-functions) have become entangled.*" A brilliant rebuttal to Einstein, Podolsky, & Rosen. (May, 1935). [See above.]

Summary

The probability relations which can occur between two separated physical systems are discussed, on the assumption that their state is known by a representative in common. The *two families* of observables, relating to the first and to the second system respectively, are linked by at least *one match* between two definite members, one of either family. The word *match* is short for stating that the *values* of the two observables in question determine each other uniquely and therefore (since the actual labelling is irrelevant) can be taken to be *equal.* In general, there is but one match, but there can be more. If, in addition to the first match, there is a second one between *canonical conjugates* [quantities which are related by definition such that one is the Fourier transform of another] of the first mates, then there are infinitely many matches, every function of the first canonical pair matching with the same function of the second canonical pair. Thus, there is a complete one-to-one correspondence between *those* two branches (of the two families of observables) which relate to the two degrees of freedom in question. If there *are* no others, the one-to-one correspondence persists as time advances, but the observables of the first system (say) change their mates in the way that the latter, i.e. the observables of the second system, undergo a certain continuous contact-transformation.

---

1. When two systems, of which we know the states by their respective representatives, enter into temporary physical interaction due to known forces between them, and when

after a time of mutual influence, the systems separate again, then they can no longer be described in the same way as before, viz. by endowing each of them with a representative of its own. *I would not call that one but rather the characteristic trait of quantum mechanics*, the one that enforces its entire departure from classical lines of thought. *By the interaction the two representatives (or ψ-functions) have become entangled.* To disentangle them we must gather further information by experiment, although we knew as much as anybody could possibly know about all that happened. Of either system, taken separately, all previous knowledge may be entirely lost, leaving us but one privilege: to restrict the experiments to one only of the two systems. After re-establishing one representative by observation, the other one can be inferred simultaneously. In what follows the whole of this procedure will be called *the disentanglement*. Its sinister importance is due to its being involved in every measuring process and therefore forming the basis of the quantum theory of measurement, threatening us thereby with at least a *regressus in infinitum*, since it will be noticed that the procedure itself involves measurement.

Another way of expressing the peculiar situation is: the best possible knowledge of a *whole* does not necessarily include the best possible knowledge of all its *parts*, even though they may be entirely separated and therefore virtually capable of being "best possibly known", i.e. of possessing, each of them, a representative of its own. The lack of knowledge is by no means due to the *interaction* being insufficiently known—at least not in the way that it could possibly be known more completely—*it is due to the interaction itself.*

Attention has recently*

> * Einstein, A., Podolsky, B. & Rosen, N. (May, 1935). Can Quantum-Mechanical Description of Physical Reality Be Considered Complete? *Phys. Rev.*, 47, 10, 777-80 [See above].

been called to the obvious but very disconcerting fact that even though we restrict the disentangling measurements to *one* system, the representative obtained for the *other* system is by no means independent of the particular choice of observations which we select for that purpose and which by the way are entirely arbitrary. It is rather discomforting that the theory should allow a system to be steered or piloted into one or the other type of state at the experimenter's mercy in spite of his having no access to it. *This paper does not aim at a solution of the paradox, it rather adds to it,* if possible. A hint as regards the presumed obstacle will be found at the end.

2. To begin with I wish to establish a simple theorem, which makes it very obvious that the phenomenon in question is a quite general one; that *it is the rule and not the exception. The representative arrived at for one system depends on the program of observations to be taken with the other one*. It is necessary to envisage the dependence on the *program*. For

since one device only can be carried out in every individual case and since, moreover, we cannot tell the result (because after all we are not actually experimenting, but sitting at our desk), there seems to be a certain liberty for presuming that perhaps, after all, there is always or at least in most cases a result possible, which is also possible when other devices are followed, and that perhaps it is this that actually would turn up.

Let x and y stand for all the *coordinates* of the first and second systems respectively and $\Psi(x,y)$ for the normalized representative of the *state* of the composed system, *when the two have separated again, after the interaction has taken place*. What constitutes the *entanglement* is that Ψ is not a product of a function of x and a function of y. Now suppose that we perform on the second system certain observations in consequence of which its representative, at the moment in which *disentanglement* is reached, is sure to turn up as one out of the known complete set of *normalized orthogonal functions* $f_n(y)$. Then, provided that the variables which we have measured all commute, we have to develop $\Psi(x,y)$ into a series with respect to the f_n,

$$\Psi(x,y) = \Sigma_n\, c_n g_n(x)\, f_n(y), \qquad (1)$$

in order to come to know the representative of the other system. When the readings on the y-system point to $f_k(y)$, we have to adopt $g_k(x)$ as the representative of the x-system. The c_k have been introduced in order to assume that the g_k are normalized, i.e. that

$$\int g_k{}^*(x)\, g_k(x)\, dx = 1. \qquad (2)$$

Of course, $|\, c_k\,|^2$ is the probability of that particular case occurring. The equations

$$c_k g_k(x) = \int f_k{}^*(y)\, \Psi(x,y)\, dy \qquad (3)$$

together with (2) determine the c's and the g's, apart from an irrelevant phase-factor in every g and its reciprocal in the corresponding c *and* apart from the possible indeterminateness of g, should the integral for some values of k vanish identically in x.

There is no reason for the g_k to be orthogonal to each other. Let us ask *when* they are, i.e. *how* must the f_k be chosen for that purpose? The condition evidently is

$$c_k{}^* c_l\, \delta_{kl} = \int dx \int dy \int dy'\, f_k(y')\, \Psi^*(x,y')\, f_l{}^*(y)\, \Psi(x,y). \qquad (4)$$

This amounts to saying that, for every k, the function

$$u_k(y) = \int dx \int dy'\, f_k(y')\, \Psi^*(x,y')\, \Psi(x,y) \qquad (5)$$

is to be orthogonal to all the $f_l(y)$, with the possible exception of $f_k(y)$. Hence $u_k(y)$ must be a numerical multiple of $f_k(y)$. From (4), with $l = k$, it is seen that the numerical multiplier is $|\, c_k\,|^2$. We have therefore

$$|\, c_k\,|^2\, f_k(y) = \int dx \int dy'\, f_k(y')\, \Psi^*(x,y')\, \Psi(x,y). \qquad (6)$$

Introducing the function

$$K(y,y') = \int dx\ \Psi^*(x,y')\ \Psi(x,y), \qquad (7)$$

which has *Hermitian symmetry*, we see from (6) that the reciprocals of the $|\ c_k\ |^2$ and the functions $f_k(y)$ are required to be the *eigenvalues* and a system of *eigenfunctions* respectively of the *homogeneous integral equation*

$$f(y) = \lambda \int K(y,y')\ f(y')\ dy'. \qquad (8)$$

Provided that the integral in (7) converges, so that K is defined, a complete solution of (8) exists. (It is convenient for our purposes, in order to be concerned with complete sets only, to include the functions, orthogonal to K, as *eigenfunctions* belonging to $A = \infty$, at variance with the custom of mathematicians.) By using this set for the development (1) one easily satisfies oneself that all requirements are fulfilled, in particular that the λ_k^{-1} are all non-negative and that their sum *is* unity.

The general case is evidently that all the λ_k^{-1} are different from one another, except maybe for an arbitrary set of them vanishing. Then the *relevant* $f_k(y)$ are uniquely determined and so are the $g_k(x)$. Hence there is always one and as a rule only one development of $\Psi(x,y)$ of the type which might suitably be called "*biorthogonal*†".

† The whole mathematical treatment is familiar to mathematicians in dealing with an "*unsymmetrical kernel*" $\Psi(x,y)$. See Courant-Hilbert, *Methoden der mathematischen Physik*, 2nd edition, p. 134.

Whenever (and of course only when) the *eigenfunctions* of a *program* to be carried out on the y-system include the relevant functions $f_k(y)$, or the *eigenfunctions* properly speaking of (8), the program will lead to the *biorthogonal development* and imply the relevant $g_k(x)$ as the other set. Now if for an arbitrarily fixed *program* of measurements on the y-system the representative arrived at for the x-system was the same *in all individual cases*, the same $g_k(x)$ would have to turn up (and even with the same probabilities) as in the *biorthogonal development*; for in two infinite series of repetitions *ab ovo* of one and of the other *program* respectively every possible result occurs according to its due probability. Hence the relevant functions $g_k(x)$ would have to be implied whatever program is carried out. But since, of course, they also determine the *biorthogonal development* uniquely and thereby require the relevant $f_k(y)$ as the other set, these would have to be included in the *eigenfunctions* of every program which cannot be, since the latter are, by principle, an entirely arbitrary complete orthogonal set. Hence the non-invariance is proved*.

* In order to adapt this proof to the case when the *biorthogonal development* is not unique, just replace the *biorthogonal development* by a particular one, on which you fix your attention.

There must, of course, be cases in which the *biorthogonal development* refers to a continuous variable (or set of commuting variables), an integral replacing the series (1); and also mixed cases. In our present treatment they would be indicated by the integral (7) diverging and would therefore require a separate treatment, on which I shall not enter here.

The *biorthogonal development* is the one to give us true insight into the *entanglement*. If there are no coincidences among the $| c_k |^2$ (excluding also the case, that more than one of them vanish) the relevant f_k's form a well-determined and complete set and so do the g_k's. Then one can say that the *entanglement* consists in that one and only one observable (or set of commuting observables) of one system is uniquely determined by a definite observable (or set of commuting observables) of the other system. *This is the general case*. We shall now turn to the opposite extreme, which is the *Einstein-Podolsky-Rosen case*. It could be characterized by all the $| ck |^2$ being equal and all possible developments being *biorthogonal*. Every observable (or set, etc.) of one system is determined by an observable (or set, etc.) of the other one. But the mere fact, that the equality of the $| ck |^2$ prevents their sum from being normalized to unity, shows us that very improper representatives (in fact much more so than Dirac's δ, δ', δ'', ...) are involved in this case, making it advisable to deal with it on slightly different lines.

3. For simplicity's sake we suppose each of the two systems to have one degree of freedom only. Let the q-numbers x_1, p_1 and x_2, p_2 denote coordinate and momentum of the first and of the second system respectively. The existence of further degrees of freedom would not affect the considerations of this section except for slight alterations in the wording; but for section 4 to hold it would have to be assumed, that within each of the two systems the degree of freedom which we investigate has its Hamiltonian separated from the rest.

The two systems are of course supposed not to interact with each other. The *entanglement* is to be such that the two *commuting* observables

$$x = x_1 - x_2, \qquad p = p_1 + p_2, \tag{9}$$

which we choose to represent the state of the *composed* system, have definite numerical values, say x' and p' respectively, which we suppose to be known. The representative Ψ of the *composed system* is a function of the *eigenvalues* of x and p, which involves x' and p' as parameters and vanishes everywhere except in that point where the former are equal to the latter. It is not a δ-function though and can hardly be written explicitly. According to our assumptions Ψ must have the properties

$$x\,\Psi = x'\,\Psi \text{ and } p\,\Psi = p'\,\Psi. \tag{10}$$

We shall use no others.

From (9) the variable x can be observed by observing x_1 and x_2 separately, because the latter commute. The difference of the observed values, x'_1 and x'_2 say, must be equal to x':

$$x'_1 - x'_2 = x'. \qquad (11)$$

Hence x'_1 can be predicted from x'_2 and *vice versa.* Similarly

$$p'_1 - p'_2 = p'. \qquad (12)$$

so that the result of measuring p_1 serves to predict the result for p_2 and vice versa. But, of course, every *one* of the *four* observations in question, when actually performed, *disentangles* the systems, furnishing each of them with an independent representative of its own. A second observation, whatever it is and on whichever system it is executed, produces no further change in the representative of the *other* system.

Yet since I can predict *either* x'_1 *or* p'_1 without interfering with system No. 1 and since system No. 1, like a scholar in examination, cannot possibly know which of the two questions I am going to ask it first: it so seems that our scholar is prepared to give the right answer to the *first* question he is asked, *anyhow*. Therefore, he must know both answers; which is an amazing knowledge, quite irrespective of the fact that after having given his first answer our scholar is invariably so disconcerted or tired out, that all the following answers are "wrong".

Thus far the results of the paper quoted above. Now I wish to point out that system No. 1 (say) has further knowledge. It does not only know these two answers but a vast number of others, and that with no mnemotechnical help whatsoever, at least with none that we know of.

Let us consider an *Hermitian operator* referring to the first system and given as a "well-ordered" analytic function of the observables x_1 and p_1:

$$F(x_1, p_1), \qquad (13)$$

which we suppose not to contain the $\sqrt{-1}$ *explicitly*. It is an observable of system No. 1.

We shall prove that its value is equal to the value of the following observable of system No. 2

$$F(x_2 + x', p' - p_2), \qquad (14)$$

so that the result of either observation can be predicted from the other one. That is not trivial, because the equations x = x' and p = p' do not hold, except in the form (10), that is to say they are not identities.

The proof will be produced, if we can show that the difference of the two operators, when applied to Ψ, gives zero:

$$\{F(x_2 + x', p' - p_2) - F(x_1, p_1)\}\Psi = 0. \qquad (15)$$

Using (9), we may write this in the form

$$\{F(x_1 + x' - x, p_1 + p' - p) - F(x_1, p_1)\}\Psi = 0. \qquad (16)$$

To prove it we observe that any operator which *ends* on its right with a factor $x' - x$ or $p' - p$ reduces Ψ to zero, from (10). Additive terms of this type can therefore be dropped within the curved bracket. Now fix the attention on one of the power products in the *minuendus* [a quantity or number to be to be lessened or diminished]. Its last factor, either $x_1 + x' - x$ or $p_1 + p' - p$, can be replaced by x_1 or p_1, as the case may be, and then this x_1 or p_1 commutes with the rest of the power product and can be removed to its extreme left. The second step consists in applying a similar treatment to the factor ($x_1 + x' - x$ or $p_1 + p' - p$, as the case may be), which has *now* become the last; but this one cannot safely be displaced to the *extreme* left but only to the second place, counting from the left. This procedure is continued until we are left with a power product which differs from the original one in that x_1 and p_1 have replaced $x_1 + x' - x$ and $p_1 + p' - p$ respectively and also that the order of factors has been reversed. But F, owing to its Hermiticity and to the further condition that *it should not contain* $\sqrt{-1}$ *explicitly*, must contain the "*reversed*" power product too. Hence after applying the same treatment to all of them, we are left with $F(x_1, p_1)$, which cancels with the *subtrahendus* [a quantity or number to be subtracted from another], *and the statement is proved.*

If F contains the $\sqrt{-1}$ explicitly, we could replace it by $(x_1p_1 - p_1x_1)/\hbar$. Then the prescription (14) would apply without corollary. It would turn the operator just mentioned into $(x_1p_1 - p_1x_1)/\hbar$ which now can be replaced by $-\sqrt{-1}$. From this follows the corollary to prescription (14), that an explicit $\sqrt{-1}$ has to change sign.

By this theorem all observables are placed on the same footing. Our system, in its virgin state, must know the answers to all of them. One might presume that it avails itself at least of a suggestive mnemotechnical device, viz. that the answer prepared for the variable $F(x_1, p_1)$ is simply $F(x'_1, p'_1)$, if x'_1 and p'_1 are those prepared for x_1 and p_1 respectively. *But this is not so*. For consider, e.g., the *series* of observables

$$F(x_1, p_1, b) = 1/b\, p_1^2 + b\, x_1^2,$$

where b is to be a positive *c-number* parameter. With every value for b we are confronted with a new observable, to which an answer must be pending. Moreover, the answer must be, irrespective of b, an odd integral multiple of ħ (though not necessarily independent of b). This shows plainly that all these answers cannot conform to the results which would be obtained by inserting into the expression the same pair of *c-numbers*, p'_1 and x'_1; and which, by the way, are simultaneously accessible to experiment in every individual case, one by direct observation, the other one by inference from an observation on the other system.

Our complete lack of insight into the relationship between the different answers in *one* system is all the more bewildering, since we have proved, on the other hand, that the one-to-one correspondence between the answers of the two systems necessarily extends to *all* pairs of observables whenever it holds for two of them.

4. If equations (10) are assumed to hold at time zero, the equations of motion determine what becomes of them as time proceeds. Let the Hamiltonian of the composed system be

$$H = H_1(x_1, p_1) + H_2(x_2, p_2) \tag{17}$$

and let it not contain the time explicitly. We shall use what Dirac calls a *Heisenberg representation*; then every variable at time t is a function of the variables at time zero, e.g.

$$x_{1t} = e^{itH1/\hbar}\, x_1\, e^{-itH1/\hbar}. \tag{18}$$

From prescription (14), including the corollary, we can find out what observable of system No. 2 is equivalent to x_{1t}; we call it $[x_{1t}]_2$,

$$[x_{1t}]_2 = e^{-itH1(x2 + x', p' - p2)/\hbar}\, (x_2 + x')\, e^{itH1(x2 + x', p' - p2)/\hbar}. \tag{19}$$

This equation, by its form, indicates the observation on No. 2 at *time zero*, which would serve to predict the coordinate of No. 1 at time t. Solving two equations, similar to (18) for x_2 and p_2, we get

$$x_2 = e^{-itH2/\hbar}\, x_{2t}\, e^{itH2/\hbar} \tag{20}$$

and similarly for p_2. Of course, H_2 has now to be thought of as written with the arguments x_{2t} and p_{2t}; which does not affect its form, since it is a constant of the motion. With these expressions replacing x_2 and p_2 in (19), the exponentials with H_2 cancel in the interior, leaving just one in front and in the rear. So the final result is

$$[x_{1t}]_2 = e^{-itH2/\hbar}\, e^{-itH1/\hbar}\, (x_{2t} + x')\, e^{itH1/\hbar}\, e^{itH2/\hbar}, \tag{21}$$

where H_1, and H_2 are precisely the functions of equation (17), but written with the arguments

$x_{2t} + x'$ and $p' - p_{2t}$ for H_1,
x_{2t} and p_{2t} for H_2.

This rather complicated function of x_{2t} and p_{2t} is *that* observable of system No. 2 which is equivalent to x_{1t}. Though we have deduced it by means of a *Heisenberg representation*, the functional connection is of course exactly the same for what Dirac calls the *Schrodinger representatives*. That is to say, we can take x_{2t} and p_{2t} to have the general meaning of x_2 and p_2 of the preceding section. Regarded as operators they then do not involve the notion of time but work on a Ψ-function, which develops according to the *wave equation*. This consideration applies to every moment of time. It is therefore correct to say, that the variable which in No. 2 is equivalent to the coordinate in No. 1 undergoes a continuous unitary or *contact transformation* as time goes on. The transformation is of course the same for every observable, so that we need not write out the formulae for $[p_{1t}]_2$ or for an arbitrary $[F(x_{1t}, p_{1t})]_2$. It is noteworthy that the two exponentials of which the transformation is composed may not be amalgamated, because H_1 and H_2, considering the arguments with which they are written, do not in general commute.

All this is moderately trivial. But it is necessary to consider it least one should believe that the antinomies [real or apparent mutual incompatibility of two notions] could be solved by suggesting or proving that some of the observations must take a certain minimum time. Provided that they *relate* to a definite moment, this will not help us. It cannot be argued that, before the results are reached, the situation to which they refer has passed away. A prediction for time zero does not dissolve into nought as time goes on, but simply transforms into the prediction of another observable. And any desired observable can be predicted for time t by making a suitable observation at time zero on the other system.

When at time zero a certain observable of system No. 1, say x_1, is inferred from observing x_2, I am forced to assign to system No. 1 a representative that makes the observable x_1 precise and tells nothing about its canonically conjugate, *although* I safely infer that system No. 1 *does* know quite a definite (as opposed to a haphazard) answer for the canonical conjugate as well, the only difference being that I know the one while I am ignorant of the other. Now this paradoxical situation is not confined to time zero and could not, therefore, be avoided by my satisfying myself that the result of observing x_2 cannot be known before a certain time has elapsed. From the moment I come to know the result I should be faced with exactly the same situation, only the pair of *canonically conjugate* observables that is involved changes with time.

The paradox would be shaken, though, if an observation did not relate to a definite moment. But this would make the present interpretation of quantum mechanics meaningless, because at present the *objects* of its predictions are considered to be the results of measurements for definite moments of time.

Bell, J. S. † (1964). On the Einstein-Podolsky-Rosen Paradox*.

Physics, 1, 195-200; https://journals.aps.org/ppf/pdf/10.1103/PhysicsPhysiqueFizika.1.195.

Department of Physics, University of Wisconsin, Madison, Wisconsin.

* Work supported in part by the U.S. Atomic Energy Commission.

† On leave of absence from SLAC and CERN.

Received November, 4 1964.

In this paper, John Stewart Bell, a physicist from Northern Ireland, building upon the Einstein–Podolsky–Rosen paradox, determined that *quantum mechanics was incompatible with local hidden-variable theories,* given some basic assumptions about the nature of measurement, subsequently known as *Bell's theorem*. "*Local*" here referred to the principle of locality, the idea that a particle can only be influenced by its immediate surroundings, and that interactions mediated by physical fields could not propagate faster than the speed of light. "*Hidden variables*" were putative properties of quantum particles that were not included in quantum theory but nevertheless affected the outcome of experiments. Bell deduced that if measurements are performed independently on the two separated particles of an *entangled* pair, then the assumption that the outcomes depended upon *hidden variables* within each half implied a mathematical constraint on how the outcomes on the two measurements were correlated. This constraint was subsequently called a *Bell inequality*. Bell then showed that quantum physics predicted correlations that violated this inequality. The first rudimentary experiment designed to test *Bell's theorem* was performed in 1972 by John Clauser and Stuart Freedman.

I. *Introduction*

The paradox of Einstein, Podolsky and Rosen was advanced as an argument that quantum mechanics could not be a complete theory but should be supplemented by additional variables. These additional variables were to restore to the theory *causality* and *locality*[2].

[2] Einstein, in (1949). *Albert Einstein, Philosopher Scientist*, edited by P. A. Schilp. Library of Living Philosophers, Evanston, Illinois, p. 85: "But on one supposition we should, in my opinion, absolutely hold fast: the real factual situation of the system S_2 is independent of what is done with the system s1, which is spatially separated from the former".

In this note that idea will be formulated mathematically and shown to be incompatible with the statistical predictions of quantum mechanics. It is the requirement of *locality*, or more precisely that the result of a measurement on one system be unaffected by operations on a distant system with which it has interacted in the past, that creates the essential difficulty. There have been attempts to show that even without such a separability or *locality* requirement no "*hidden variable*" interpretation of quantum mechanics is possible. These attempts have been examined elsewhere and found wanting. Moreover, a *hidden variable* interpretation of elementary quantum theory has been explicitly constructed. That particular interpretation has indeed a grossly non-local structure. This is characteristic, according to the result to be proved here, of any such theory which reproduces exactly the quantum mechanical predictions.

II. *Formulation*

With the example advocated by Bohm and Aharonov, the EPR argument is the following. Consider a pair of *spin one-half* particles formed somehow in the singlet *spin state* and moving freely in opposite directions. Measurements can be made, say by Stern-Gerlach magnets, on selected components of the spins $\vec{\sigma}_1$ and $\vec{\sigma}_2$. If measurement of the component $\vec{\sigma}_1 \cdot \vec{a}$, where $\vec{a}$ is some unit vector, yields the value + 1 then, according to quantum mechanics, measurement of $\vec{\sigma}_2 \cdot \vec{a}$ must yield the value –1 and vice versa.
Now we make the hypothesis [2], and it seems one at least worth considering, that if the two measurements are made at places remote from one another the orientation of one magnet does not influence the result obtained with the other. Since we can predict in advance the result of measuring any chosen component of $\vec{\sigma}_2$, by previously measuring the same component of $\vec{\sigma}_1$, it follows that the result of any such measurement must actually be predetermined. Since the initial quantum mechanical wave function does not determine the result of an individual measurement, this predetermination implies the possibility of a more complete specification of the state.

Let this more complete specification be effected by means of parameters λ. It is a matter of indifference in the following whether λ denotes a single variable or a set, or even a set of functions, and whether the variables are discrete or continuous. However, we write as if A were a single continuous parameter. The result A of measuring $\vec{\sigma}_1 \cdot \vec{a}$ is then determined by $\vec{a}$ and λ, and the result B of measuring $\vec{\sigma}_2 \cdot \vec{b}$ in the same instance is determined by $\vec{b}$ and λ, and

$$A(\vec{a}, \lambda) = \pm 1, \quad B(\vec{b}, \lambda) = \pm 1. \tag{1}$$

The vital assumption [2] is that the result B for particle 2 does not depend on the setting $\vec{a}$, of the magnet for particle 1, nor A on $\vec{b}$.

If p(A) is the probability distribution of λ then the expectation value of the product of the two components $\vec{\sigma}_1 \cdot \vec{a}$ and $\vec{\sigma}_2 \cdot \vec{b}$ is

$$P(\vec{a}, \vec{b}) = \int d\lambda \rho(\lambda)\, A(\vec{a}, \lambda)\, B(\vec{b}, \lambda) \quad (2)$$

This should equal the quantum mechanical expectation value, which for the *singlet state* is

$$< \vec{\sigma}_1 . \vec{a}\ \vec{\sigma}_2 . \vec{b} > = - \vec{a} . \vec{b}. \quad (3)$$

But it will be shown that this is not possible.

Some might prefer a formulation in which the hidden variables fall into two sets, with A dependent on one and B on the other; this possibility is contained in the above, since A stands for any number of variables and the dependences thereon of A and B are unrestricted. *In a complete physical theory of the type envisaged by Einstein, the hidden variables would have dynamical significance and laws of motion*; our λ can then be thought of as initial values of these variables at some suitable instant.

III. ***Illustration***

The proof of the main result is quite simple. Before giving it, however, a number of illustrations may serve to put it in perspective.

…

IV. ***Contradiction***

The main result will now be proved. Because p is a normalized probability distribution,

$$\int d\lambda \rho(\lambda) = 1, \quad (12)$$

and because of the properties (1), P in (2) cannot be less than – 1. It can reach – 1 at $\vec{a} = \vec{b}$ only if

$$A(\vec{a}, \lambda) = B(\vec{b}, \lambda) \quad (13)$$

except at a set of points λ of zero probability. Assuming this, (2)

$$[P(\vec{a}, \vec{b}) = \int d\lambda \rho(\lambda)\, A(\vec{a}, \lambda)\, B(\vec{b}, \lambda) \quad (2)]$$

can be rewritten

$$P(\vec{a}, \vec{b}) = - \int d\lambda \rho(\lambda)\, A(\vec{a}, \lambda)\, A(\vec{b}, \lambda). \quad (14)$$

It follows that c is another unit vector

$$P(\vec{a}, \vec{b}) - P(\vec{a}, \vec{c}) = - \int d\lambda \rho(\lambda)\, [A(\vec{a}, \lambda)\, A(\vec{b}, \lambda) - A(\vec{a}, \lambda)\, A(\vec{c}, \lambda)]$$
$$= \int d\lambda \rho(\lambda)\, [A(\vec{a}, \lambda)\, A(\vec{b}, \lambda)\, [A(\vec{b}, \lambda)\, A(\vec{c}, \lambda) - 1]$$

using (1), whence

$$| P(\vec{a}, \vec{b}) - P(\vec{a}, \vec{c}) | \leq \int d\lambda \rho(\lambda) [1 - A(\vec{b}, \lambda) A(\vec{c}, \lambda)].$$

The second term on the right is $P(\vec{b}, \vec{c})$, whence

$$1 + P(\vec{b}, \vec{c}) \geq | P(\vec{a}, \vec{b}) - P(\vec{a}, \vec{c}) |. \quad (15)$$

Unless P is constant, the right-hand side is in general of order $| \vec{b} - \vec{c}) |$ for small $| \vec{b} - \vec{c}) |$. Thus $P(\vec{b}, \vec{c})$ cannot be stationary at the minimum value (-1 at $\vec{b} = \vec{c}$) and cannot equal the quantum mechanical value (3).

$$[< \vec{\sigma_1} . \vec{a} \; \vec{\sigma_2} . \vec{b} > = - \vec{a} . \vec{b}. \quad (3)]$$

Nor can the quantum mechanical correlation (3) be arbitrarily closely approximated by the form (2).

$$[P(\vec{a}, \vec{b}) = \int d\lambda \rho(\lambda) A(\vec{a}, \lambda) B(\vec{b}, \lambda) \quad (2)]$$

The formal proof of this may be set out as follows. …

…

V. *Generalization*

The example considered above has the advantage that it requires little imagination to envisage the measurements involved actually being made. In a more formal way, assuming[7] that any Hermitian operator with a complete set of *eigenstates* is an "observable", the result is easily extended to other systems.

[7] Dirac, P. A. M. (1947). *The Principles of Quantum Mechanics*, (3rd Ed.) p. 37. The Clarendon Press, Oxford.

If the two systems have state spaces of dimensionality greater than 2 we can always consider two dimensional subspaces and define, in their direct product, operators $\vec{\sigma_1}$ and $\vec{\sigma_2}$ formally analogous to those used above and which are zero for states outside the product subspace. Then for at least one quantum mechanical state, the "*singlet*" state in the combined subspaces, the statistical predictions of quantum mechanics are incompatible with separable predetermination.

VI. *Conclusion*

In a theory in which parameters are added to quantum mechanics to determine the results of individual measurements, without changing the statistical predictions, there must be a mechanism whereby the setting of one measuring device can influence the reading of another instrument, however remote. Moreover, the signal involved must propagate instantaneously, so that *such a theory could not be Lorentz invariant*.

Of course, the situation is different if the quantum mechanical predictions are of limited validity. Conceivably they might apply only to experiments in which the settings of the instruments are made sufficiently in advance to allow them to reach some mutual rapport by exchange of signals with velocity less than or equal to that of light. In that connection, experiments of the type proposed by Bohm and Aharonov[6], in which the settings are changed during the flight of the particles, are crucial.

[6] Bohm, D. & Aharonov, Y. (1957). *Phys. Rev.*, 108, 1070.

John Francis Clauser (born December 1, 1942).

Clauser is an American theoretical and experimental physicist known for contributions to the foundations of quantum mechanics, in particular the Clauser–Horne–Shimony–Holt inequality.

Clauser was born in Pasadena, California. His father, Francis H. Clauser, was a professor of aeronautical engineering who founded and chaired the aeronautics department at Johns Hopkins University. His mother, Catharine McMillan, was the humanities librarian at Caltech and sister of 1951 Nobel Prize in Chemistry laureate Edwin McMillan.

He received a BSc in physics from Caltech in 1964, and MA in physics in 1966 and PhD in physics in 1969 from Columbia University.

In 1969, while still a graduate student at Columbia University, Clauser, along with Michael Horne, Abner Shimony, and Richard Holt, transformed Bell's 1964 mathematical theorem into a very specific experimental prediction via what is now called the Clauser–Horne–Shimony–Holt (CHSH) inequality. In 1972, when he was a postdoctoral researcher at the University of California Berkeley and Lawrence Berkeley National Laboratory, Clauser and graduate student Stuart Freedman were the first to prove experimentally that two widely separated particles (about 10 feet apart) can be entangled.

John Clauser standing with his second *quantum entanglement* experiment at UC Berkeley in 1976. Credit: University of California Graphic Arts / Lawrence Berkeley Laboratory.

Fifty years later, Clauser was awarded the 2022 Nobel Prize in Physics, jointly with Alain Aspect and Anton Zeilinger "for experiments with *entangled photons*, establishing the violation of Bell inequalities and pioneering quantum information science".

First Experimental Proof That Quantum Entanglement Is Real* (1972).

* Based on article at https://scitechdaily.com/first-experimental-proof-that-quantum-entanglement-is-real/California Institute of Technology, October 9, 2022.

"When scientists, including Albert Einstein and Erwin Schrödinger, first discovered the phenomenon of *entanglement* in the 1930s, they were perplexed. Disturbingly, *entanglement* required two separated particles to remain connected without being in direct contact. In fact, Einstein famously called entanglement "*spooky action at a distance*," because the particles seemed to be communicating faster than the speed of light.

To explain the bizarre implications of *entanglement*, Einstein, along with Boris Podolsky and Nathan Rosen (EPR), argued that "*hidden variables*" should be added to quantum mechanics. These could be used to explain *entanglement*, and to restore "*locality*" and "*causality*" to the behavior of the particles. *Locality* states that objects are only influenced by their immediate surroundings. Causality states that an effect cannot occur before its cause, and that causal signaling cannot propagate faster than light speed. Niels Bohr famously disputed EPR's argument, while Schrödinger and Wendell Furry, in response to EPR, independently hypothesized that *entanglement* vanishes with wide-particle separation.

Unfortunately, at the time, no experimental evidence for or against *quantum entanglement* of *widely separated particles* was available. Experiments have since proven that *entanglement* is very real and fundamental to nature. Furthermore, quantum mechanics has now been proven to work, not only at very short distances but also at very great distances."
...

"The very first of these experiments was proposed and executed by Caltech alumnus John Clauser (BS '64) in 1969 and 1972, respectively. His findings are based on *Bell's theorem*, devised by CERN theorist John Bell. In 1964, Bell ironically proved that EPR's argument actually led to the opposite conclusion from what EPR had originally intended to show. Bell demonstrated that *quantum entanglement* is, in fact, incompatible with EPR's notion of *locality* and *causality*."

Freedman, S. J. & Clauser, J. F. (April, 1972). Experimental Test of Local Hidden-Variable Theories*

Physical Review Letters, 28, 14, 938–41; https://journals.aps.org/prl/pdf/10.1103/PhysRevLett.28.938.

Department of Physics and Lawrence Berkeley Laboratory, University of California, Berkeley, California.

* Work supported by the U.S. Atomic Energy Commission.

Received February 4, 1972.

"In the present work we measured the correlation in linear polarization of two *photons* emitted in an atomic cascade. The decaying atoms were viewed by two symmetrically placed optical systems, each consisting of two lenses, a wavelength filter, a rotatable and removable polarizer, and a single-photon detector. … We made the following assumptions for any *local hidden-variable* theory: (1) The two *photons* propagate as separated localized particles. (2) A binary selection process occurs for each *photon* at each polarizer (transmission or no-transmission). This selection does not depend upon the orientation of the distant polarizer. In addition, we made the following assumption to allow a comparison of the *generalization* of *Bell's inequality* without experiment: (3) All *photons* incident on a detector have a probability of detection that is independent of whether or not the *photon* has passed through a polarizer. … It has been shown by this generalization of *Bell's inequality* that the existence of *local hidden variables* imposes restrictions on this correlation in conflict with the predictions of quantum mechanics. Our data, in agreement with quantum mechanics, violate these restrictions to high statistical accuracy, *thus providing strong evidence against local hidden-variable theories*."

***Abstrac*t**

We have measured the linear polarization correlation of the *photons* emitted in an atomic cascade of calcium. It has been shown by a generalization of *Bell's inequality* that the existence of *local hidden variables* imposes restrictions on this correlation in conflict with the predictions of quantum mechanics. Our data, in agreement with quantum mechanics, violate these restrictions to high statistical accuracy, *thus providing strong evidence against local hidden-variable theories*.

Since quantum mechanics was first developed, there have been repeated suggestions that its statistical features possibly might be described by an underlying deterministic substructure. Such features, then, arise because a *quantum state* represents a statistical ensemble of "*hidden-variable states*". Proofs by von Neumann and others, demonstrating the impossibility of a *hidden-variable* substructure consistent with quantum mechanics, rely on various assumptions concerning the character of the *hidden variables*. Bell has argued that these assumptions are unduly restrictive. However, by considering an idealized case of two spatially separated but quantum-mechanically correlated systems, he was able to show that any *hidden-variable* theory satisfying only the natural assumption of "*locality*" also leads to predictions ("*Bell's inequality*") in conflict with quantum mechanics.

Bell's proof was extended to realizable systems by Clauser, Horne, Shimony, and Holt, who also pointed out that their generalization of Bell's inequality can be tested experimentally, thus testing all *local hidden-variable theories*, but that existing experimental results were insufficient for this purpose. *This Letter reports the results of an experiment which are sufficiently precise to rule out local hidden-variable theories with high statistical accuracy.*

In the present work we measured the correlation in linear polarization of two *photons* γ_1 and γ_2 emitted in a $J = 0 \rightarrow J = 1 \rightarrow J = 0$ atomic cascade. The decaying atoms were viewed by two symmetrically placed optical systems, each consisting of two lenses, a wavelength filter, a rotatable and removable polarizer, and a single-photon detector (see Fig. 1).

...

Fig. 1. Schematic diagram of apparatus and associated electronics. Scalers (not shown) monitored the outputs of the discriminators and coincidence circuits during each 100-sec count period. The contents of the scalers and the experimental configuration were recorded on paper tape and analyzed on an IBM 1620-II computer.

The following quantities were measured: $R(\varphi)$, the coincidence rate for two-photon detection, as a function of the angle φ between the planes of linear polarization defined by the orientation of the inserted polarizers; R_1, the coincidence rate with polarizer 2 removed; R_2, the coincidence rate with polarizer 1 removed; R_0, the coincidence rate with both polarizers removed. Quantum mechanics predicts that $R(\varphi)$ and R, are related as follows:

$$R(\varphi)/R_0 = \tfrac{1}{4} (\varepsilon_M^1 + \varepsilon_m^1)(\varepsilon_M^2 + \varepsilon_m^2) + \tfrac{1}{4} (\varepsilon_M^1 - \varepsilon_m^1) (\varepsilon_M^2 - \varepsilon_m^2)F_1(\theta) \cos 2\varphi \quad (1a)$$

while

$$R_1/R_0 = \tfrac{1}{2} (\varepsilon_M^1 + \varepsilon_m^1), \quad (1b)$$

and

$$R_2/R_0 = \tfrac{1}{2} (\varepsilon_M^2 + \varepsilon_m^2). \quad (1c)$$

Here ε_M^i (ε_m^i) is the transmittance of the ith polarizer for light polarized parallel (perpendicular) to the polarizer axis, and $F_1(\theta)$ is a function of the half-angle θ subtended by the primary lenses. It represents a depolarization due to *noncollinearity* of the two *photons*, and approaches unity for infinitesimal detector solid angles. [For this experiment, $\theta = 30^o$, and $F_1(30^o) = 0.99$.]

We make the following assumptions for any *local hidden-variable* theory: (1) The two *photons* propagate as separated localized particles. (2) A binary selection process occurs for each *photon* at each polarizer (transmission or no-transmission). This selection does not depend upon the orientation of the distant polarizer. In addition, we make the following assumption to allow a comparison of the generalization of *Bell's inequality* without experiment: (3) All *photons* incident on a detector have a probability of detection that is independent of whether or not the *photon* has passed through a polarizer.

The above assumptions constrain the coincidence rates by the following inequalities:

$$-1 \leq \Delta(\varphi) \leq 0, \qquad (2)$$

where

$$\Delta(\varphi) = 3R(\varphi)/R_0 - R(3\varphi)/R_0 - (R_1 + R_2)/R_0.$$

For sufficiently small detector solid angles and highly efficient polarizers, *these inequalities (2) are not satisfied by the quantum-mechanical prediction (1) for a range of values of* φ. Maximum violations occur at $\varphi = 22\ \frac{1}{2}^o$ $[\Delta(\varphi) > 0]$ and $\varphi = 67\ \frac{1}{2}^o$ $[\Delta(\varphi) < -1]$. At these angles of maximum violation, inequalities (2) can be combined into the simpler and more convenient expression

$$\delta = |\, R(22\ \tfrac{1}{2}^o)/R_0 - R(67\ \tfrac{1}{2}^o)/R_0 \,| - \tfrac{1}{4} \leq 0, \qquad (3)$$

which does not involve R_1 or R_2.

…

The results of the measurements of the correlation $R(\varphi)/R_0$, corresponding to a total integration time of ~200 h, are shown in Fig. 3. … Using the values at 22 ½° and 67 ½°, we obtain $\delta = 0.050 \pm 0.008$ *in clear violation of inequality* (3). Furthermore, we observe no evidence for a deviation from the predictions of quantum mechanics, calculated from the measured polarizer efficiencies and solid angles, and shown as the solid curve in Fig. 3.

…

We consider these results to be strong evidence against local hidden-variable theories.

The Nobel Prize in Physics 2022, Press release, October 4, 2022.

"The Royal Swedish Academy of Sciences has decided to award the Nobel Prize in Physics 2022 to

Alain Aspect, Institut d'Optique Graduate School – Université Paris-Saclay and École Polytechnique, Palaiseau, France

John F. Clauser, J. F. Clauser & Assoc., Walnut Creek, CA, USA

Anton Zeilinger, University of Vienna, Austria

"for experiments with *entangled photons*, establishing the violation of Bell inequalities and pioneering quantum information science".

Entangled states – from theory to technology.

Alain Aspect, John Clauser and Anton Zeilinger have each conducted groundbreaking experiments using *entangled quantum states*, where two particles behave like a single unit even when they are separated. Their results have cleared the way for new technology based upon quantum information.

The ineffable effects of quantum mechanics are starting to find applications. There is now a large field of research that includes quantum computers, quantum networks and secure quantum encrypted communication.

One key factor in this development is how quantum mechanics allows two or more particles to exist in what is called an *entangled state*. What happens to one of the particles in an *entangled pair* determines what happens to the other particle, even if they are far apart.

For a long time, the question was whether the correlation was because the particles in an *entangled pair* contained *hidden variables*, instructions that tell them which result they should give in an experiment. In the 1960s, John Stewart Bell developed the mathematical inequality that is named after him. This states that if there are *hidden variables*, the correlation between the results of a large number of measurements will never exceed a certain value. However, quantum mechanics predicts that a certain type of experiment will violate *Bell's inequality*, thus resulting in a stronger correlation than would otherwise be possible.

John Clauser developed John Bell's ideas, leading to a practical experiment. When he took the measurements, they supported quantum mechanics by clearly violating a *Bell inequality*. This means that quantum mechanics cannot be replaced by a theory that uses *hidden variables*.

Some loopholes remained after John Clauser's experiment. Alain Aspect developed the setup, using it in a way that closed an important loophole. He was able to switch the measurement settings after an *entangled pair* had left its source, so the setting that existed when they were emitted could not affect the result.

Using refined tools and long series of experiments, Anton Zeilinger started to use *entangled quantum states*. Among other things, his research group has demonstrated a phenomenon called *quantum teleportation*, which makes it possible to move a *quantum state* from one particle to one at a distance.

"It has become increasingly clear that a new kind of quantum technology is emerging. We can see that the laureates' work with *entangled states* is of great importance, even beyond the fundamental questions about the interpretation of quantum mechanics," says Anders Irbäck, Chair of the Nobel Committee for Physics.

Experimenting with Bell inequalities

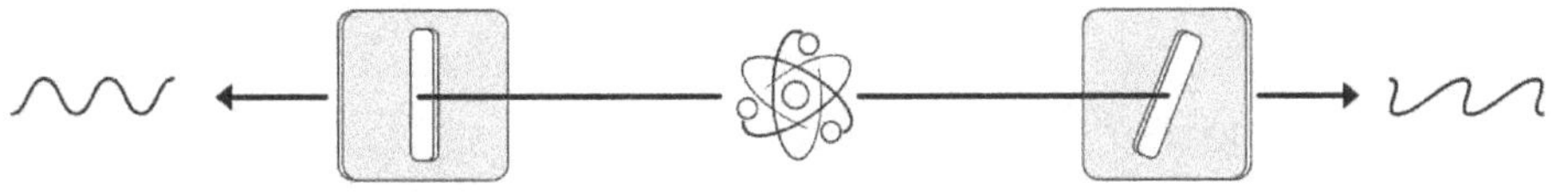

John Clauser used calcium atoms that could emit *entangled photons* after he had illuminated them with a special light. He set up a filter on either side to measure the photons' polarization. After a series of measurements, he was able to show they violated a *Bell inequality*.

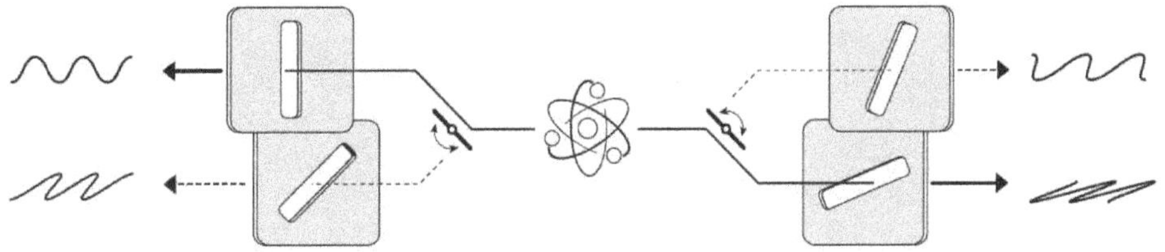

Alain Aspect developed this experiment, using a new way of exciting the atoms so they emitted *entangled photons* at a higher rate. He could also switch between different settings, so the system would not contain any advance information that could affect the results.

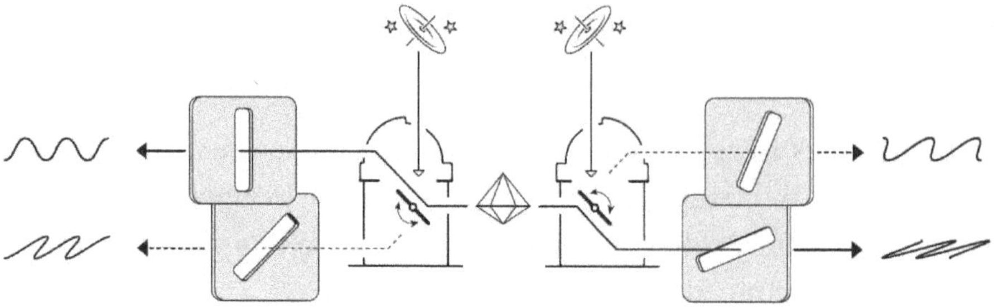

Anton Zeilinger later conducted more tests of *Bell inequalities*. He created *entangled pairs of photons* by shining a laser on a special crystal, and used random numbers to shift between measurement settings. One experiment used signals from distant galaxies to control the filters and ensure the signals could not affect each other."

"John F. Clauser
The Nobel Prize in Physics 2022
Prize share: 1/3:
One of the most remarkable traits of quantum mechanics is that it allows two or more particles to exist in what is called an entangled state. What happens to one of the particles in an entangled pair determines what happens to the other particle, even if they are far apart. In 1972, John Clauser conducted groundbreaking experiments using entangled light particles, photons. This and other experiments confirm that quantum mechanics is correct and pave the way for quantum computers, quantum networks and quantum encrypted communication." [NobelPrize.org.https://www.nobelprize.org/prizes/physics/2022/clauser/facts/].

Experiments testing macroscopic quantum superpositions must be slow. (September, 2015). Mari, A.[1], De Palma, D.[1,2] & Giovannetti, V. [1.]

(Addition: June 23, 2025).

Nature, Scientific Reports, 6, 22777 (2016); arXiv:1509.02408v1[quant-ph]

[1] NEST, Scuola Normale Superiore & Istituto Nanoscienze-CNR, I-56126 Pisa, Italy.

[2] INFN, Pisa, Italy.

Abstract

We consider a thought experiment where the preparation of a macroscopically massive or charged particle in a ***quantum superposition*** and the associated dynamics of a distant test particle apparently allow for ***superluminal communication***. We give a solution to the paradox which is based on the following fundamental principle: any ***local experiment***, discriminating a coherent superposition from an incoherent statistical mixture, necessarily requires a minimum time proportional to the mass (or charge) of the system.

For a charged particle, we consider two examples of such experiments, and show that they are both consistent with the previous limitation. In the first, the measurement requires to accelerate the charge, that can ***entangle*** with the emitted photons. In the second, the limitation can be ascribed to the quantum vacuum fluctuations of the electromagnetic field.

On the other hand, ***when applied to massive particles our result provides indirect evidence for the existence of gravitational vacuum fluctuations and for the possibility of entangling a particle with quantum gravitational radiation***.

I. INTRODUCTION

The existence of ***coherent superpositions*** is a fundamental postulate of quantum mechanics but, apparently, implies very counterintuitive consequences when extended to macroscopic systems. This problem, already pointed out since the beginning of quantum theory through the famous Schrodinger cat paradox, has been the subject of a large scientific debate which is still open and very active.

Nowadays there is no doubt about the existence of ***quantum superpositions***. Indeed, *this effect has been demonstrated in a number of experiments involving microscopic systems* (photons, electrons, neutrons, atoms, molecules, etc.). However, at least in principle, the standard theory of quantum mechanics is valid at any scale and does not put any limit on the size of the system: if you can delocalize a molecule then nothing should forbid you to delocalize a cat, apart from technical difficulties. Such difficulties are usually associated with the impossibility of isolating the system from its environment, because it is well known that *any weak interaction changing the state of the environment is sufficient to destroy the initial coherence of the system*.

In this work we are interested in the ideal situation in which we have a ***macroscopic mass*** or a ***macroscopic charge*** perfectly isolated from the environment and prepared ***in a quantum superposition of two spatially separated states***. Without using any speculative theory of quantum gravity or sophisticated tools of quantum field theory, we propose a simple thought experiment based on particles interacting via semiclassical forces. Surprisingly a simple consistency argument with ***relativistic causality*** is enough to obtain *a fundamental result which, being related to gravitational and electric fields, indirectly tells us something about quantum gravity and quantum field theory*.

> [***Relativistic causality*** is the hypothesis that observed preferred orders of physical processes are not disturbed by transformation between inertial frames. The principle of relativity of simultaneity states that the cause must precede its effect according to all inertial observers. This is equivalent to the statement that the cause and its effect are separated by a time-like interval, and the effect belongs to the future of its cause.]

The result is the following: assuming that a ***macroscopic mass*** m is prepared in a ***superposition*** of two states separated by a distance d, then any experiment discriminating the coherent superposition from a classical incoherent mixture requires a ***minimum time*** $T \propto md$, proportional to the ***mass*** and the ***separation distance***. Analogously for a ***quantum superposition of a macroscopic charge*** q, such minimum time is proportional to the associated ***electric dipole*** $T \propto qd$. In a nutshell, experiments testing macroscopic superpositions are possible in principle, but ***they need to be slow***. For common experiments involving systems below the Planck mass and the Planck charge this limitation is irrelevant, however such time can become very important at macroscopic scales.

As an extreme example, if the center of mass of the Earth were in a ***quantum superposition*** with a separation distance of one micrometer, according to our result one would need a time equal to the age of the universe in order to distinguish this state from a classical statistical mixture. Clearly this limitation suggests that at sufficiently macroscopic scales quantum mechanics can be safely replaced by classical statistical mechanics without noticing the difference.

The fact that large gravitational or electromagnetic fields can be a limitation for the observation of quantum superpositions is not a new idea. In the past decades, several models of spontaneous localization have been proposed which, going beyond the standard theory of quantum mechanics, postulate the existence a ***gravity*** induced collapse at macroscopic scales. Remaining within the domain of standard quantum mechanics, the loss of coherence in interference experiments due to the emission of electromagnetic radiation has been already studied in the literature. Similarly, the interaction of a massive particle with gravitational waves and the dephasing effect of time dilation on internal degrees of freedom have been considered as possible origins of quantum decoherence.

For what concerns our thought experiment, a similar setup can be found in the literature where the interference pattern of an electron passing through a double slit is destroyed by a distant measurement of its electric field. This thought experiment can be traced back to Bohr as quoted in, was discussed by Hardy interviewed in and appears as an exercise in the book by Aharanov and Rohrlich. Moreover, *recently different experiments involving interacting test particles have been proposed in order to discriminate the quantum nature of the gravitational field from a potentially classical description*, [in March 2016 version published in *Nature*: "while some limitations that ***relativistic causality*** imposes to the possible measurements in quantum field theory have been investigated in Benincasa D. M. T., Borsten L., Buck M. & Dowker F. Quantum information processing and relativistic quantum fields. *Class. Quantum Grav.*, 31, 075007 (2014)"].

The original contribution of our work *is that,* ***imposing the consistency with relativistic causality****, our thought experiment allows the derivation of a fundamental minimum time which is valid for any possible experiment involving macroscopic superpositions*. In this sense our bounds represent universal limitations having a role analogous to the Heisenberg uncertainty principle in quantum mechanics. For this reason, while our results could be observable in advanced and specific experimental setups, their main contribution is probably a better understanding of the theory of quantum mechanics at macroscopic scales. ***For charged particles*** we propose two different measurements for testing the coherence. The first requires to accelerate the charge, and our bound on the discrimination time is due to the ***entanglement*** with the ***emitted photons***. In the second, the bound can be instead ascribed to the presence of the vacuum fluctuations of the ***electromagnetic field***. On the other hand, we also find an equivalent bound associated to ***quantum superposition of large masses***. What is the origin of this limitation? The analogy suggests that ***the validity of our bound could be interpreted as indirect evidence for the existence of quantum fluctuations of the gravitational field, and of quantum gravitational radiation.***

This work is structured in the following way: in **Section II** we propose our thought experiment which suggests a minimum discrimination time for any macroscopic quantum superposition. In **Section III** we derive a quantitative bound. Finally in **Sections IV** and **V** we check the consistency of our results with an explicit analysis of a charged particle interacting with the electromagnetic field. Here we propose two different measurements, and show that in both cases they are able to check the coherence only if their duration satisfies our fundamental limit.

II. THOUGHT EXPERIMENT Consider the thought experiment represented in Fig. 1, and described by the following protocol. The protocol can be equivalently applied to quantum superpositions of large masses or large charges. …

…

…

www.ingramcontent.com/pod-product-compliance
Lightning Source LLC
Chambersburg PA
CBHW061323190326
41458CB00011B/3872